# Lecture Notes in Physics

For information about Vols. 1–172, please contact your bookseller or Springer-Verlag.

# Lecture Notes in Physics

## 238

# The Free-Lagrange Method

Proceedings of the First International Conference
on Free-Lagrange Methods, Held at Hilton Head Island,
South Carolina, March 4–6, 1985

Edited by M. J. Fritts, W. P. Crowley and H. Trease

Springer-Verlag
Berlin Heidelberg GmbH

**Editors**

Martin J. Fritts
Science Applications International Corporation
Annapolis, MD 21401, USA

W. Patrick Crowley
Lawrence Livermore National Laboratory
Livermore, CA 94550, USA

Harold Trease
Los Alamos National Laboratory
Los Alamos, NM 87545, USA

ISBN 978-3-540-15992-6

Library of Congress Cataloging in Publication Data. International Conference on Free-Lagrange
Methods (1st: 1985: Hilton Head Island, S.C.) The Free-Lagrange method. (Lecture notes in
physics; 238) 1. Hydrodynamics—Congresses. 2. Lagrangian functions—Congresses. 3. Numeri-
cal analysis—Congresses. I. Fritts, M. J. (Martin J.), 1943-. II. Crowley, William Patrick. III. Trease,
H. (Harold), 1955-. IV. Title. V. Series.
QC150.I57 1985 515'.55 85-26154
ISBN 978-3-540-15992-6      ISNB 978-3-540-39697-0 (eBook)
DOI 10.1007/978-3-540-39697-0

2153/3140-543210

# PREFACE

This volume is devoted to the Proceedings of the First International Conference on Free-Lagrange Methods which was held at Hilton Head Island, South Carolina, March 4 - 6, 1985. The Conference was attended by 39 scientists representing five countries (the People's Republic of China, England, France, the United States of America, and Wales).

Fourteen papers are contained in the Proceedings, thirteen of which were presented during the Conference. These Proceedings assemble together, for the first time, discussions of the basic Free-Lagrange Method (FLM) and a range of recent implementations and extensions, presented by researchers and by the original developers. The Conference was designed to allow in-depth coverage for each of the primary topics. Two round-table discussions amplify a number of crucial points, and the references included with the papers are extensive enough to allow interested readers access to material which is diffused over several different research communities. The Proceedings therefore provide a basic introduction to FLM for researchers and students using computational fluid dynamics techniques.

Two separate round-table discussions attended by all Conference participants were simultaneously recorded on magnetic tape and by a court reporter. Nevertheless, the final transcripts contained many ambiguities both in context and in speaker identity. Regretfully, a few comments were completely unintelligible and these were necessarily deleted. To those contributors whose statements were deleted or misinterpreted we apologize. We sincerely hope to have represented the spirit, if not the exact text, of the discussions.

Many people were responsible for the success of the Conference and should be acknowledged for their contributions. First of all we thank the authors for their papers, and everyone who attended the Conference for their interest and comments. We wish to thank in particular the staff of Sea Pines Plantation, Hilton Head Island, for providing us with a congenial atmosphere in which to hold the Conference. Also, we appreciate the secretarial help of Margaret Dixon of LLNL before, during, and after the Conference and of Lynn Trease for typing the final draft of the edited transcript. Special thanks to Margaret York for providing us with an idyllic setting, an undeserved but greatly appreciated menu, and in general an outstanding atmosphere in which to edit these Proceedings. Support from LANL, LLNL, and SAI is gratefully acknowledged. Finally, we acknowledge Springer-Verlag, particularly Mr. Jeffrey Robbins, for their help in publishing these Proceedings.

> "I guess the hindalex form is real easy, you know?"
> "...If you can't follow the vertex, work six more years on it."
> From the unedited discussions.

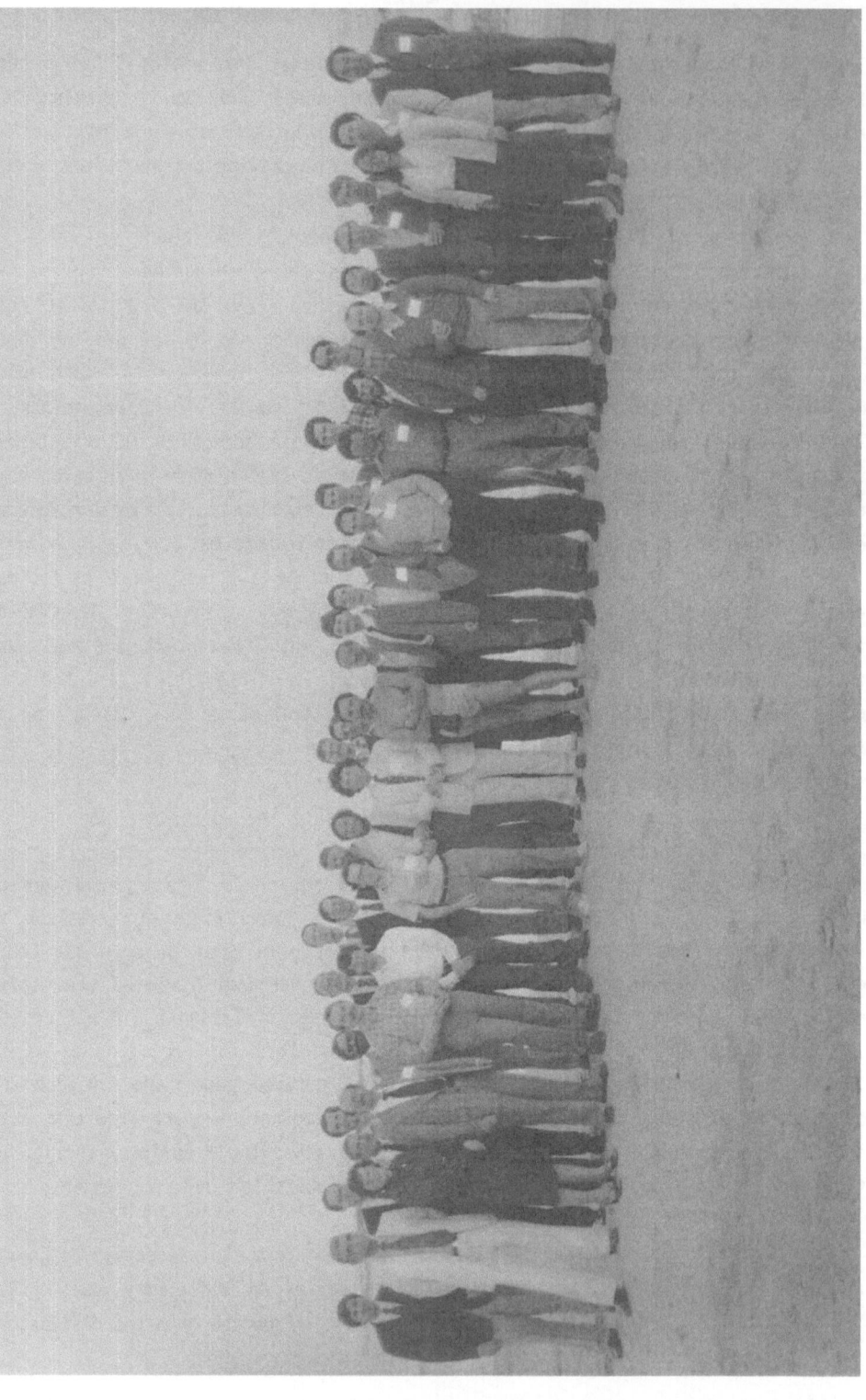

The numerical methods available to computational physicists are constantly being extended by the steadily increasing power of computing machinery. Larger and faster machines permit more extensive calculations to be performed more quickly. In addition, they open up entirely new research areas to computation through the development of radically different techniques which simply could not be attempted on earlier machines. The Free-Lagrange Method (FLM) is one such technique which has benefited from advances in computing power.

The FLM is a means of extending the advantages of the Lagrangian formulation of the fluid dynamics equations to a wider set of problems by freeing the numerical implementation of the method from the restrictions of a fixed connectivity mesh. The method has been developed under a number of different programs over several generations of computers. The purpose of the Conference on the FLM, the subject of these Proceedings, is to introduce the FLM as a legitimate, developed numerical tool to a broader community of workers and students using computational methods.

The FLM is designed to solve the evolution equations for fluids which are characterized by significant relative fluid motions. Calculations which use Eulerian methods lose accuracy through the introduction of numerical diffusion. Conversely, Lagrangian simulations of highly sheared flows are subject to large mesh distortions and mesh tangling which quickly degrade the accuracy of the solutions. The FLM retains the accuracy of the Lagrangian formulation by restructuring the mesh to obtain more reasonable difference templates. The types of fluid problems that this method is intended to solve are those which are critically dependent upon an accurate definition of interfaces between fluid types and which contain a significant component of shear; e.g., hydrodynamic instabilities, jet formation, bubble collapse, thin shell deformations, and asymmetric collisions.

Numerical representations of the Lagrangian form of the fluid flow equations approximate a continuous fluid region by a discrete set of points which are carried by the advective velocity. The connectivity between points is stored in the same manner as in Eulerian codes, through implicit indexing based on the relative positions of the points in the fluid. In the FLM, the connectivity of the points is arbitrary as well. A connectivity matrix is stored which contains a list of nearest neighbors that a point will interact with on a hydrodynamic basis. This connectivity matrix is dynamic in the sense that the list of neighbors change as the Free-Lagrange points follow the (Lagrangian) motion of the fluid. The initialization and maintenance of the nearest neighbor lists are separate subtechnologies associated with the FLM. These two functions are significant factors which may limit the speed of certain Free-Lagrange algorithms, even with the present supercomputers. Hardware improvements in gather/scatter operations are expected to reduce or eliminate this factor.

The computational overhead of the FLM is mainly due to the indirect addressing associated with the arbitrary and continually changing lists of nearest neighbors. This implies that the data structures that support Free-Lagrange algorithms must be designed and optimized for both speed and memory storage efficiency. The techniques that are used to maintain the connectivity matrix fall into three different classes, which also characterize the types of meshes that underlie different algorithms. These meshes, which will be referred to throughout the Proceedings, are classified as follows; the Voronoi mesh, having n-sided polygonal or polyhedral cells about each mesh point; triangular and tetrahedral meshes, and in particular the Delaunay mesh, which use mesh points as triangle or tetrahedron vertices; and mixed meshes, usually a combination of triangles and quadrilaterals having corners defined by mesh points. The Voronoi and Delaunay meshes are duals of each other, i.e., they both have the same connectivity and one can be constructed from the other.

The arbitrarily connected mesh gives the FLM its power over fixed connectivity meshes. Mesh points can truly follow the Lagrangian motion of the fluid without the loss in accuracy attended by mesh distortion, and without the code crashes associated with mesh tangling. The mesh connectivity can be tailored to suit the type of problems and the partial differential equation operators to be approximated. A continuously updated definition of nearest neighbors determines which mesh connections are used in evaluating the differenced form of those operators. Local mesh modifications can be made without worrying about destroying the global connectivity of the mesh, of having to introduce special templates to accommodate blended or interpolated meshes. Mesh points can be added or deleted at arbitrary locations, thereby increasing or decreasing the local resolution. The robustness of the FLM is generally much greater than standard Lagrangian methods because the code can automatically rezone itself, according to some predefined set of criteria, during the evolution of a problem.

The flexibility of the Free-Lagrangian technique is evident in the range of problems for which it has been used. Several early attempts were made in the 60s to eliminate the restrictions of fixed connectivity meshes for fluid dynamics problems: for example, the PANACEA code by Howard Kolsky at LANL, and the PAF, "Particle-And-Force", algorithm by Frank Harlow at LANL. The method was improved and extended by Pat Crowley in the 70s, resulting in the first convincing demonstrations of the Free-Lagrange Method. Further contributions to the still immature technology were made by Martin Fritts and Jay Boris at the Naval Research Laboratory and by Charles Peskin at the Courant Institute in the late 70s. During this period the method came into use for a number of different applications. These included studies of classic hydrodynamic instabilities, heat conduction problems, high-temperature plasma physics, water wave/structure interactions, tokamak modelling, and a substantial collection of compressible and incompressible hydrodynamics problems.

These Proceedings make no attempt to collect and classify these earlier algorithms and codes, but instead focus on current extensions of the method which are in use at several institutions in a large variety of applications. The extent of the development of the method and the range of applications are amply represented in presentations and discussions.

The order of the papers in these Proceedings will be the same as the order in which they were presented at the Conference. The Conference was organized to provoke the greatest amount of discussion. In most cases papers are grouped according to subject matter, to juxtapose similar work developed by different researchers. Unfortunately, it was not practical to attempt to record the very vigorous and animated discussions which followed every talk. However, some flavor of these discussions can be obtained from the round-table sessions, which were recorded by a court reporter and included here, in an edited form, in the same chronological order as in the Conference.

The first two papers in the first session serve as a general introduction to the Free-Lagrange method and the use of unstructured grids. Pat Crowley's paper provides an overview of the evolution of a highly optimized algorithm for calculating two-dimensional compressible fluid flows on a Free-Lagrange mesh composed of a combination of triangles and quadrilaterals. This paper also discusses the logic used in the local mesh optimization scheme that is used to maintain the nearest neighbor connectivity matrix. In the next paper, Gordon Erlebacher discusses the fundamentals of difference approximations which are based on unstructured triangular meshes. Error estimates are given for approximations to several operators using a number of different cell shapes.

The next two papers by Jeffrey Augenbaum and Christoph Börgers discuss various aspects of solving differential equations on grids constructed of Voronoi polygons. The Voronoi gridding technique, which is also discussed in several subsequent papers, is used to systematically tessellate a region of space containing an arbitrary distribution of points. The resulting mesh is unique and completely spans a region of space with polygons in two dimensions and polyhedrons in three dimensions. Jeffrey Augenbaum applies the method to the solution of the shallow water equations on a sphere, and Christoph Börgers discusses its application to the solution of the Navier-Stokes equations on a periodic domain.

The last paper to be presented in the first session presents work done by Peter Eltgroth on a method of mapping the difference form of the hydrodynamic equations onto a multi-processing computing environment. This type of method holds great promise in accelerating the real-time throughput of hydrodynamic calculations.

The first two papers of the second session of the Conference deal with the setup and maintenance of the nearest neighbor connectivity matrix for three-dimensional calculations. The paper by Martin Fritts outlines an efficient scheme for constructing and maintaining a tetrahedral mesh, the Delaunay mesh, in three dimensions. In the following paper, Harold Trease describes a similar algorithm for constructing the Voronoi mesh in three dimensions. Trease's paper also describes the solution of fluid flow equations in three dimensions on the Voronoi mesh.

The last two papers of the second session, presented by Jay Boris and Robert Cooper, describe methods of optimizing data structures for Free-Lagrange codes to maximize computer speed and data storage. The paper by Jay Boris introduces an efficient, vectorizable three-dimensional nearest neighbor algorithm and data structure. The presentation by Cooper describes the data structure and the code optimization techniques that have been implemented in Crowley's Free-Lagrangian algorithm.

The last part of the second session was devoted to a round-table discussion concerning various questions addressed to the authors of the papers in the first two sessions. A large part of this discussion centered around the applicability of the Voronoi mesh, as applied in hydrodynamic calculations, and questions on the formal accuracy of the method.

On the third day of the Conference four papers were presented. The first paper was given by Peter Eiseman, and it discusses the use of adaptive mesh techniques as applied to triangular grids. The following paper, by Rainald Löhner, presents a review of finite element methods applied to unstructured, stationary grids.

The two final papers, by Arthur Armstrong and Robert Clark, present two contrasting techniques for solving the compressible fluid equations by using Free-Lagrangian techniques. Arthur Armstrong discusses the solution of the integral form of the fluid flow equations on a Voronoi mesh. Robert Clark, on the other hand, solves the differential form of the equations. In his formulation, a mesh is defined by Delaunay triangulation, but is used only to calculate finite-difference templates.

The Conference closed with another round-table discussion. The emphasis in the second discussion was on questions addressed to the authors of the papers in the third session and to the problem of conservation properties of the various algorithms.

# CONTENTS

FREE-LAGRANGE METHODS FOR COMPRESSIBLE HYDRODYNAMICS IN TWO SPACE DIMENSIONS

W. P. Crowley
Lawrence Livermore National Laboratory, University of California
P.O. Box 808
Livermore, CA 94550

ABSTRACT

Since 1970 a research and development program in Free-Lagrange methods has been active at Livermore. The initial steps were taken with incompressible flows for simplicity. Since then the effort has been concentrated on compressible flows with shocks in two space dimensions and time. In general, the line integral method has been used to evaluate derivatives and the artificial viscosity method has been used to deal with shocks.

Basically, two Free-Lagrange formulations for compressible flows in two space dimensions and time have been tested and both will be described. In method one, all prognostic quantities were node centered and staggered in time. The artificial viscosity was zone centered. One mesh reconnection philosophy was that the mesh should be optimized so that nearest neighbors were connected together. Another was that vertex angles should tend toward equality. In method one, all mesh elements were triangles.

In method two, both quadrilateral and triangular mesh elements are permitted. The mesh variables are staggered in space and time as suggested originally by Richtmyer and von Neumann. The mesh reconnection strategy is entirely different in method two. In contrast to the global strategy of nearest neighbors, we now have a more local strategy that reconnects in order to keep the integration time step above a user chosen threshold. An additional strategy reconnects in the vicinity of large relative fluid motions.

Mesh reconnection consists of two parts: (1) the tools that permit nodes to be merged and quads to be split into triangles etc. and; (2) the strategy that dictates how and when to use the tools. Both tools and strategies change with time in a continuing effort to expand the capabilities of the method. New ideas are continually being tried and evaluated. The successful ones stay in the code, and in some sense its intelligence increases with time.

Examples of mesh optimization tools and strategies and of sample problems will be given.

## 1. Introduction

The Free Lagrange effort at Livermore has been active since 1970 and has moved sequentially in three different directions. The initial effort was confined to incompressible flows [1]. During this time basic ideas were explored such as initial mesh construction and mesh reconnection during the evolution of a problem. It was soon realized that initial mesh generation and mesh reconnection were separable problems. In this paper we will assume that a mesh exists and concentrate on mesh reconnection.

Another problem that must be faced involves the centering of prognostic variables in space and time. In the initial incompressible effort and in the first compressible effort all prognostic variables were node centered. This has the potential advantage that mesh reconnection will not change the values of prognostic variables - that is, mesh reconnection does not introduce diffusion. Node centering has difficulties when it comes to multi-material problems. The notion of material interfaces, if done correctly, introduces complexities. In the second effort, material interfaces were handled in a rather crude way and this led to somewhat crude results.

The third and current effort solves the material interface problem by recentering the variables according to the traditional Lagrangian centering of Richtmyer and von Neumann [2]. Energy and mass are zone centered - position, velocity and acceleration are node centered. With this approach, material interfaces are well defined; they fall on zone edges (along lines connecting nodes). The disadvantage is that mesh reconnections now introduce a mixing of adjacent zone attributes and thus diffusion occurs. However since mesh reconnection is now more local than global, it is thought that the effect of diffusion is small.

Shock waves are an important aspect of compressible flows. In these formulations they are automatically taken care of by an artificial viscosity [2,3].

Spatial derivatives are done with the Line Integral method [4].

In the third effort the mesh reconnection philosophy changed along with the centering of variables. In the first two efforts, mesh reconnection was a global affair with the goal in mind of connecting nearest neighbors. In the third effort the strategy is threefold: (1) to accommodate shearing motions; (2) to keep the integration time step up and; (3) to prevent boomerang and bowtie zone topologies.

Since all these formulations use an explicit scheme for time differencing, the integration time step is an important consideration.

The remainder of this paper is divided into six sections. Section 2 gives the equations of motion. Section 3 describes the line integral method with particular application to the momentum equation. In section 4 the first compressible effort will be described. In section 5 the current effort will be described and in section 6 mesh reconnection and optimization will be described. An interactive

graphics capability is quite important to the development of these methods.
Comments on interactive graphics are in section 7. Results from a sample test
problem are described in section 8.

## 2. Equations of Motion

In traditional Lagrangian codes, each node has an invariant set of neighbors and
the spatial difference terms usually involve the original coordinates and a Jacobian
that transforms from orginal to current coordinates. In many cases the mesh elements
(zones) are quadrilaterals.

In this formulation, the mesh elements are both triangles and quadrilaterals and
the number of neighbors of each node may change with time. It is more convenient to
express derivatives in terms of current coordinates; Pomraning [5] refers to this
formulation as the Modified Eulerian and to the traditional formulation as the
Lagrangian.

The equations of motion are

1. $\dfrac{d\vec{u}}{dt} = -\dfrac{1}{\rho}(\text{grad } p + \text{div } \bar{\bar{q}})$          Momentum

2. $\dfrac{d\rho V}{dt} = 0$          Mass

3. $\dfrac{d\varepsilon}{dt} = -\dfrac{p}{\rho}\text{div}(\vec{u}) - \dfrac{\bar{\bar{q}}}{\rho}\text{grad}(\vec{u})$          Specific Internal energy

To these are appended an equation of state

$$p = p(\varepsilon, \rho)$$

a kinematic equation

$$\dfrac{d\vec{x}}{dt} = \vec{u}$$

and an equation for the artificial viscosity

$$q = \begin{cases} \rho\Delta u(\Delta u - C_1 C) & \text{if } \Delta u < 0 \\ \\ 0 & \text{otherwise.} \end{cases}$$

The artificial viscosity actually used is somewhat more complex in 2D - different
forms are required by quadrilateral and by triangular zones. The tensor $\bar{\bar{q}}$ used here
is in a state of continual development. In addition, an optional $qk\ell$ [6] may be used
to inhibit nonphysical small scale motions in quads.

The equations are solved in a cylindrical geometry with

$\vec{x}$ : $(r,z)$

$\vec{u}$ : $(\frac{dr}{dt}, \frac{dz}{dt})$

## 3. The Line Integral Method

Green's Theorem in the plane states

$$\iint (\frac{\partial P}{\partial x} + \frac{\partial Q}{\partial y}) \, dx \, dy = \oint (P \, dy - Q \, dx)$$

where P and Q are arbitrary functions of x and y. Taking first P to be zero and then Q we have

$$\frac{\partial P}{\partial x} = \frac{\oint p \, dy}{\oint x \, dy}$$

$$\frac{\partial P}{\partial y} = \frac{- \oint p \, dx}{\oint x \, dy}$$

where $\frac{\partial P}{\partial x}$ and $\frac{\partial P}{\partial y}$ are mean values over the area $\oint x \, dy$. In finite difference form we have

$$\oint p \, dy \approx \sum P_{\ell-1/2}(y_\ell - y_{\ell-1})/2 \tag{3.1}$$

The path of integration connects the mid points of the edges – it does not include points in the zones because $P_{\ell-1/2}$ is assumed constant in a zone.

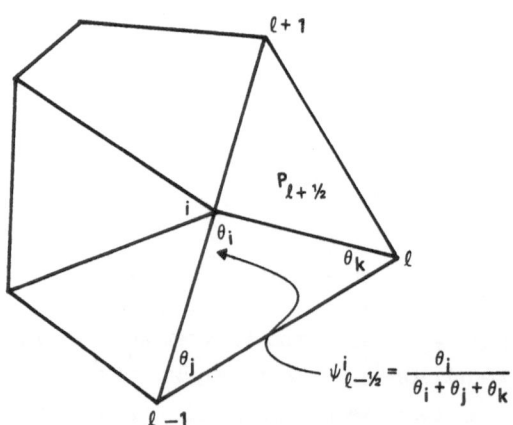

In the momentum equation we need the gradient multiplied by the reciprocal of the density. The denominator becomes

$$\rho \oint x\ dy \approx \sum_{\ell} \rho_{\ell-1/2}\ \psi^i_{\ell-1/2}\ A_{\ell-1/2} \tag{3.2}$$

where $A_{\ell-1/2}$ is the area of zone $\ell-1/2$ and where $\psi^i_{\ell-1/2}$ is the ratio of angle $(\ell-1,i,\ell)$ to the sum of all angles in zone $\ell-1/2$. The quantity $\psi^i_{\ell-1/2}\ A_{\ell-1/2}$ is thus the fractional contribution of zone $\ell-1/2$ to node i.

The r-component of acceleration is the ratio of equations (3.1) and (3.2) and similarly for the z-component.

The acceleration computed above is in general not centered at node i, but rather at some other point, say $a_i$. The r-component of acceleration at node i is

$$\dot{u}_i = \dot{u}_{a_i} + (\text{grad } \dot{u}) \cdot (\vec{x}_i - \vec{x}_{a_i}) .$$

and similarly for the z-component. A similar correction for velocity rather than acceleration was suggested by Margolin and Nichols [7]. Here the term grad $\dot{u}$ is evaluated by a line integral ratio over a path including the neighbors of node i; the path follows the points $a_\ell$.

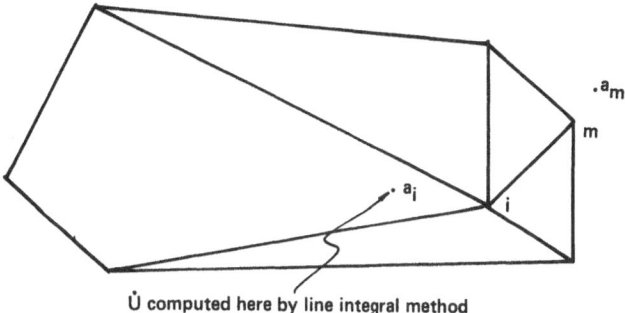

$\dot{U}$ computed here by line integral method

Work on this correction term is in progress now. At first glance, the point $a_i$ should be at the average centroid of the zones surrounding node i.

## 4. The Node-Centered Compressible Formulation

In this formulation all mesh elements are triangles and all prognostic variables are point-centered quantities. However, at the beginning of each cycle and after optimization, temporary zone-centered energies and masses are computed by a mapping process. Once these quantities are computed, the traditional Lagrangian centered difference equations [2] are used. At the end of the cycle, the point-centered quantities have been advanced in time, and all zone-centered information is forgotten. The mapping process will be described here.

First it is necessary to introduce some notation and some conventions. There are NPTOT mesh points and each point is connected to its nearest neighbors to give triangular mesh elements. Each point has NUM neighbors where

$$\text{NUM} \geq 2 \text{ for boundary points}$$

$$\text{NUM} \geq 3 \text{ for interior points.}$$

The three vertices i, j, and k of each triangle are ordered in a counterclockwise sense in our right-handed (x,y) coordinate plane, Fig. la. Since j and k are two of the neighbors of i, i is a neighbor of j and of k.

The complete neighborhood of point i consists of several triangles. For example, a typical interior point is sketched in Fig. lb: point i has five neighbors and is surrounded by five triangles. The first neighbor of this point may be any of its neighbors, but the others must follow in counterclockwise order. A typical boundary point is sketched in Fig. lc. There, point i has four neighbors and three triangles. Interior points have NUM neighbors and NUM neighboring triangles while boundary points have NUM neighbors and NUM-1 neighboring triangles. We are more particular about the organization of boundary neighborhoods. The first and last neighbors of boundary points are also boundary points, and the counterclockwise ordering of the neighbors means that a path from the first to the last neighbor lies within the boundaries of the problem.

In referring to the neighbors of a point, we may use the local index $\ell$ where

$$1 \leq \ell \leq \text{NUM} .$$

Using this, we may refer to a triangle by its "proper" name (i,j,k) or by its first name and two indices (i,$\ell$,$\ell$+1). It is then natural to denote point quantities

$$f^i \text{ or } f_i$$

and zone quantities

$$f_{\ell+1/2} .$$

When time is introduced, we will use an additional superscript, and denote a mixed quantity at time level n

$$(f^i_{\ell+1/2})^n$$

or a point quantity at time level n

$$(f_i)^n$$

In general, the time index is suppressed implying that latest available values are used.

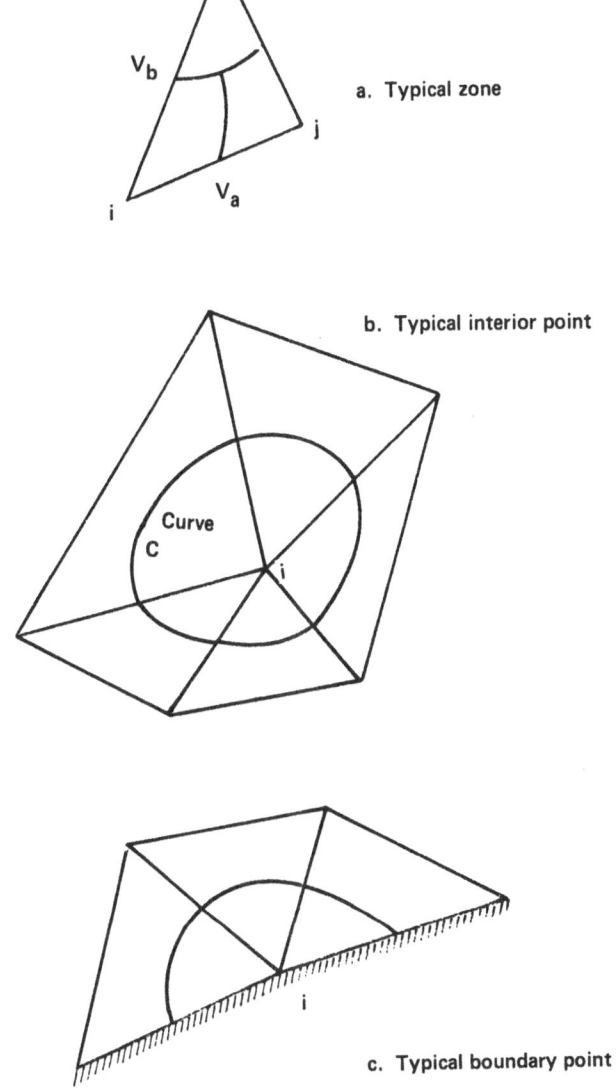

a. Typical zone

b. Typical interior point

Curve
C

c. Typical boundary point

**REPRESENTATIVE GEOMETRIC SKETCHES**

Figure 1

Consider the interior mesh point i in Fig. 1b.  It is surrounded by five triangles each of which has a definite area and volume.  The curve C defines the area of the secondary mesh element surrounding the point i.  That is, we subdivide each triangle into three parts associating one part with each vertex of the triangle.  If $A_{\ell+1/2}$ is the area of triangle (i,j,k) (Fig. 1a) then

$$a^i_{\ell+1/2} = \psi^i_{\ell+1/2} \, A_{\ell+1/2}$$

is the area that point i and zone $\ell+1/2$ have in common. The product $\pi\psi^i_{\ell+1/2}$ is the angle in radians between side $\overline{IJ}$ and $\overline{IK}$.

An indefinite summation notation is used and is to be interpreted in the following manner. The zone area is given by

$$A_{\ell+1/2} = \sum_{ijk} a^i_{\ell+1/2}$$

and the sum is over the three vertices i, j, and k associated with zone $\ell+1/2$. The point area is given by

$$A^i = \sum_{\ell} a^i_{\ell+1/2}$$

and the sum is over all zones $\ell+1/2$, ($\ell=1$, $L-1$) surrounding a point where, if NUM is the number of neighbors for point i,

$$L = \begin{cases} \text{NUM for interior zones} \\ \\ \text{NUM-1 for boundary zones} \end{cases}$$

Similarly, there is a volume associated with each zone and with each point

$$S_{\ell+1/2} = \bar{r}_{\ell+1/2} \, A_{\ell+1/2} = \left(\frac{r^i + r^j + r^k}{3}\right) A_{\ell+1/2}$$

$$s^i_{\ell+1/2} = \bar{r}^i_{\ell+1/2} \, a^i_{\ell+1/2}$$

$$s^i = \sum_{\ell} s^i_{\ell+1/2}$$

where S is the volume of revolution divided by $2\pi$ and where

$$\bar{r}^i_{\ell+1/2} = \begin{cases} (\Omega_{\ell+1/2} \, r^i + \zeta) \, \bar{r}_{\ell+1/2} & \text{(cylindrical geometry)} \\ \\ 1 & \text{(plane geometry)} \end{cases}$$

and

$$\Omega_{\ell+1/2} = \frac{1 - \zeta}{\displaystyle\sum_{ijk} r^i \, \psi^i_{\ell+1/2}}$$

The constant $\zeta$ is taken to be 7/12 to give the correct volumes for equilateral triangles. We define the geometric factor

$$h^i_{\ell+1/2} = \bar{r}^i_{\ell+1/2}\ \psi^i_{\ell+1/2}$$

and the mapping factor

$$f^i_{\ell+1/2} = h^i_{\ell+1/2} \Big/ \sum_\ell h^i_{\ell+1/2} \ .$$

The mass contribution from point i to zone $\ell+1/2$ is (Fig. 2)

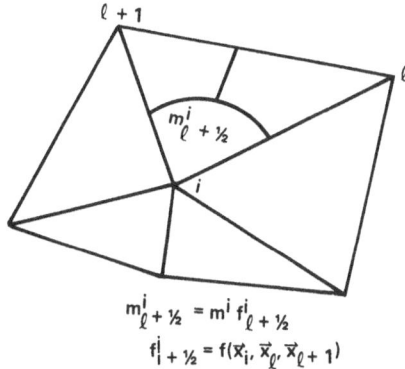

$$m^i_{\ell+1/2} = m^i\ f^i_{\ell+1/2}$$

$$f^i_{i+1/2} = f(\bar{x}_i, \bar{x}_\ell, \bar{x}_{\ell+1})$$

Figure 2

$$m^i_{\ell+1/2} = m^i\ f^i_{\ell+1/2}$$

and the mass in zone $\ell+1/2$ is the sum of the three vertex contributions

$$m_{\ell+1/2} = \sum_{ijk} m^i_{\ell+1/2}\ .$$

The specific internal energy in zone $\ell+1/2$ is

$$\varepsilon_{\ell+1/2} = \frac{1}{m_{\ell+1/2}} \sum_{ijk} e^i\ m^i_{\ell+1/2}\ .$$

and the density is

$$\rho_{\ell+1/2} = m_{\ell+1/2} / S_{\ell+1/2}\ .$$

Once the energy and density have been determined, the pressure is determined through the equation of state $p = p(\varepsilon, \rho)$.

These mapping equations hold for boundary zones (Fig. 1c) as well as interior zones.

The mapping described here is conservative in that the sum of $m^i$ or $\epsilon^i$ over all the points is identical to the sum of $m_{\ell+1/2}$ or $\epsilon_{\ell+1/2}$ over all the zones.

Materials, material properties, and equations of state (EOS) are associated with points. When zone quantities are constructed, it is necessary to associate an EOS with the zone. If the three vertex points are all the same material then the zone is of that material. That is, the EOS of the zone is the same as the EOS of the points. If all vertex points do not have the same EOS, then the EOS of the zone is that of the material contributing the most mass to the zone.

At the beginning of a cycle we have

$$x_i^n, \; y_i^n, \; u_i^{n-1/2}, \; w_i^{n-1/2}, \; \epsilon_i^n, \; \rho_i^n, \; m_i \; .$$

First zone centered energies, densities and pressures are computed

$$\epsilon_{\ell+1/2}^n, \; \rho_{\ell+1/2}^n, \; p_{\ell+1/2}^n$$

and a zone centered artificial viscosity is constructed

$$q_{\ell+1/2}^n = q(\rho_{\ell+1/2}^n, \; \Delta\vec{u}_{\ell+1/2}^{n-1/2}) \; .$$

From these quantities a pressure gradient can be calculated, and this gives the two components $(\dot{u}_i^n, \; \dot{w}_i^n)$ of the acceleration vector by the line integral method.

The acceleration provides an increment in the velocities

$$\vec{u}_i^{n+1/2} = \vec{u}_i^{n-1/2} + \dot{\vec{u}}_i^n \; \Delta t^n$$

and this results in an advanced value for the coordinates

$$\vec{x}_i^{n+1} = \vec{x}_i^n + \vec{u}_i^{n+1/2} \; \Delta t^{n+1/2}$$

and thus a new value for the volume of each zone

$$S_{\ell+1/2}^{n+1} = \bar{r}_{\ell+1/2}^{n+1} \; A_{\ell+1/2}^{n+1} \; .$$

The zonal energy change is then computed based upon the work term which includes both p and q, using a time centered pressure, $p^{n+1/2}$. Time centering this term is important to conserve energy. For general equations of state it is done with one iteration involving the EOS and the energy equation.

Energy changes are mapped back to the node quantities by defining the spatial sum

$$(md\epsilon)_i^n = \sum_{\ell} (m_{\ell+1/2} \; \epsilon_{\ell+1/2} \; h_{\ell+1/2}^i / \bar{r}_{\ell+1/2})^n$$

Then

$$\epsilon_i^{n+1} = \epsilon_i^n + [(md\epsilon)_i^{n+1} - (md\epsilon)_i^n]/m_i$$

Care must be taken in updating the nodal density. If it is done with a strict volume change, or by a simple mapping from zone to node, then the density may change due to mesh optimization. This is undesirable because such changes will cause pressure changes which will cause mesh motion. We want the density to change only due to fluid motions and this is done with the notion of pseudo volumes.

Let $m_i$ and $S_i^n$ be the nodal mass, and pseudo-volume of node i at the beginning of a cycle. At problem generation time $J_i^n = S_i^n$.

First we reconnect the mesh and then we compute $J_i^n$, the actual volume, after mesh reconnection but before hydro. After hydro, the actual volume is $J_i^{n+1}$ and the pseudo volume is

$$S_i^{n+1} = S_i^n + J_i^{n+1} - J_i^n$$

The new density is $\rho_i^{n+1} = \rho_i^n S_i^n/S_i^{n+1}$. With this approach the mass of a node remains constant and the density changes only due to fluid motions.


## 5. Staggered Mesh Compressible Formulation

In this formulation mesh elements are a mixture of quads and triangles and the prognostic variables are staggered in space and time as suggested by von Neumann and Richtmyer [2]. The variables are

$$\vec{x}_j^n, \; \dot{\vec{u}}_j^n, \; \vec{u}_j^{n+1/2}, \; \rho_z^n, \; \epsilon_z^n, \; EOS_z^n, \; Q_z^{n+1/2}$$

where a subscript z designates a zonal quantity.

The calculation is advanced in time as follows:

1. Mesh optimization

2. $Q^{n-1/2} (\rho^n, \Delta u^{n-1/2})$

3. $\dot{\vec{u}}_j^n$

3a. Apply boundary conditions

4. $\vec{u}_j^{n+1/2}$

5. $x_j^{n+1} \rightarrow S_z^{n+1}$

6. $\rho^{n+1} = \rho^n S^n/S^{n+1}$

7a. $\quad \delta\tilde{\varepsilon}^{n+1} = \varepsilon(p^n,\ Q^{n-1/2},\ u^{n+1/2})$

7b. $\quad \tilde{p}^{n+1} = p(\tilde{\varepsilon}^{n+1},\ \rho^{n+1})$

7c. $\quad \delta\varepsilon^{n+1} = \varepsilon(p^n,\ \tilde{p}^{n+1},\ Q^{n-1/2},\ u^{n+1/2})$

8. $\quad \varepsilon^{n+1} = \varepsilon^n + \delta\varepsilon^{n+1}$

$\Delta t$

We start with a mesh that consists primarily of quads. This has the positive aspect of reducing the number of zones by a factor of two and thus of reducing memory requirements and of speeding up the calculation by a factor near two.

Experimentally we have found that in calculating the growth of Meshkov-Richtmyer instabilities, triangular zones tend to stiffen the mesh and to reduce the growth rate significantly. On the other hand, quads can tangle with resulting bowties and boomerangs which may result in negative volumes.

As a problem runs and the fluid distorts, mesh optimization tends to introduce triangular zones as will be discussed in section 6.

## 6. Mesh Reconnection and Optimization

With this Free-Lagrange algorithm, the nodes are reconnected subject only to a few constraints. Mainly we require that the area of each zone be positive and that the connectivity be symmetric. The first concern is that any small area of the physical domain of the problem be covered uniquely by zones. The symmetry concern helps to conserve momentum.

The mesh may be reconnected in two modes: (1) by an interactive graphics code that operates on a restart dump and attempts to reconnect as directed by a human; (2) automatically by the main code at execution time. Comments on the first mode of operation will be found in section 7. The second mode requires a predetermined strategy that is determined by a set of user controlled parameters. We develop this strategy and the options it employs through experimenting with both modes 1 and 2. Each new problem brings with it new challenges for the optimization strategy.

We learn how to run complex problems by a trial and error procedure - by trying out new and different mesh optimization options for example. Older options are rarely discarded in favor of new options - they are reprioritized relative to the new options in terms of when and if they are used. The prioritized collection of mesh reconnection options is called an optimization strategy. Thus the number of mesh optimization options increases with age as does the complexity of the mesh optimization strategy. This growth permits more and more complex fluid motions to be treated automatically by the code as time passes. The analogy with a growing child is impossible to avoid.

Different mesh situations may call for different options and the user can influence the optimization priorities with a set of parameters.

Mesh optimization involves reconnecting nodes together in order to enhance the calculation in some sense. In the first and second versions of the code the primary strategy had to do with connecting nearest neighbors together. In the third version we are more concerned with keeping $\Delta t$ up. In some sense we've moved from a global to a local strategy.

In this algorithm, mesh optimization or reconnection takes place for three reasons: (1) we try to prevent the mesh from getting into unsuitable topologies such as bowties and boomerangs (a non-problem with triangular mesh elements); (2) when $\Delta t$ falls below a user defined threshold, we change the mesh locally to improve $\Delta t$; (3) mesh reconnection takes place along material interfaces when one material wants to slide along another.

In terms of optimizing $\Delta t$ and accommodating shearing motions this manifestation of the Free-Lagrange method may be classified as an adaptive mesh scheme.

A simple example of mesh optimization is based on the notion of nearest neighbors. We want each node to be coupled only to those nodes close to it because in some sense this improves the accuracy. Assume all mesh elements are triangles and that a mesh exists. Figure 3a shows the relation between mesh elements $ik\ell$ and $ijk$ - they have a common side $ik$ which is a diagonal of quadrilateral $ijk\ell$. Their connections to the rest of the mesh are not shown. Since every quad has two diagonals it is possible to do a local reconnection by breaking connection $ik$ and making connection $j\ell$ as shown in Fig. 3b. That is, we disconnect one diagonal and connect the other. Nodes $i$ and $k$ each lose a neighbor while nodes $j$ and $\ell$ each gain a neighbor. The number of nodes and zones are conserved. Both new zones are allowed since their areas are positive. This reconnection is called a "FLIP".

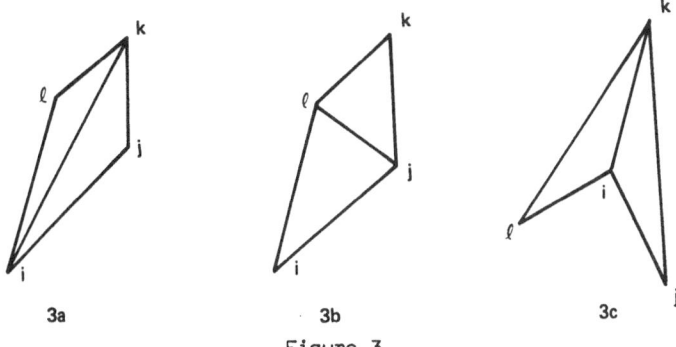

3a                          3b                          3c

Figure 3

Considering Fig. 3c, we see that a constraint on the FLIP option is necessary. Distance $j\ell$ is less than distance $ik$ but a FLIP would result in zone $ij\ell$ having a negative area. The FLIP could possibly still take place if the next layer of neighbors were considered, but the loss of simplicity for a small gain makes this extension unattractive.  Other constraints can also be applied - for example, we want the area of the smallest new zone to be greater than some fraction of the sum of the two areas (typically 20%).

The FLIP operation is local but the notion of a nearest neighbor mesh is global.  We construct a nearest neighbor mesh by sweeping over all mesh connections and considering a FLIP for each one.  This is an iterative process and we continue making mesh passes until we find that no flips have occurred.  We then have a nearest neighbor mesh.

The FLIP option can be used to optimize the mesh to states other than nearest neighbor.  For example we might want to generate a mesh in which the variations in zone area is minimized and so flips would make/break connections so that the difference in the two new areas was less than the difference in the two old areas. Equilateral triangles minimize truncation errors and so FLIP could be made to minimize the angle variation in the new zones.

With a triangular mesh, the FLIP option was quite simple as illustrated at the top of Fig. 4.  The introduction of quads means more flexibility in maximizing Δt with FLIP as is seen in the remainder of Fig. 4.

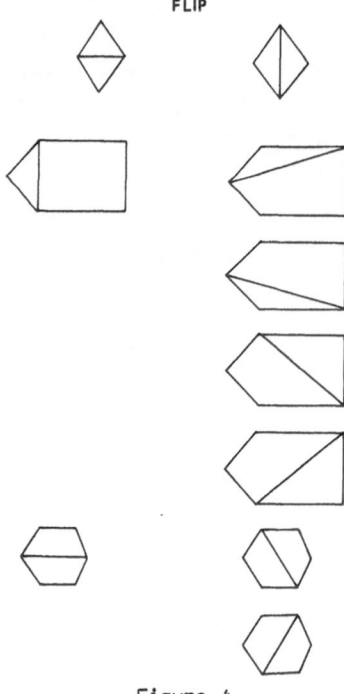

Figure 4

The FLIP option may be considered a tool and the mesh loop inside an iteration may be considered a strategy for accomplishing some goal (creating almost equilateral triangles). Another strategy for using FLIP comes into play when we have large relative fluid motions (shearing motions for example).

When one fluid moves relative to another along a material interface, the mesh is reconnected by flips along the interface. This SLIP-FLIP operation is triggered when an angle (with its vertex node on the interface) exceeds 95°. Figure 5A shows a slip motion with SLIP-FLIP inhibited and Fig. 5B shows the same motion with SLIP-FLIP operational. In Fig. 5 the left material has moved upwards since the initial time when both fluids were at the same height.

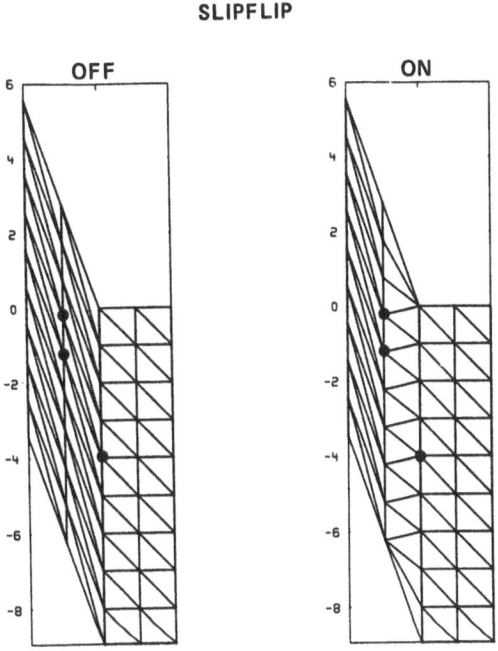

**SLIPFLIP**

Figure 5

Compressible hydro codes using an explicit time centering scheme have a $\Delta t$ limitation based on the Courant condition,

$$\Delta t < \min\ (L/C)$$

where C is the sound speed and L is the altitude of the triangle. As zones compress and distort L may decrease and C may increase to the point where economics dictates that the problem cannot continue due to a small $\Delta t$. At this point the problem dies or we invent ways to increase $\Delta t$ by increasing L (C is not at our disposal). The primary tools for increasing $\Delta t$ are FLIP, MERGE, NYM and ZAZ.

The MERGE option is used to merge two nodes of a triangle together when $\Delta t$ in that zone falls below a user defined threshold as seen in Fig. 6. If nodes A and B are both interior nodes, the MERGE action destroys two zones and one node. A new

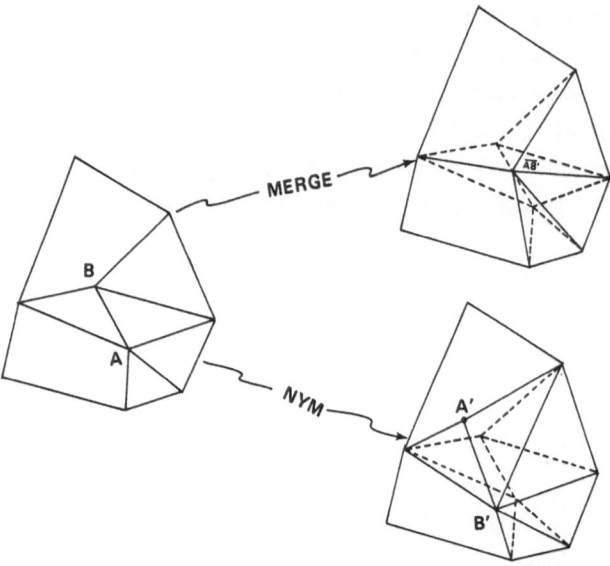

Figure 6

option, NYM (after Peter Roberts) tries to improve Δt by moving the nodes away from each other. The idea here is to increase Δt by increasing the critical length in the zone controlling Δt. Zones are not destroyed and it appears that NYM is an improvement over MERGE. As with other mesh optimization actions the price to be paid for NYM is a small amount of local diffusion because of the necessary overlay calculation that conserves mass and energy. The current strategy is to try NYM first and if it fails (e.g., it can't improve Δt enough) then to resort to MERGE.

The ZAZ option (Fig. 7) destroys a zone by moving the node opposite the longest edge over to the longest edge. This destroys one zone, but creates a new one by subdividing the zone on the opposite side of the longest edge. Since ZAZ subdivides a zone it may in fact cause Δt to decrease. In the overall Δt optimization strategy, it is therefore used only when all other tools have failed. In practice it may lower Δt, but it also loosens up the mesh structure so that MERGE or another option may fix the problem the following cycle.

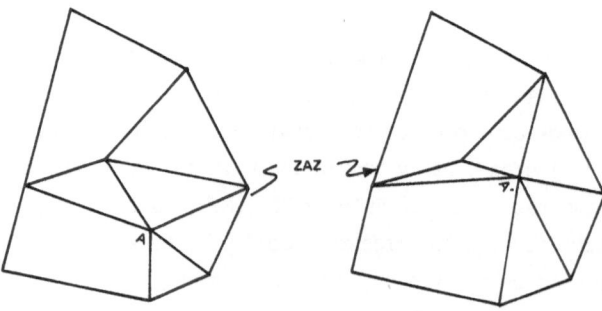

Figure 7

If a quad turns into a boomerang it is split into two triangles (Fig. 8). Similarly, if we project that a quad will bowtie in 3 cycles the quad is split into two triangles. We try to avoid SPLIT by doing a NYM operation first. If the NYM fails to fix the problem (possible because of geometric constraints on node motion) then we resort to SPLIT.

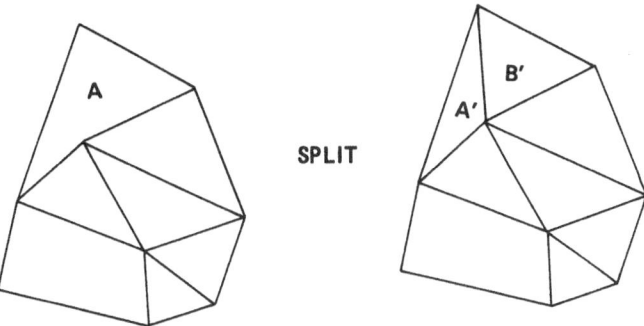

Figure 8

As a problem evolves, part of the mesh may become poorly resolved because nodes move away from each other - consider for example the growth of a perturbation on an unstable interface. The CREAP option (Fig. 9) subdivides zone edges (CREATE POINTS) when they become longer than a given value. Currently CREAP operates only on interfaces and boundaries.

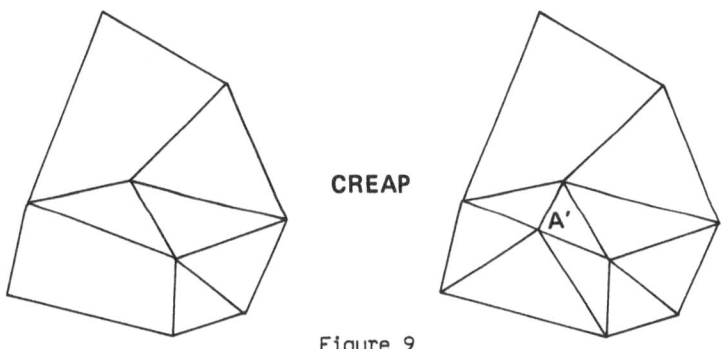

Figure 9

Two simplified examples of mesh strategy are shown in Figs. 10 and 11. In Fig. 10 we see a "prevent" strategy in which we are attempting to anticipate the occurrence of an undesirable situation (a bowtie here) and to prevent this situation from developing. In Fig. 11 we attempt to keep the integration time step up by first trying to NYM and then, if that fails, to merge zones whose current $\Delta t$ falls below user defined thresholds. The complete mesh optimization strategy including SLIP-FLIP checks, checks on geometry and interfaces and so on is much more complex than these simple examples.

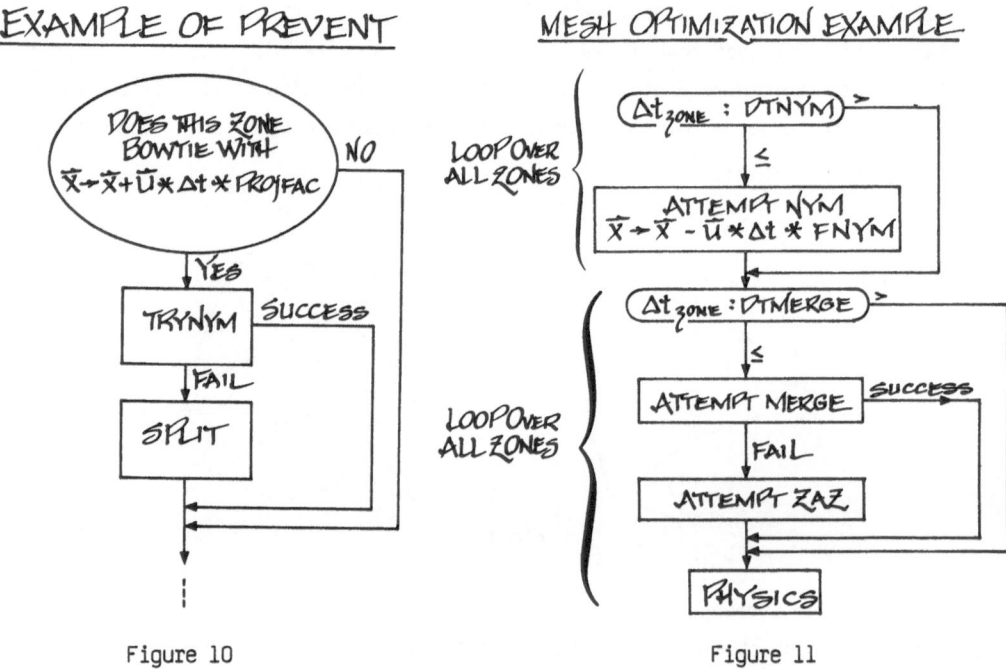

EXAMPLE OF PREVENT

DOES THIS ZONE
BOWTIE WITH
$\bar{x} \to \bar{x} + \bar{u} * \Delta t * PROJFAC$

NO

YES

TRYNYM | SUCCESS

FAIL

SPLIT

Figure 10

MESH OPTIMIZATION EXAMPLE

LOOP OVER
ALL ZONES

$\Delta t_{zone} : DTNYM$

$\leq$

ATTEMPT NYM
$\bar{x} \to \bar{x} - \bar{u} * \Delta t * FNYM$

LOOP OVER
ALL ZONES

$\Delta t_{zone} : DTMERGE$

$\leq$

ATTEMPT MERGE | SUCCESS

FAIL

ATTEMPT ZAZ

PHYSICS

Figure 11

## 7.  Interactive Graphics (IG)

As a problem runs it changes its shape and it is of some interest to the user to study this shape change.  We can specify ahead of time that the hydro code produce graphics output as it runs but many times we do not know ahead of time what we want to see.  As the problem runs we therefore leave a trail of restart dumps and we can use IG on these and get snapshots of any part of the problem.

Another important aspect of IG is that it allows humans to reconnect the mesh. Sometimes during the evolution of a problem, the mesh gets into a configuration that cannot be improved by automatic reconnections.  This happens because the reconnection algorithm is not smart enough or because the wrong optimization strategy has been chosen by the user.  In these situations, the user can fix the problem with IG because the human has a more global view of the mesh than does the reconnection algorithm.  With more global information the human can cause mesh reconnections to happen that the hydro code has not been taught to consider.  Sometimes this leads to an improvement in the hydro code; usually it results in a broken problem being fixed and being able to continue.

In the early development stages of the hydro code, the reconnection algorithm had very little intelligence.  By using IG on a number of different problems, humans were able over a period of time to figure out how best to reconnect the mesh and were able to teach much of this to the hydro code in terms of expanding the mesh reconnection algorithm.

## 8. Results from Meshkov-Richtmyer Instability Growth

An example of instability growth will illustrate the staggered mesh scheme. In Fig. 12 we see the evolution of a perturbation on an interface that is accelerated by several shocks. In this Meshkov-Richtmyer situation the interface is unstable to shocks in both directions. The jump in perturbation velocity is proportional to the jump in fluid velocity due to the shock.

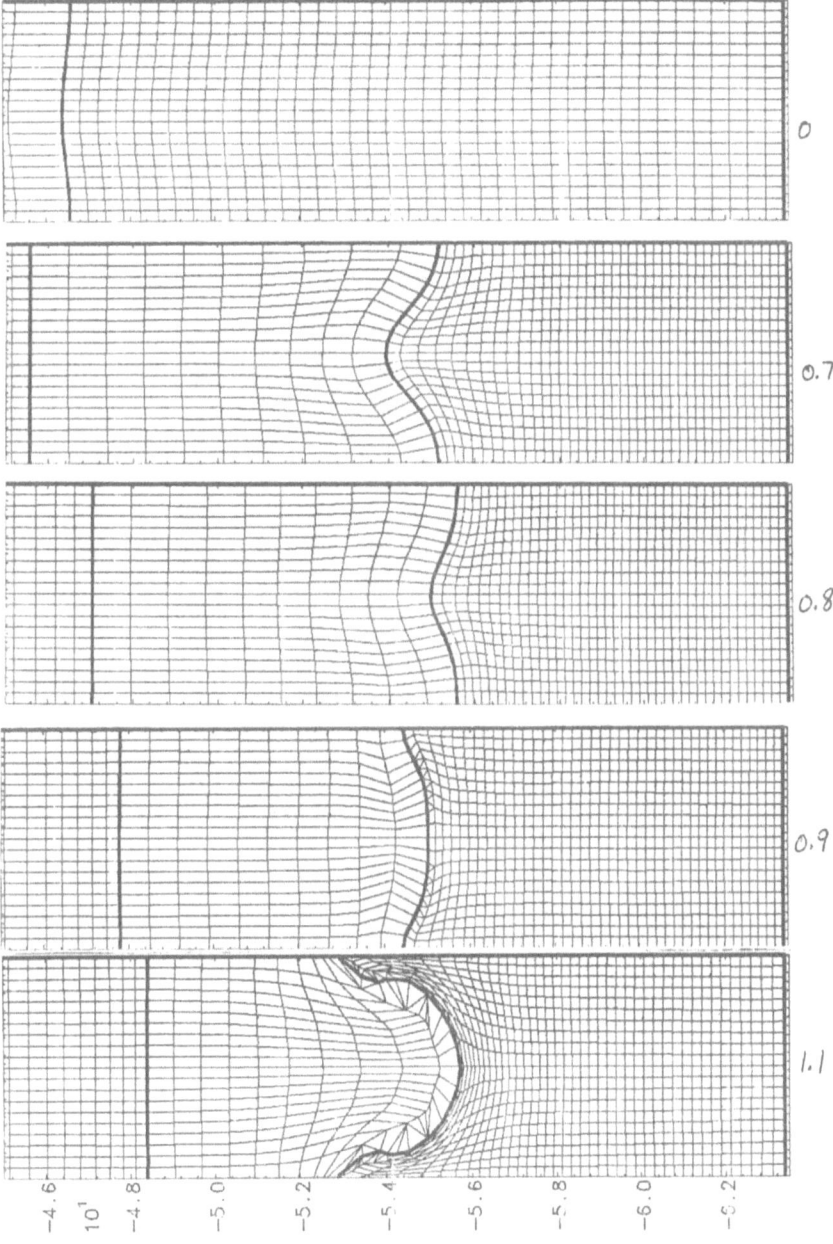

Figure 12

The initial shock travels from left to right and the interface separates light (left) from heavy (right) fluids. Only part of the total mesh is shown. The initial shock reflects off the right edge of the problem and reaches the interface about t = 0.7. At this time the reflected shock passes through the interface. Since this shock is from heavy to light, the perturbation goes through a phase change while continuing to grow.

Figure 13 shows the mesh around the perturbation at the final time. As the perturbation grows, resolution is lost along the interface. The CREAP option is being used to extend the calculation to later times.

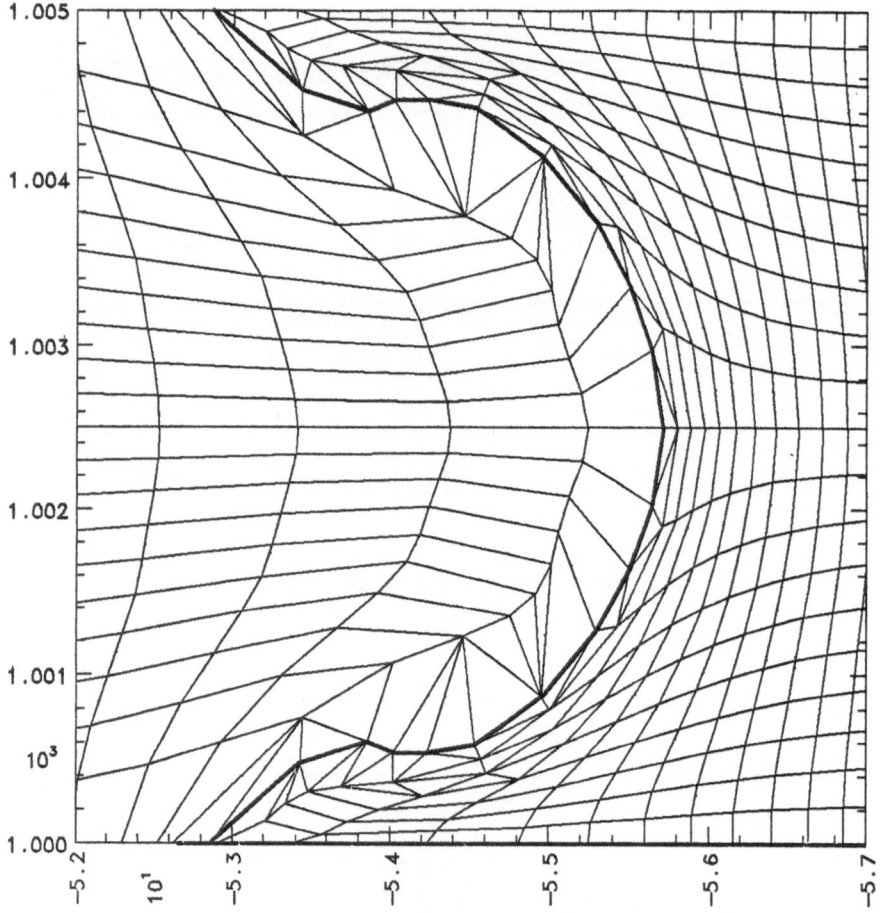

Figure 13

References

1.  Crowley, W. P. "Flag:  A Free-Lagrange Method for Numerically Similating
    Hydrodynamic Flows in Two Dimensions", Proceedings of the Second International
    Conference on Numberical Methods in Fluid Dynamics, Springer-Verlag, pp. 37-43,
    (1970).

2.  von Neumann, J. and Richtmyer, R. D., "A Method for the Numerical Calculation of
    Hydrodynamic Shocks:, J. App. Phys., vol. 21, pp. 232-257 (1950).

3.  White, J. W. "A New Form of Artificial Viscosity for Elastic Solids", J. Comp.
    Phys., vol. 16, pp. 119-126 (1974).

4.  Noh, W. F., "CEL:  A Time Dependent, Two-Space-Dimensional, Coupled
    Eulerian-Lagrange Code", Methods in Computational Physics, vol. 3, pp. 117-179
    (1963).

5.  Pomraning, J., Radiation Hydrodynamics, Pergamon Press, p. 232 (1973).

6.  Maenchen, G. and Sack, S. "The Tensor Code", Methods in Computational Physics,
    vol. 3, pp. 181-210 (1963).

7.  Margolin, L. G. and Nichols, B. D. "Momentum Control Volumes for Finite
    Difference Codes", Proceedings:  Third International Conference on Numerical
    methods in Laminar and Turbulent Flow, 1983.  Also, LANL manuscript LA-UR-83-524.

Acknowledgment

This project has profited over the years by the efforts of a number of people.  I
would especially like to thank T. Rudy, M. Uyemura, R. Williams, R. Cooper,
D. Braddy, and M. Johnson for their invaluable aid and assistance.

This work was performed under the auspices of the U.S. Department of Energy by
the Lawrence Livermore National Laboratory under contract No. W-7405-Eng-48.

# FINITE DIFFERENCE OPERATORS ON UNSTRUCTURED TRIANGULAR MESHES

G. Erlebacher
NASA Langley Research Center
Hampton, Va 23665

The solution to partial differential equations (PDE's) on increasingly complicated two-dimensional geometries has for a long time held the interest of the computational fluid dynamics community. One of the major difficulties encountered is the discretization of the PDE's on the grid underlying the physical domain in such a way that the resulting numerical scheme is stable and has good convergence properties. To this end, much effort is currently being devoted to constructing grids which simplify the discretization process, without sacrificing their ability to cover general domains and without compromising the convergence and stability of the resulting numerical schemes. These considerations have led to research on zonal gridding[1], global cartesian gridding,[2] and embedded grids[3] just to name a few. These techniques are generally applied to curvilinear grids for which a vast store of numerical knowledge has been accumulated over the last decade.

An alternative to one or more curvilinear meshes is to triangulate the physical domain. This approach has the advantage that any two-dimensional geometrical configuration, however complicated, can be gridded with well-behaved, nearly equilateral triangles[4]. Both structured and unstructured triangular grids can be used. They are distinguished by the fact that whereas an interior node on the former is always common to six edges, there is no such restriction on the latter. Of course, structured triangular grids retain some of the regular features of curvilinear grids.

Once the physical domain is triangulated, one must derive the discrete form of the PDE's, or of the individual operators. Finite-volume techniques are ideally suited to discretize hyperbolic systems of PDE's (notably the Euler equations) conservatively[5]. The method is not limited to triangles or quadrilaterals, which leaves the possibi-

lity open for grids with mixed cell types. Integrated forms of the
conservation laws are evaluated over a triangle area or an appropriate
cell area surrounding a triangle node (depending on whether the
variables are evaluated at triangle centroids or at the nodes).
Conservation of energy and mass impose additional constraints on the
difference formulas[6].

An alternative to treating integral conservation laws is to
discretize each operator of the differential form of the equations
separately. This can also be accomplished with the help of finite-
volume techniques, or by using a variational principle. For example,
the latter method can be used to effectively generate discrete forms
of the Poisson operator[4]. One of its chief assets is that only operators of lower order than the original must be discretized, which is
clearly a simpler task.

In this work, a new form of the Laplace operator is derived which
is shown to be second-order accurate when calculated at triangle nodes
if the nodes have the property that their corresponding cells possess
a high degree of symmetry. On the other hand, it is shown that the
Laplace operator directly derived from Stokes' integral theorem is
only zeroth-order accurate on these same cells. The reason the zero
order approximation has been successfully used in the past is that the
error most probably arises from the discretization of the cell area,
which often multiplies the equation that is being solved, thereby
cancelling the error.

The other major topic of this paper is a discussion of the construction of gradient and Laplace operators on unstructured triangular
grids, mainly from the variational point of view. Two discrete representations of the Laplacian are compared to each other, and conclusions are drawn regarding their accuracy on a pointwise basis.
Finally, geometric relations valid on arbitrary polygons are given for
completeness in the appendix. These were derived in the course of a
truncation error analysis.

TRIANGULAR GRIDS

Two-dimensional triangular grids have been used for a number of
years, mostly within the context of finite-element theory[6-8], because
the formulas they lead to are simpler and easier to handle than the
ones derived on curvilinear quadrilaterals. They allow the treatment
of many different types of geometries with equal ease. On the other
hand, the corresponding triangular gridded finite-difference tech-

niques are still quite new, and as yet, not well documented. The purpose of this paper is to partially remedy this situation by presenting several methods for deriving difference operators on 2-D triangular grids.

The paper begins with a review of the notation used, followed by a derivation of a series of relationships on rectilinear, non-orthogonal coordinate systems. These operators are then compared with the corresponding finite-difference operators on triangular grids. This comparison provides a basis for estimation of maximum bounds on truncation errors.

## Triangles

A triplet of non-colinear points in the plane defines a triangle. The coordinates of the 3 vertices are referenced by their x,y coordinates in a cartesian system x,y,z where the unit vector $\hat{z}$ is normal to the x,y plane. These points (also called nodes, or vertices) are referred to by radius vectors subscripted with lowercase letters i,j,k, numbered counterclockwise as one moves around the triangle's perimeter. The difference between two such vectors, $(\vec{r}_j - \vec{r}_i)$ is denoted by $\vec{r}_{i,j}$ . Triangles are indicated by capitalized superscripts. When summing over the vertices of a triangle B, the symbol $\sum\limits_{i(B)}$ is used. Unit vectors are recognized by the presence of a carat. Consistent with this notation, the outward unit normal vector to a triangle edge i,i+1 is

$$\hat{n}_{i,i+1} = \frac{\vec{r}_{i,i+1}}{r_{i,i+1}} \times \hat{z} = \hat{r}_{i,i+1} \times \hat{z}$$

where $r_{i,i+1}$ is the length of $\vec{r}_{i,i+1}$ . The normal vector is outward because of the counterclockwise vertex ordering and positive z direction. Occasionally $r_{i+1/2}$ is substituted for $r_{i,i+1}$ when the meaning is clear from the context. Otherwise, the original convention is assumed. Each node is connected to a set of nodes which form a corresponding polygon. Its nodes are numbered in a counterclockwise order (fig. 1). The center node is generally labelled c, and for simplicity, expressions such as $f_i - f_c$ are replaced by $F_i$. Here, $f_i$ is the value of the analytical function f (typically $c^2$) to which the operators are applied.

Cells

Fundamental to the construction of discrete operators on unstructured grids is the concept of a cell which satisfies the following
three conditions:

1. Each grid point has a cell associated with it.

2. A cell contains one and only one grid point.

3. An arbitrary point in the computational domain is either included in a single cell, or lies on a cell boundary.

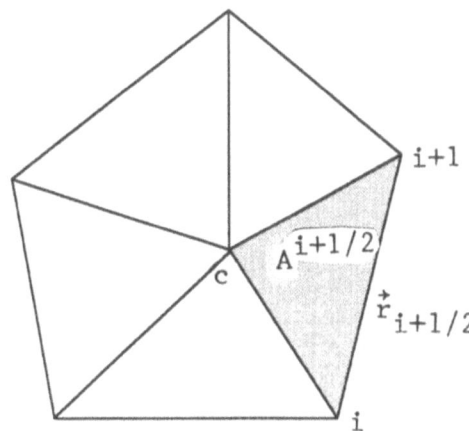

Fig. 1: Polygon surrounding node c

The third condition immediately implies that the domain of intersection between two cells is at most an edge, which guarantees that
cells do not overlap. Thus, the area of the computational domain is
just the sum of the cell areas. Moreover, since each cell is associated with a unique (interior or boundary) mesh point, it follows
that the sum can be indexed by the mesh points or simply by the
corresponding mesh point index. The computational domain can also be
interpreted as a fluid modeled by a finite number of particles
(lattice points) of finite volume (cell areas).

Consider now the construction of a cell from a given triangular
lattice. Two approaches are commonly referred to: the Voronoi cell,

and the median based cell. The Voronoi cell (fig. 2) associated with a node c is by definition the collection of points that lie closer to c than to any other node[9]. By virtue of its definition, it is evident that this construction is unique. The major advantage of this type of cell is that its volume varies continuously as the mesh nodes move across the physical domain. Furthermore, they are inherently convex, which guarantees that they always contain their centroid. This can have an impact on certain adaptive techniques[10]. The associated triangular mesh is called the Delaunay mesh. An alternative to the Voronoi cells are the median-based cells which are used throughout this text. The construction of such a cell is made clear in fig. 3. Medians to the triangles that have c as one of their nodes,

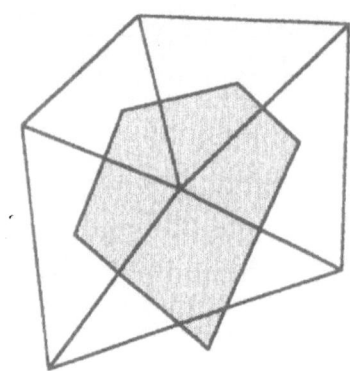

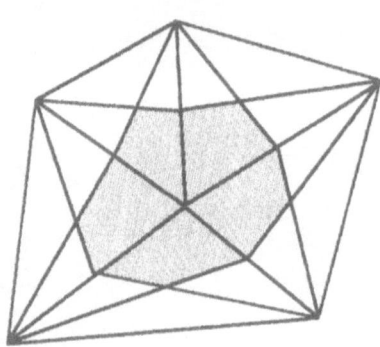

Fig. 2: Voronoi cell                    Fig. 3: Median-based cell

define the perimeter of the cell associated with c. It is also easily shown that the cell area is given by

$$A_c = \frac{1}{3} \sum_{B(c)} A^B \qquad (2)$$

The cells just defined interlock like a jigsaw puzzle and can partition every bounded planar region.

## Non-Orthogonal Rectilinear Coordinate Systems

A cartesian coordinate system is not always the most convenient system in which to work. For example, in order to estimate truncation errors of operators on triangular meshes, Taylor expansions of their finite-difference representations are required. On a cartesian grid, an expansion in the x,y directions would explicitly involve the various angles in the triangle. However, simpler formulas result if

the expansion is done along two preferred directions. From symmetry
considerations, the most natural point about which to expand is the
triangle's centroid. The independent variables used are, therefore,
chosen to be the distances along 2 of the triangle's 3 medians. In
general, consider a positive oriented coordinate system $\hat{u}_1$, $\hat{u}_2$,
along two arbitrary directions such that $\hat{u}_1 \cdot \hat{u}_2 = 0$ (fig. 4). It is
desirable to derive an analytical expression for the gradient operator
in these coordinates. Since the gradient operator is two-dimensional,
it can be expressed as a linear combination of the independent unit
operators $\partial_1$, $\partial_2$, where the partial derivatives are respectively
taken along the $u_1$ and $u_2$ axes. A simple calculation leads to the
formula

$$\nabla = \hat{z} \times (-\hat{u}_1 \partial_2 + \hat{u}_2 \partial_1) / \sin\theta \tag{3}$$

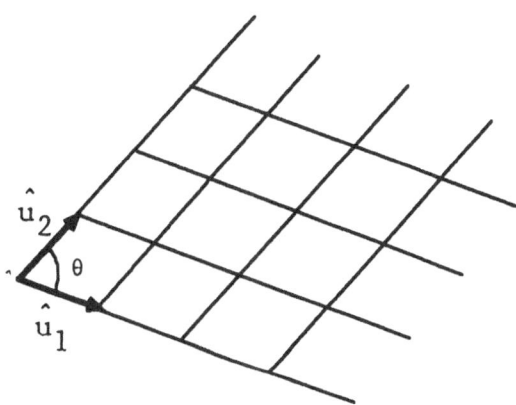

Fig. 4. Non-orthogonal rectilinear basis

Since the angle between coordinate lines is constant over the entire
domain, the Laplace operator is simply the scalar product of the gradient operator with itself, which immediately yields the formula

$$\nabla^2 = \frac{1}{\sin^2\theta} (\partial_{11} - 2 \cos\theta \, \partial_{12} + \partial_{22}) \tag{4}$$

This clearly reduces to the standard form on a cartesian grid in the
limit $\theta = \frac{\pi}{2}$.

The next step is to derive an exact formula for the gradient operator at a cell center c (fig. 5). The expression above is obviously still exact if one works in a basis defined along any two directions, but we wish to take all the neighboring vertices of c into account. To do so, equation (3) is written for all consecutive pairs of nodes $(1,2),(2,3),(3,4),\ldots(n-1,n)$:

$$(\nabla_E)_c^{i+1/2} = \hat{z} \times (-\hat{r}_{c,i}\ \partial_{i+1} + \hat{r}_{c,i+1}\ \partial_i) / \sin\theta_{i+1/2} \tag{5}$$

where $(\nabla_E)_c^{i+1/2}$ is the gradient calculated at c, in the rectilinear

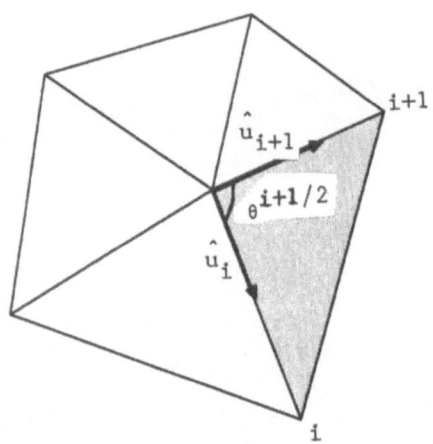

Fig. 5. Non-orthogonal basis superimposed on
a triangular mesh

basis $\vec{r}_{c,i},\ \vec{r}_{c,i+1}$ . The gradient operator can be written as a weighted sum of these functions if the weighting coefficients sum up to unity. The coefficients are chosen to be proportional to the triangle areas, which leads to

$$\nabla = \frac{1}{6}\ A_c\ \hat{z} \times \sum_{i(c)} \vec{r}^i\ (\vec{r}_{c,i} \cdot \nabla) \tag{6}$$

where the notation $\vec{r}^i = \vec{r}_{i-1,i+1}$ has been adopted. The expression above is consistent with the intuitive requirement that triangles with zero area should not contribute to the gradient. Note that other choices for the coefficients are possible. For example, Crowley[11] reports successful computations when the coefficients are proportional to the product of the triangle area and the angle $\theta_{i+1/2}$ .

## Numerical operators

When a computational mesh has an irregular structure, it has been found that discrete approximations to a number of differential opera-tors can be obtained from integral relations such as Stokes' and Gauss' theorems. The gradient and divergence operators are respecti-vely determined by

$$\int_{\sigma} \nabla f \ da = \oint_{\partial\sigma} f \ \hat{n} \ dr \tag{7}$$

$$\int_{\sigma} \nabla.\vec{v} \ da = \oint_{\partial\sigma} \vec{v}.\hat{n} \ dl \tag{8}$$

where the notation is clear from fig. 6a.

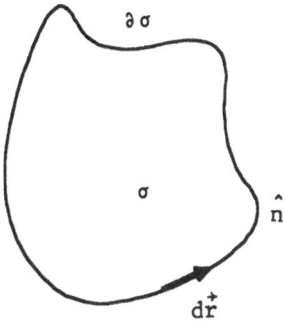

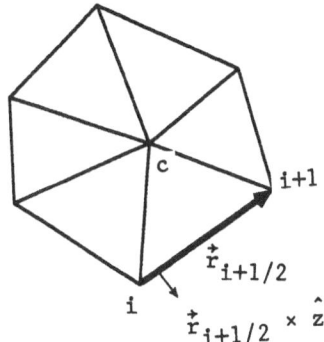

Fig. 6a.  Domain $\sigma$ with $c^1$
          boundary $\partial\sigma$ .

Fig. 6b: Polygonal boundary.

From the mean-value theorem, there exists points $p_1$, $p_2$, in $\sigma$ such that

$$\int_{\sigma} \nabla f \ da = A_{\sigma} \ \nabla f_{p_1} \tag{9}$$

$$\int_{\sigma} \nabla.\vec{v} \ da = A_{\sigma} \ \nabla.\vec{v}_{p_2} \tag{10}$$

where $A_{\sigma}$ is the area of the domain $\sigma$ , and $\nabla f_p$ is the analytical gradient calculated at the point p.  Since these relations are valid on all simply-connected domains, they must also be satisfied on domains with polygonal boundaries.  To derive finite-difference approximations to (9-10), the points $p_i$ are identified with the cell center c, and the integrals on the left-hand side are approximated with a box integration scheme. Equations (9-10) respectively become

$$\nabla f_c = \frac{1}{3A_c} \sum_{i(c)} f_{i,i+1} \hat{n}_{i,i+1} r_{i,i+1} \qquad (11)$$

$$\nabla \cdot \vec{v}_c = \frac{1}{3A_c} \sum_{i(c)} \vec{v}_{i,i+1} \cdot \hat{n}_{i,i+1} r_{i,i+1} \qquad (12)$$

which can be seen from figure 6b. First order operators can also be defined at the barycenter (not a grid point) of a triangle if $3A_c$ is replaced by the triangle area $A^B$. A knowledge of a variable U at the barycenters of a cell's triangles enables it to be calculated at the cell's center from the area weighted average

$$U_c = \frac{1}{3A_c} \sum_{B(c)} A^B U^B \qquad (13)$$

For example, with $U = \nabla f$ , an estimate for $\nabla f$ at the cell center is given by

$$3 A_c \nabla f_c = \sum_{B(c)} A^B (\nabla f)^B \qquad (14)$$

$$= \sum_{i(c)} f_{i+1/2} \vec{r}_{i+1/2} + f_{i+1} \vec{r}_{i+1,c} + f_{c,i} \vec{r}_{c,i}) \times \hat{z} \qquad (15)$$

$$= \sum_{i(c)} f_{i+1/2} \vec{r}_{i+1/2} \times \hat{z} \quad . \qquad (16)$$

Expressions (16) and (11) are identical, which justifies the use of (14) as an interpolation formula to derive a discrete representation of the Laplace operator. Other interpolation schemes have also been considered[11] where $A^B/3A_c$ is replaced by $A^B \alpha^B/(6A_c \pi)$ . Here, $\alpha^B$ is the angle centered at c within triangle B. These are not pursued further in this paper.

A rather simple form of the gradient is obtained if $\vec{r}^i = \vec{r}_{i-1,i+1}$ is substituted for $\vec{r}_{i-1,i+1}$ and when $f_{i,i+1}$ is expanded in terms of $f_i$ and $f_{i+1}$. After simplifications, The gradient becomes

$$\nabla f_c = \frac{1}{6A_c} \sum_{i(c)} f_i \vec{r}^i \times \hat{z} \quad . \qquad (17)$$

In the same manner, vectors are defined at a triangle's barycenter by the arithmetic average of its value at the 3 triangle vertices, which allows the gradient operator at c to be expressed, not as a function of the variable evaluated at the vertices, but as a function of its value at the barycenters. Starting from (16), $f_{i+1/2}$ is expanded to obtain

$$3A_c \; \nabla f_c = -\frac{1}{2} \hat{z} \times \sum_{i(c)} (f_i + f_{i+1}) \; \vec{r}_{i+1/2} \tag{18}$$

which can be simplified by observing that

$$\sum_{i(c)} \vec{r}_{i+1/2} = 0 \tag{19}$$

around a closed contour. One can therefore add $f_c$ to the expression in parenthesis in (18) to obtain

$$\nabla f_c = -\frac{1}{2A_c} \hat{z} \times \sum_{i(c)} f^{i+1/2} \; \vec{r}_{i,i+1} \tag{20}$$

where $f^{i+1/2}$ is the value of $f$ at the centroid of triangle B, with vertices $i, i+1, c$. In terms of the outward normals $\hat{n}_{i+1/2}$, the cell centered gradient reduces to

$$\nabla f_c = \frac{1}{2A_c} \sum_{i(c)} f^{i+1/2} \; \hat{n}_{i+1/2} \; r_{i+1/2} \tag{21}$$

A similar expression can also be derived for the divergence operator by multiplying both sides of equation (20) by 3/2 and replacing $\vec{v}_{i,i+1}$ by

$$\vec{v}^{i+1/2} = \frac{1}{3} (\vec{v}_i + \vec{v}_{i+1} + \vec{v}_c)$$

Equations (11) and (21) are not exactly identical since (13) is only an approximation. Nevertheless, they do agree to first order. These same integral techniques are employed to derive a discrete representation for the Laplace operator:

$$\oint_\sigma \nabla^2 f \; da = \oint_{\partial\sigma} (\nabla f) \; \hat{n} \; dl \tag{22}$$

where $\hat{n} \; dl = d\vec{l} \times \hat{z}$ and with an assumed knowledge of the gradient operator. The domain of integration of (22) is the (hatched) cell area in fig. 3. The approximate formula

$$A_c \; \nabla^2 f_c = \frac{1}{2} \sum_{i(c)} (\nabla f)^{i+1/2} \; (\vec{r}_{i+1/2} \times \hat{z}) \tag{23}$$

is then immediately deduced from the mean value theorem. One then obtains for the Laplacian,

$$\nabla^2 f_c = - \frac{1}{4A_c A^{i+1/2}} \sum_{i(c)} \vec{r}_{i+1/2} \cdot [\vec{r}_{i+1,c} \, F_i + \vec{r}_{c,i} \, F_{i+1}] \tag{24}$$

where $F_i = f_i - f_c$. This formula is well known, but it is found to be only zeroth order accurate on most cell shapes. Nonetheless it is often used to solve elliptic problems when a stationary solution is sought. The error is primarily a result of the approximation of the line integral in (22). In the next section, a more accurate discrete representation of the Laplace operator is derived, based on a variational principle. The difference between the two formulations will be seen to originate from alternate partitionings of the physical domain. The generalization of (24) to the operator $\nabla \cdot (\sigma \nabla)$ is immediate:

$$\nabla \cdot (\sigma \nabla f)_c = - \frac{1}{4A_c A^{i+1/2}} \sum_{i(c)} \sigma^{i+1/2} \, r_{i,i+1} \cdot$$
$$(\vec{r}_{i+1,c} \, F_i + \vec{r}_{c,i} \, F_{i+1}) \tag{25}$$

where $\sigma^{i+1/2}$ is a spatial function evaluated at the barycenter of triangle $i+1/2$.


## Variational Principles

In the previous section, an expression for the discrete Laplace operator was derived with the help of integral theorems. Unfortunately, as will be shown later, it is only zeroth order accurate on many cell geometries. The objective here is to derive an improved, more accurate version of the discrete Laplace operator. Rather than resort to the methods presented in the previous section, variational methods are chosen for their flexibility and ease of use[12-13]. The method is mesh independent, invariant with respect to a change of variables, and relies on the discretization of an operator of lower order, which is one of its chief assets. Some authors find the variational approach a useful tool to develop finite-difference approximations to systems of equations which can be derived from a minimization principle. Discrete natural boundary conditions can also be deduced from this approach in a natural way. Even though some equations do not possess a minimization principle, it still makes sense to separate out the differential operators that do, and to derive some finite-difference formulas for them.

Formulation

A map $I(f)$ from the space of $C^2$ functions to the real line is called a functional. Given a set of functions that all satisfy identical boundary conditions, which one(s) make $I(f)$ stationary with respect to arbitrary perturbations of f? To answer this question, an arbitrary perturbation $\delta f$ is applied to f:

$$f \rightarrow f + \lambda \, \delta f \tag{26}$$

where $\lambda$ is a scalar parameter. $I(f)$ then becomes $I(f + \lambda \, \delta f)$, and the functions that satisfy

$$\left. \frac{dI(\lambda)}{d\lambda} \right|_{\lambda=0} = 0 \tag{27}$$

make the functional stationary. In what follows, only functionals of the form

$$I(f) = \int_\sigma L(f, \nabla f, \vec{r}) \, d\vec{r} \tag{28}$$

integrated over a fixed two-dimensional domain $\sigma$ are considered. All the concepts described below can be extended to the treatment of moving boundaries, natural boundary conditions, and different types of Lagrangian functions. The objective is to calculate which function(s) among those with identical boundary conditions on $\partial\sigma$ make $I(f)$ stationary with respect to perturbations of f. When written out, the derivative of $I(f)$ with respect to $\lambda$ becomes

$$\frac{dI(\lambda)}{d\lambda} = \frac{d}{d\lambda} \int_\sigma L(f + \lambda \, \delta f, \nabla f + \lambda \, \nabla \delta f, \vec{r}) \, d\vec{r} \quad, \tag{29}$$

and the first variation of I is simply

$$\delta I = \left. \frac{dI(\lambda)}{d\lambda} \right|_{\lambda=0} = \int_\sigma \left( \frac{\partial L}{\partial f} - \frac{d}{d\vec{r}} \frac{\partial L}{\partial \nabla f} \right) \delta f \, d\vec{r} \tag{30}$$

where the boundary terms drop out since $\delta f$ vanishes on $\partial\sigma$. The functions that make the functional stationary must therefore satisfy the Euler-Lagrange equations. For notational simplicity, let $\Lambda(f)$ denote the Euler-Lagrange operator acting on f. Equation (30) then becomes

$$\delta I = \int_\sigma \Lambda(f) \, \delta f \, d\vec{r} \tag{31}$$

The inherent flexibility of this approach lies in the fact that the accuracy of the discrete Euler-Lagrange operator, $\Lambda_h$ , is the result of two approximations: $L_h$ and $I_h$. The former is equivalent to the requirement that the calculation be based on a good representation of the lower order differential operator. This will be addressed in this section. On the other hand, the latter is directly affected by the integration scheme employed to approximate the functional. Of course, more accurate schemes will lead to more complicated formulas for $\Lambda_h$ . In the following, we confine ourselves to a box integration scheme. It is interesting to note that while $I(f)$ may be approximated on a mesh $M(f_i, A_i)$, its first variation can be evaluated on a second mesh $M'(f_j, A_j)$, after $L_h$ has been appropriately interpolated onto it. All the formulas remain invariant. This property is often used to express differential operators in various coordinate systems. Examples of mesh switching are given in the next section when the formalism is applied to triangular grids. The discrete representation of (31) is obtained as follows. First, the functional I is made discrete:

$$I_h(f) = \sum_{i(c)} L_{h_i}(f) A_i \quad . \tag{32}$$

Next, f is perturbed according to (29) and $\delta I_h$ is calculated on the mesh M:

$$\delta I_h = \sum_{i(c)} \frac{\partial L_{h_i}(f+\lambda\delta f)}{\partial \lambda}\Big|_{\lambda=0} A_i \tag{33}$$

$$= \sum_{j(c)} \left( \sum_{i(c)} \frac{\partial L_{h_i} A_i}{\partial f_j A_j} \right) \delta f_j A_j \tag{34}$$

where the derivatives are evaluated on the mesh M'. Comparing equations (31) and (34), $\Lambda_h(f)$: becomes

$$\Lambda_{h_j}(f) = \sum_{i(c)} \frac{\partial L_{h_i} A_i}{\partial f_j A_j} \tag{35}$$

at each nodal point j on the mesh M'. When these results are specialized to the functional

$$I(f) = -\frac{1}{2} \int_\sigma (\nabla f)^2 \, d\vec{r} \quad , \tag{36}$$

it is easy to verify that its first variation is

$$\delta I = \int_{\sigma} \nabla^2 f \, \delta f \, d\vec{r} \tag{37}$$

if $\delta f = 0$ on $\partial \sigma$. Equation (37) direcly leads, after the identification $\Lambda_{h_j} = \nabla_h^2 f_j$, to the discrete approximation

$$\nabla_h^2 f_i = - \frac{1}{2A_i} \sum_{j(i)} \frac{\partial (\nabla_h f)^2_i}{\partial f_j} A_j \tag{38}$$

where the sum extends over all nodes j which partake in the definition of $\nabla_h f_i$. As expected, the discrete representation of the Laplacian is solely dependent on a knowledge of the discrete gradient. By merely switching from one discrete gradient formulation to another, on various sub-meshes, a variational approach immediately results in a variety of discrete approximations. Two versions of the discrete Laplacian on arbitrary triangular meshes are now presented.

## Triangular Grids

A discrete representation of the Laplacian of a scalar f is now derived from two variational principles that differ from each other in the definition of the mesh M over which the gradient is approximated. The grid points of M' at which the Lapacian is evaluated are the same in both cases; it is defined by triangle vertices. The domain is then partitioned into median-based cells of area $A_c$.

## Version 1

A typical cell from the mesh M is illustrated as a hatched region in fig. 1. The nodes on the mesh M, defined by the triangle centroids, lead to a functional $I_h(f)$ of the form

$$I_h(f) = - \frac{1}{2} \sum_{B(\sigma)} (\nabla_h f^B)^2 A^B \tag{39}$$

which, after substitution of the discrete gradient operator (18) becomes

$$I_h(f) = - \frac{1}{2} \sum_{B(\sigma)} \frac{1}{A^B} \sum_{\substack{i(B) \\ j(B)}} (f_{i+1/2} \, f_{j+1/2} \, \vec{r}_{i+1/2}) \cdot \vec{r}_{j+1/2} \tag{40}$$

on the mesh M'. Rearranging the terms, and noting that

$$f_{i,i+1} = (f_i + f_{i+1})/2 \quad ,$$

it follows that

$$I_h(f) = -\frac{1}{8} \sum_{B(\sigma)} \frac{1}{A^B} f_i \, f_j \, \vec{r}^i \cdot \vec{r}^j \tag{41}$$

which is a quadratic expression in the $f_i f_k$. The coefficients of $\delta f_c$ in the first variation

$$\delta I_h(f) = \sum_{c(\sigma)} \frac{\partial I_h}{\partial f_c} \, \delta f_c \tag{42}$$

are identified with the right-hand side of equation (35) after substitution of $\nabla_h^2 f_j$ for $\Lambda_{h_j}$ , which leads to

$$\nabla_h^2 f_c = \frac{1}{A_c} \frac{\partial I_h}{\partial f_c} \tag{43}$$

for the approximate Laplace operator. The derivative of $I_h$ with respect to $f_c$ must now be evaluated. From (41)

$$\frac{\partial I_h}{\partial f_c} = \frac{1}{4} \sum_{B(\sigma)} \frac{\vec{r}_{c-1,c+1}}{A^B} \cdot \sum_{j(B)} f_j \, \vec{r}_{j-1,j+1} \tag{44}$$

where all superscripts refer to the triangle B with vertices c,i,i+1 (the last two lie on the polygon's boundary) and area $A^{i+1/2}$ (fig. 1). Expand the sum over the triangle's vertices in (44) yields

$$\nabla_h^2 f_c = -\frac{1}{4A_c} \sum_{B(c)} \frac{\vec{r}_{i+1/2}}{A^{i+1/2}} \cdot (f_c \vec{r}_{i+1/2} + f_i \, \vec{r}_{i+1,c} + f_{i+1} \vec{r}_{c,i}) \tag{45}$$

which is identical to the formulas derived from integral theorems in the previous section. Henceforth, this form of the Laplacian is referred to as FL.

In a later section, it will be shown that this representation of the Laplace operator is zeroth order accurate on most cell shapes except for those that tile planar domains in a regular fashion. Usually a scalar coefficient multiplies a first order accurate expression. This is probably due to the approximation of the cell area in the integration scheme. Previous authors, when solving time-independent elliptic problems involving the Laplace operator, missed this error, assuming that the expression was first order accurate.

Moreover, Fritts[7] has indicated that on average, the discrete algorithm is slightly less than second order accurate. A second formula, second-order accurate on well-behaved cells, is now derived.

## Version 2

The mesh M' is unchanged (triangle vertices), whereas the mesh M is now the union of the vertices of M' and the midpoints of all its edges as illustrated in fig. 7, which depicts a cell of mesh M as a hatched area identified with node k. Let k be an arbitrary vertex from M with a cell area of $A_{kM}$ where the suffix M is present as a reminder of the mesh currently in use. The discrete functional is

$$I_h(f) = -\frac{1}{2} \sum_{k(\sigma)} (\nabla f_k)^2 A_{kM} \qquad (46)$$

Expanding the gradients, the functional becomes

$$I_h(f) = -\frac{1}{2} \sum_{k(\sigma)} \frac{1}{9A_{kM}} \sum_{\substack{i(k) \\ j(k)}} f_{i,i+1} \, f_{j,j+1} \, \vec{r}_{i+1/2} \cdot \vec{r}_{j+1/2} \qquad (47)$$

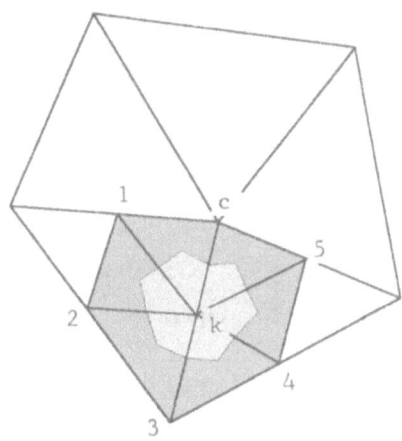

Fig. 7: Mesh M (node k) and mesh M' (node c)

Using the following relation true at all the interior nodes k,

$$\sum_{i(k)} f_{i,i+1} \vec{r}_{i,i+1} = \frac{1}{2} \sum_{i(k)} f_i \, \vec{r}^i \qquad (48)$$

to express (47) as a quadratic function of $f_i f_j$, results in the simplified expression

$$I_h(f) = -\frac{1}{72} \sum_k \frac{1}{A_{k^M}} \sum_{i,j} f_i \, f_j \, \vec{r}^{\,i} \cdot \vec{r}^{\,j} \tag{49}$$

formally identical to (41). Equation (49) is now differentiated with respect to $f_c$ to obtain

$$\frac{\partial I_h}{\partial f_c} = -\frac{1}{36} \sum_{k(c)} \frac{1}{A_{k^M}} \, \vec{r}^{\,c} \cdot \sum_{\substack{i(k)\\j(k)}} f_i \, \vec{r}^{\,i} \, f_j \, \vec{r}^{\,j} \tag{50}$$

where $\vec{r}^{\,c}$ is defined in figure 8. Note that the sum only includes the polygons k connected to node c, for these are the only ones which give a non-zero contribution to (50). All that remains to be done now is to interpolate all the variables in (50) onto the grid M'. For

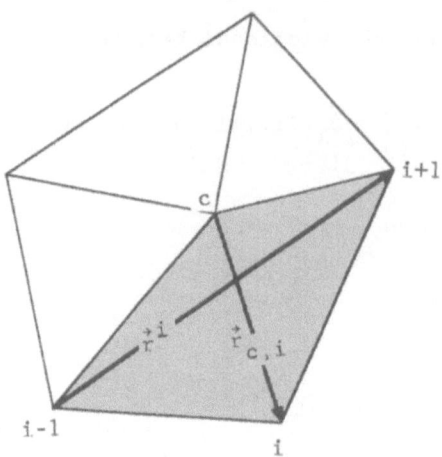

Fig. 8: Notation for equation (50)

maximum clarity, the sum over i is expanded, noting that there are always six terms as can be seen from figure 7:

$$f_i \vec{r}^{\,i} = f^c \vec{r}_{5,1} + f_1 \vec{r}_{c,2} + f_2 \vec{r}_{1,3} + f_3 \vec{r}_{2,4} + f_4 \vec{r}_{3,5} + f_5 \vec{r}_{4,1} \quad .$$

Every term in (51) on the fine mesh M is now interpolated onto the coarser mesh M' by assuming that the value of f at an edge midpoint is the average of its value at the extremities. Equation (51) becomes after some straightforward algebra

$$f_i \vec{r}^i = \frac{3}{4} \left[ (f_i - f_c) \vec{r}^i - (f_{i+1} - f_{i-1}) \vec{r}_{c,i} \right] \quad . \tag{52}$$

The area of the quadrilateral c,i-1,i,i+1 is denoted by $A^i$, (fig. 8) and $A_{kM}$ is expressed as a function of $A^i$:

$$A_{kM} = \frac{A^i}{4} \quad . \tag{53}$$

Substitution of (52) and (53) into (50) leads to

$$\frac{\partial I_h}{\partial f_c} = \frac{1}{24} \sum_{i(c)} \frac{\vec{r}^i}{A^i} \cdot \left[ (f_i - f_c) \vec{r}^i - (f_{i+1} - f_{i-1}) \vec{r}_{c,i} \right] \tag{54}$$

where $\vec{r}^c$ has been replaced by $-\frac{\vec{r}^i}{2}$ .

Finally equation (54), inserted into (43) for the Laplacian gives the desired result

$$\nabla^2_h f_c = \frac{1}{6A_c} \sum_{i(c)} \frac{\vec{r}^i}{A^i} \cdot \left[ (f_i - f_c) \vec{r}^i - (f_{i+1} - f_{i-1}) \vec{r}_{c,i} \right] \tag{55}$$

This form of the Laplacian is henceforth referred to as GL. In a later section, the accuracy of GL and FL will be compared to one another on various cell shapes, and it will be concluded that GL is often more accurate than FL on a pointwise basis.

The generalization of (55) to approximate $\nabla \cdot (\sigma \nabla f)$ is straightforward:

$$\nabla \cdot (\sigma \nabla f) = \frac{1}{6A_c} \sum_{i(c)} \frac{\sigma^i \vec{r}}{A^i} \cdot \left[ (f_i - f_c) \vec{r}_i - (f_{i+1} - f_{i-1}) \vec{r}_{c,i} \right] \tag{56}$$

where $\sigma^i = \frac{\sigma_i + \sigma_c}{2}$ .

Formulas for the second partial derivatives with respect to x and y may also be derived from appropriate functionals[4].

This section treated interior points only. If natural boundary conditions must be calculated, the procedure is identical. The function f is perturbed by an increment $\lambda \delta f$ among the set of all functions differentiable to second order. The first variation of the functional is calculated, this time including the boundary terms. It is then made discrete. At the same time, the functional itself (same as previously) is made discrete, and differentiated with respect to the nodal values of the function. The finite-difference formulas corresponding to the natural boundary conditions are then determined from equation (35).

## Cell Geometry

In this section, the explicit expressions for GL and FL are compared to one another, first at triangle centroids, and then on more general cell shapes. It is demonstrated that the the best linear approximation to the Laplace operator estimated at the centroid of arbitrary triangles is obtained when they are equilateral, and that even then, only GL provides first order accuracy. A discrete representation of an operator L is considered $n^{th}$ order accurate if

$$L_h(f) = L_E(f) + O(h^n) \tag{57}$$

where h is a typical scale length, for example the average length of a triangle edge.

## Arbitrary Triangular cell

From basic geometry, it is well known that all the $A^{i+1/2}$ in an arbitrary triangle B with centroid c are equal to each other. This simplifies the formulas considerably. It is easy to apply formulas (45) and (55) to this cell (which respectively correspond to FL and GL). They reduce to the single formula

$$\nabla_h^2 f_c = \alpha_i \frac{(\sum_i \vec{r}_{c,i}^2)}{4 A_c^2} \sum_j F_j \tag{58}$$

where $A_c = A^B/3$ is the cell area. The scale factors that correspond to GL and FL are denoted by $\alpha_i$, i=FL,GL. They are given by $\alpha_{GL} = \frac{1}{3}$ and $\alpha_{FL} = 1$. This difference in values immediately implies that FL and GL cannot simultaneously be first order accurate on any given triangle.

Under what conditions (if any) does there exist a linear finite-difference formula in the $f_i$ of higher accuracy than zero on a triangle? The most general linear expression for $\nabla_h^2 f_c$ as a function of f at the triangle barycenter c and nodes i is

$$\nabla_h^2 f_c = \beta_c f_c + \sum_{i=1}^{3} \beta_i f_i \tag{59}$$

where $\beta_i$, $\beta_c$ are 4 undetermined constants. A Taylor expansion of $f_i$ about $f_c$ at node c up to second order leads to

$$f_i = f_c + \vec{r}_{c,i} \cdot \nabla f_c + \frac{1}{2} \vec{r}_{c,i} \vec{r}_{c,i} : \nabla \nabla f + O(r^2_{c,i}) \tag{60}$$

Equation (59) is required to be exact for constant functions. Furthermore, (59) and (60) together imply that

$$\beta_c = -\sum_i \beta_i \qquad \cdot \qquad \overline{\phantom{-----}} \tag{61}$$

Substituting (59) into (60), and choosing a function f linear in x,y, all the terms of second and higher order derivatives drop out and there remains

$$\sum_i \beta_c \vec{r}_{c,i} \cdot \nabla f = 0 \tag{62}$$

which must be satisfied for all linear functions f. Expressing $\vec{r}_{c,3}$ as the linear combination

$$\vec{r}_{c,3} = -(\vec{r}_{c,1} + \vec{r}_{c,2}) \tag{63}$$

and substituting this into (62) implies that

$$(\beta_1 - \beta_3) \partial_1 f + (\beta_2 - \beta_3) \partial_2 f = 0 \tag{64}$$

where the partials are taken along the corresponding vectors $\vec{r}_{c,i}$. The derivatives are acting along two independent directions; therefore their coefficients must vanish. As a consequence,

$$\beta_1 = \beta_2 = \beta_3 = \beta \qquad \cdot \tag{65}$$

With this simplification (59) reduces to

$$\nabla^2_h f_c = \beta \sum_i F_i \tag{66}$$

which is structurally identical to (57). The $\beta_i$ have now been chosen to cancel out the zeroth and first order terms in the Taylor expansion of (59) for all possible functions. This leaves the second order term, plus a third order remainder:

$$\nabla_h^2 f_c = \frac{\beta}{2} \left( \sum_{i(c)} r_{c,i}^2 \; \partial_{ii} f_c \right) + \beta \; O(r_{c,i}^3) \quad . \tag{67}$$

After expressing $\partial_{33}$ as a function of $\partial_{11}$, $\partial_{12}$, and $\partial_{22}$ and substituting the result in the expression for the discrete Laplace operator, (66) becomes

$$\nabla_h^2 f_c = \beta \; (r_{c,1}^2 \partial_{11} + r_{c,2}^2 \partial_{22} + \vec{r}_{c,1} \cdot \vec{r}_{c,2} \; \partial_{1,2}) \; f_c + \beta \; O(r_{c,i}^3) \tag{68}$$

which approximates the exact Lapacian to first order at node c if $r_{c,i} = r_{c,2}$, $-2 \cos\theta = 1$, and $\beta = 4/(3h^2)$ where $h = \vec{r}_{c,i}$ (see (4)). The first two conditions imply that the triangle must be equilateral. Finally, the discrete representation of the Laplacian is found by substituting $\beta$ back into (67),

$$\nabla_h^2 f_c = \frac{4}{3h^2} \; \sum_i F_i \quad . \tag{69}$$

Therefore, FL gives a zeroth order accuracy, whereas GL is first order accurate, on equilateral triangles. This is easily seen after expressing h as a function of $A_c$. On polygons with more than 4 edges, a sufficient number of degrees of freedom (>6) are available to theoretically insure the existence of a first order accurate discrete Laplace operator. In the next section, it is shown that GL is zeroth order accurate on asymmetrical cells, and second order on highly symmetric ones. It is conjectured that when solving a time-dependent equation that involves the Laplace operator, that these zero order errors average out and that the actual error is closer to second order. This is borne out by work done by Fritts[6]. Finally, note that if the triangle is not equilateral, (14) and (68) cannot be identified with each other, and zeroth order accuracy is the best that can be expected from any linear formula.


More general shapes

Table I lists several cell shapes, along with the corresponding discrete formulas derived from GL and FL. The discrete formulas are equated to the exact Laplacian operator acting upon the function f with the addition of an error term of order n. Several facts emerge from a close inspection of table I. First of all, there are cells (i.e. cell 7) for which both GL and FL are simultaneously zeroth order accurate. Accordingly, they do not produce the correct values when applied to a function f as the cell size decreases to zero. Second,

Cell Geometries with discrete GL and FL

## Cell 1

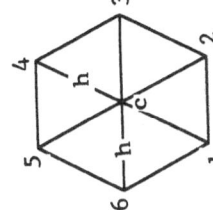

$$\nabla^2_h f^{FL}_c = \nabla^2_h f^{GL}_c = \frac{2}{3h^2} \sum_{i(c)} F_i = \nabla^2_E f_c + O(h^2)$$

## Cell 2

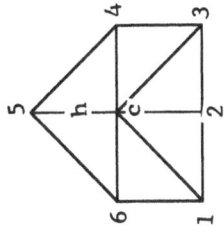

$$\nabla^2_h f^{FL}_c = \frac{1}{h^2}(F_2 + F_4 + F_5 + F_6) = \nabla^2_E f_c + O(h^2)$$

$$\nabla^2_h f^{GL}_c = \frac{5}{6h^2}[F_4 + F_6 + F_1 + F_3 + 4 F_2 + 6 F_5] = \nabla^2_E f_c + O(h)$$

Table I

## Cell 3

$$\nabla^2_h f_c \; FL = \frac{3}{4h^2} \sum_i F_i = \nabla^2_E f_c + O(1)$$

$$\nabla^2_h f_c \; GL = \frac{1}{h^2} \sum_i F_i = \nabla^2_E f_c + O(h^2)$$

## Cell 4

$$\nabla^2_h f_c \; FL = \frac{3}{4h^2} \sum_{even\ i} F_i = \nabla^2_E f_c + O(1)$$

$$\nabla^2_h f_c \; GL = \frac{1}{4h^2} \left( 2 \sum_{odd\ i} + \sum_{even\ i} \right) F_i = \nabla^2_E f_c + O(h^2)$$

## Cell 5

$$\nabla^2_E f_c \; FL = \frac{6}{5h^2} \sum_i F_i = \nabla^2_E f_c + O(1)$$

$$\nabla^2_h f_c \; GL = \frac{1}{h^2} \sum_i F_i = \nabla^2_E f_c + O(h^2)$$

Table I (cont.)

Cell 6

$$\nabla^2_h f^{FL}_c = \frac{6}{h+k} \left[ \frac{F_1 + F_2}{k} + \frac{F_3 + F_4}{h} \right] = \nabla^2_E f_c + O(1)$$

$$\nabla^2_h f^{GL}_c = \frac{2}{h+k} \left[ \frac{F_1 + F_2}{k} + \frac{F_3 + F_4}{h} \right] = \nabla^2_E f_c + O(h)$$

Cell 7

$$\nabla^2_h f^{FL}_c = \frac{1}{2h^2} \left[ F_1 + F_2 + F_3 + F_5 \right] = \nabla^2_E f_c + O(1)$$

$$\nabla^2_h f^{GL}_c = \frac{1}{18h^2} \left[ 13(F_3 + F_5) + 9(F_1 + F_2) + 18 F_4 \right] = \nabla^2_E f_c + O(1)$$

Table I (cont.)

FL is only second order accurate on cells 1 and 2 which share the interesting property that they can both tile the plane. This fact has been verified on every cell shape possessing this property that has been tried to date[4]. However this has not been formally demonstrated. This property is not shared by GL since it is only first order accurate on cell 2. GL produces second order truncation errors on cells 3 and 4, contrary to FL which, surprisingly enough is only zeroth order accurate. A more detailed look at the formulas reveals however that they are proportional to the exact Laplacian. Although cells 3 and 4 cannot tile the plane individually, they can do so together. This, along with the fact that the ratio of GL to FL is smaller and greater than unity on cells 3 and 4 respectively, suggests that in some sense, the truncation errors are closer to first or second order, as was pointed out by Fritts.

In the context of adaptive grids, it is very important to use accurate pointwise approximations to operators that provide information about surface topology, in order to pull nodes to the areas where they are really needed[10]. This requirement is not as severe when approximating equations which contain this operator, because they are often multiplied out by the cell area which has the effect of guaranteeing the consistency of the discrete differential equations.

## Truncation error of the gradient operator at a cell node

To date, very little is known about the accuracy of discrete operators constructed on arbitrary triangular grids, with the exception of the work done in finite-element theory[14]. Their philosophy however is somewhat different from the one adopted in finite-difference methods. In the former, basis vectors are typically defined at each mesh point, and the solution vector is linearly expanded in this basis with arbitrary coefficients that are chosen to make a functional stationary. This eliminates the need to approximate the differential operators since they can be directly applied to the original function. In contrast, finite-difference theory constructs these operators from the unknown solution at the grid points. Despite these differences, there exists a similar dependence of the truncation error bound of the gradient operator on the geometry of the cell, which brings both methodologies closer together. The motivation for a truncation error analysis is mainly to sharpen our instincts regarding the choice of an optimal cell or triangle geometry. It seems intuitively obvious that a grid composed of equilateral triangles is the best in the sense that it leads to the lowest maximum bound on

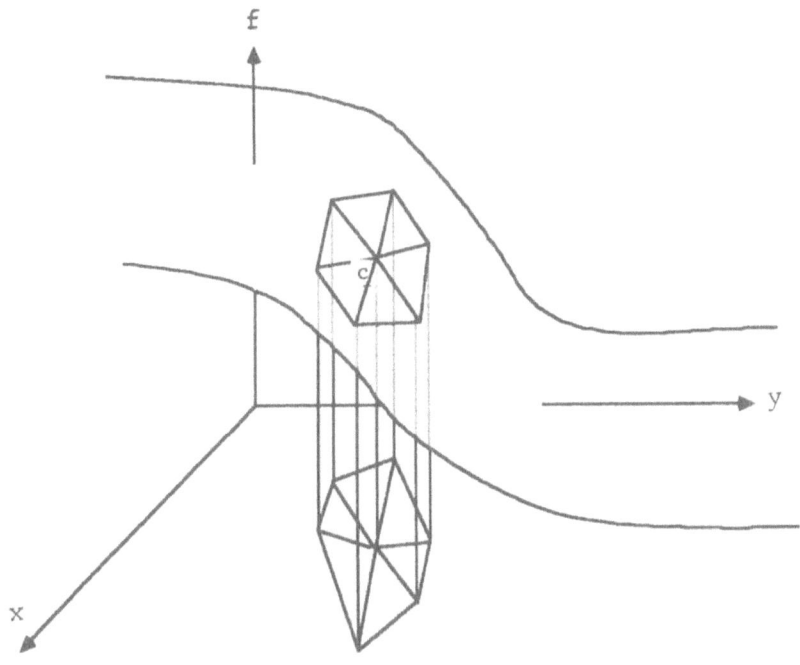

Fig. 9. The triangular grid is lifted up onto the solution surface. Well-behaved surface triangles in high gradient and curvature regions of the solution result in skewed elements when projected back onto the physical plane

truncation errors. However, note that cells built from isoceles triangles may be strong contenders for optimality, since these triangles are so closely related to the cartesian grid. Experience with curvilinear grids has proven that extreme skewness of the cells increases the inaccuracy of a computation. This is a consequence of an inverse dependence of the error on the Jacobian of the transformation between the computational and the physical domain. The calculation of truncation error estimates typically leads to formulas for maximum error bounds of the form[4]

$$\text{error} < K(f) \ \frac{h^n}{\text{minimum angle in triangle}} \tag{71}$$

where h is a typical scale length (e.g. average edge length). K is an analytic operator that acts on f. These maximum bounds are often much larger than the actual error incurred during the numerical computation. For example, if the triangle becomes very skewed, the denominator in (71) goes to zero, which magnifies the error bound. However, this does not necessarily imply that the error is large. Consider a triangular grid, not in the physical plane, but on the surface f itself (fig. 9). If f has very high curvature or gradient regions, triangles that are well structured on the surface will project down to the physical plane and produce skewed elements. But the following question must be posed: since the errors should intuitively be minimum when the surface is well resolved, shouldn't the well-behaved triangles lie on this surface, rather than on the x-y plane? The maximum error bounds will still take on a form similar to (71), but both h and the minimum triangle angle will contain information about the function f, which will probably have the effect of reducing the magnitude of K(f), and provide tighter error bounds. This speculative conclusion still requires verification.

## Conclusion

In this paper, attention has been focused on the construction of discrete representations of the gradient and Laplace operators on unstructured triangular grids. Emphasis was placed on the derivation of a discrete Laplacian which was at least first order accurate. This objective was not achieved. However, with the help of a variational principle, a new discrete approximation to the Laplacian (GL) was constructed, and shown to be at least first order accurate on simple cell shapes. The standard Laplacian (FL) was found to be zeroth order accurate on most cell shapes, except on those which tile the plane.

This substantiates, in an empirical manner, the observation that it is second order accurate in some average sense.

One of the main reasons for constructing operators which have a high order of accuracy (on a pointwise basis) is to correctly estimate the properties of the solution surface (e.g. curvature). These quantities are crucial within the context of adaptive grid strategies. Therein, nodes are dynamically moved, added, or deleted from the mesh according to the properties of the solution surface evolving in time.

Ongoing research is being focused on the construction of higher order accurate operators, as well as to sharpening the maximum bounds on truncation error estimates. The approach to the latter is based on performing a truncation error analysis on the solution surface rather than in the physical plane.

## Appendix

For the interested reader, some geometrical relationships are presented that were derived as byproducts of the truncation error analysis[4]. These derivations demonstrate some of the subtelties involved in index manipulation when working on unstructured grids. The first formula provides a relation between the lengths of the sides of an arbitrary triangle, and the length of its medians. With the notation of figure 10, let

$$I = \sum_{i(c)} r^2_{i+1/2} \quad \text{and} \quad J = \sum_{i(c)} r^2_{c,i}$$

where c is the triangle's centroid. The radius vector $\vec{r}_{i+1/2}$ is expanded along the two median vectors $\vec{r}_{c,i}$ and $\vec{r}_{c,i+1}$. I becomes

$$I = \sum_{i(c)} (\vec{r}_{i,c} + \vec{r}_{c,i+1})^2$$

$$= 2J + 2K$$

where

$$K = \sum_{i(c)} \vec{r}_{i,c} \cdot \vec{r}_{c,i+1} \quad .$$

To evaluate K as a function of J, let $\vec{r}_{c,i+1} = \vec{r}_{i,c} + \vec{r}_{i-1,c}$ which implies that

$$K = \sum_{i(c)} r^2_{i,c} + \sum_{i(c)} \vec{r}_{i,c} \cdot \vec{r}_{i-1,c}$$

This leads to the desired result

$$\sum_{i(c)} r^2_{i+1/2} = 3 \sum_{i(c)} r^2_{c,i}$$

The second formula presented is a consequence of the fact that the FL applied to linear functions is zero. Starting from FL evaluated at a

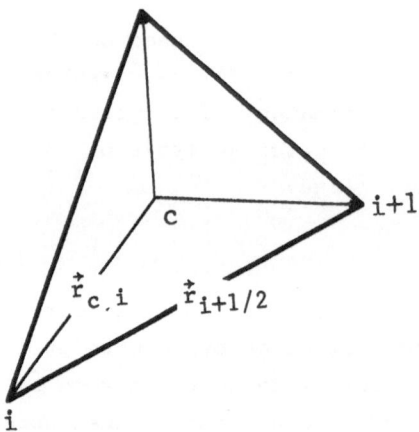

Fig. 10

cell center, the linear function values $f^i$ are Taylor expanded:

$$f_i = f_c + \vec{r}_{c,i} \cdot \nabla f_c \quad .$$

Therefore FL becomes

$$\nabla^2_h f^{FL}_c = -\frac{1}{2A_c} \sum_{i(c)} \frac{\vec{r}_{i+1/2} \cdot [\vec{r}_{i+1,c} \vec{r}_{c,i} - \vec{r}_{c,i} \vec{r}_{i+1,c}]}{(\vec{r}_{c,i} \times \vec{r}_{i+1,c}) \cdot \hat{z}} \cdot \nabla f_c \quad (A1)$$

where the triangle area $A^{i+1/2}$ has been replaced by a vector product. The numerator is a standard cross-product formula; therefore equation (A1) simplifies to

$$\nabla^2_h f^{FL}_c = -\frac{1}{2A_c} \sum_{i(c)} \frac{\vec{r}_{i+1/2} \times (\vec{r}_{c,i} \times \vec{r}_{i+1,c})}{(\vec{r}_{c,i} \times \vec{r}_{i+1,c}) \cdot \hat{z}} \cdot \nabla f_c$$

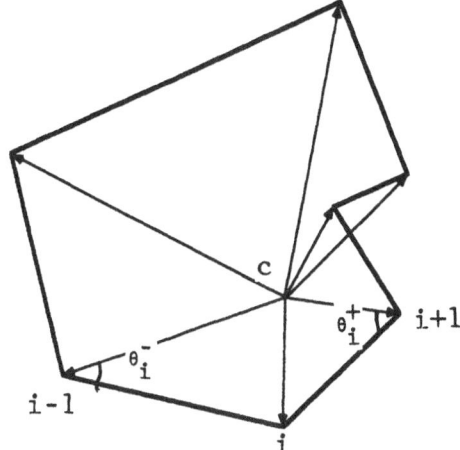

Fig. 11.

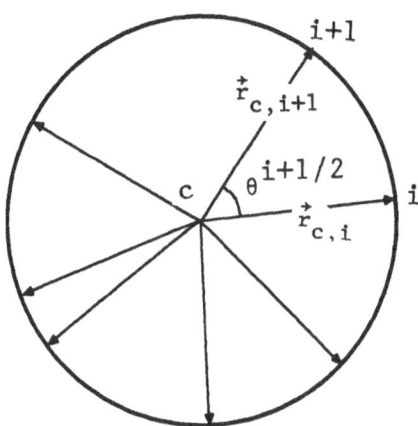

Fig. 12.

$$= - \frac{1}{2A_c} \sum_{i(c)} (\vec{r}_{i+1/2} \times \hat{z}) \cdot \nabla f_c = 0$$

which proves that the FL is exact for linear functions. Equation (A1) can be rewritten in terms of cotangents and set to zero to obtain the geometrical formula

$$\sum_{i(c)} (ctg\theta_c^+ + ctg\ \theta_c^-)\ \vec{r}_{c,i} = 0 \tag{A2}$$

which is valid on any convex or concave polygon, with c in its interior or exterior. The notation is defined in fig. 11. A particular case is obtained if the nodes i lie on a circumference of radius $r_{c,i}$ with c at its center (fig. 12). Equation (A2) then reduces to

$$\sum_{i(c)} (tg\ \frac{\theta_i^+}{2} + tg\ \frac{\theta_i^-}{2})\ \vec{r}_{c,i} = 0 \quad .$$

## References

1.  Hessenius, K. A.; and Rai, M. M.: Applications of a Conservative Zonal Scheme to Transient and Geometrically Complex Problems. AIAA paper 84-1532.

2.  Wedan, B.; and South, J. C.: A Method for Solving the Transonic Full-Potential Equation for General Configurations. AIAA paper 83-1889.

3.  Berger, M.; Gropp, W., and Oliger, J.: Grid Generation for Time-Dependent Problems: Criteria and Methods. Numerical Grid Generation Techniques, Proc. of NASA Langley Research Workshop, Va, 181-188, (1980).

4.  Erlebacher, G.: Solution Adaptive Triangular Meshes with Application to the Simulation of Plasma Equilibrium. Ph. D. Thesis, Columbia University, N.Y., (1984).

5.  Pelz, R. B. and Jameson, A.: Transonic Flow Calculations using Triangular Finite Elements. AIAA Computational Fluid Dynamics Conf., 253-260, (1983)

6.  Fritts, M. J. and Boris, J. P.: The Lagrangian Solution to transient Problems in Hydrodynamics using a Triangular Mesh. J. Comp. Phys. 31, 173-215, (1979)

7.  Miller, K. and Miller, R. N.: Moving Finite Elements I. Siam J. Num. Anal., vol 18, No.6, 1019-1032, (1981)

8.  Miller, K.: Moving Finite-Element Methods II. Siam J. Num. Anal., 18, No.6, 1033-1057, (1981)

9.  Dukowicz, J. K.: Generalized Grids: and Application of the Voronoi Mesh. Adaptive Grid Workshop, Los Alamos

10. Erlebacher, G. and Eiseman, P. R.: Adaptive Triangular Mesh Generation, AIAA-84-1607, AIAA 17th Fluid Dynamics, Plasma Dynamics, and Lasers conference.

11. Crowley, W. P., Free Lagrange Methods for Compressible Flows. Free-Lagrange Methods Conference, Hilton Head Island, SC, (1985)

12. Salzman, J.: A Variational Method for generating Multidimensional Adaptive Grids. Ph.D thesis, New York University, (1981)

13. Bauer, F., Betancourt, O., and Garabedian, P.: A Computational Methods in Plasma Physics. Springer Verlag, New York (1978)

14. Strang, G. and Fix, G.: An Analysis for the Finite-Element Methods. Prentice Hall, Englewook Cliffs, N.J., (1973)

# A LAGRANGIAN METHOD FOR THE SHALLOW WATER EQUATIONS
## BASED ON A VORONOI MESH - FLOWS ON A ROTATING SPHERE

Jeffrey M. Augenbaum

Code 611

NASA/Goddard Space Flight Center

Greenbelt, MD 20771

## Introduction

Ever since the development of modern high speed computers, most of the work on large scale hydrodynamic codes and particularly on meteorological codes have centered around those of Eulerian type. This is due to their ease in programming and higher order of accuracy. However, one of the biggest drawbacks in using Eulerian methods is the presence of the nonlinear convective terms $\underline{u} \cdot \nabla \underline{u}$ which leads to inaccurate representations of advection and discontinuities.

An alternative, and conceptually simpler, approach is to use a Lagrangian method. In such a method the fluid particles themselves are tracked and equations can be derived based on local particle interactions. The main advantage of the Lagrangian approach is that the conservation equations take their simplest form. In particular, the nonlinear convection terms $\underline{u} \cdot \nabla \underline{u}$ do not appear explicitly.

Despite this, advantage, the Lagrangian approach has a fundamental difficulty associated with large deformations generated by typical fluid motions. Fluid particles which are close together at time t=0 may become farther and farther apart as time evolves. Conversly, fluid particles that are far apart may come closer together and even collide. A computational mesh that moves with the fluid becomes increasingly distorted, and the difference approximation to derivatives become worse and worse. Thus, a grid which is continuously deforming and always linking nearest neighbors is needed. Such a grid, the Voronoi mesh has been around for a long time, but has only recently been introduced into hydrodynamic codes.

At each time step, the fluid markers find their natural neighbors. This is acomplished by assigning to each fluid marker the region of space consisting of points which are closer to that marker than to any other. Thus the fluid markers generate a natural partition of the domain into convex polygons, if the domain is two dimensional, or into convex polyhedron if it is three dimensional. The mesh so generated is known as a Voronoi mesh. (For this reason, the fluid markers will be called generating points.)

Two generating points are considered neighbors if their polyhedra have a face of nonzero area in common. As the generating points move, the polyhedra deform continuously, but a given generating point is free to lose old neighbors and aquire new ones. These changes in structure occur continuously in the following sense. Let $A_{jk}$ be the area of the face in common between polyhedra $j$ and $k$ when there is such a face. Otherwise, set $A_{jk} = 0$. Then $A_{jk}$ is a continuous function of the coordinates of the generating points. Since the fluid markers find their natural neighbors at each time step, there is no tendency for the distance between neighbors to increase with time, and the difficulty of large deformations is overcome.

While there have been several successful free Lagrangian (FLAG) codes used for gas dynamics and incompressible flows e.g. Crowley (1971), Dukowicz (1981), Fritts and Boris (1979), Peskin (1985) and Trease (1981), there has been very little application of such methods to meteorological problem. A first attempt was made by Mesinger (1971), using a set of floating points to solve the shallow water equations on a sphere. In his method all points within a fixed radius from a given point are used in computing the derivatives at that point. Mesinger compares his method to existing Eulerian schemes and reports very encouraging results. However, no mention is made of the cost of his method when large deformation occur after a number of time steps, nor is there any attempt to optimize the grid algorithm.

In Augenbaum (1982, 1984) a Lagrangian scheme is introduced for the shallow water equations based on the use of the voronoi mesh. In this paper we describe the method for flow on a sphere. At time $t = 0$ we place N fluid markers on the sphere and partition the sphere via the voronoi mesh. Specifically, the N fluid markers partition the surface of the sphere into convex spherical polygons in such a way that each marker point represents all the fluid that is closer to that marker than to any other one.

Every spherical ploygon is the base of a fluid cell. Each cell is a column of fluid consisting of a polygonal base on the sphere, sides with height H, and a spherical polygonal top at the free surface. Since the fluid in incompressible, the mass of fluid in a cell is constant throughout the evolution of the flow. By adjusting the free surface height, at each time step, to compensate for changes in surface area of the base, conservation of mass is trivally satisifed.

Equations of motion can be derived by applying Hamiltons principle of least action to the N particle fluid system (see Salmon (1983), and Wang (1984)). The discrete shallow water equations can be found by directly incorporating the shallow water assumption into the Lagrangian in the action integral. This approach can also be used to derive the continuous shallow water equations. This is described in Sections 2 and 3.

These equations involve certain difference operators $\underline{G}$ which by analogy, appear to correspond to the gradient. These difference operators couple generating points to their neighbors, and they have coefficients that depend on the coordinates of the generating points. A formula for $\underline{G}$ is derived in Sec. 4.

In order to define a useful numerical method, dissipation is put into the model, when needed, by merging fluid points when they get too close to each other. This is described in section 5, and the full numerical scheme is described in Sec. 6. Numerical examples are then presented in Sec. 7.

## 2. DISCRETE SHALLOW WATER EQUATIONS ON A NON-ROTATING SPHERE

We now describe the discrete approximation to the shallow water equations on the surface of a non-rotating sphere.  The continous equations in Eulerian form, for a rotating sphere (in spherical co-ordinates) can be found in Williamson and Browning (1973).  We recall these equations for the convenience of the reader.

### Conservation of Mass

$$\frac{\partial h}{\partial t} + \frac{u}{a \sin^2 \theta} \frac{\partial h}{\partial \phi} + \frac{v}{a \sin \theta} \frac{\partial h}{\partial \theta} + \frac{h}{a \sin^2 \theta} \frac{\partial u}{\partial \phi} + \frac{h}{a \sin \theta} \frac{\partial v}{\partial \theta} = 0 \tag{2.1a}$$

### Conservation of Momentum

$$\frac{\partial u}{\partial t} + \frac{u}{a \sin^2 \theta} \frac{\partial u}{\partial \phi} + \frac{v}{a \sin \theta} \frac{\partial u}{\partial \theta} + \frac{g}{a} \frac{\partial h}{\partial \phi} + vf = 0 \tag{2.1b}$$

$$\frac{\partial v}{\partial t} + \frac{u}{a \sin^2 \theta} \frac{\partial v}{\partial \phi} + \frac{v}{a \sin \theta} \frac{\partial v}{\partial \theta} + \frac{g \sin \theta}{a} \frac{\partial h}{\partial \theta} - \frac{(u^2 + v^2)}{a \sin^2 \theta} - uf = 0 \tag{2.1c}$$

where  θ = colatitude

  φ = longitude

  u = longitudinal velocity

  v = colatitudinal velocity

  h = height of the free surface

  g = gravity

  a = radius of the earth (assume a = 1)

  f = 2Ω cos θ = Coriolis force (Ω = angular rotation rate)

We follow the usual convention of using small letters to denote Eulerian variables and capital letters to denote Lagrangian variables.

We now derive the discrete shallow water equations in Lagrangian variables.  Note that the derivation will consider fluid particles on the surface of a sphere to be imbedded in the Euclidean space $R^3$ and therefore all norms denote the usual Euclidean norm in $R^3$ (i.e., $|\underline{X}|^2 = X^2 + Y^2 + Z^2$).

Assume now that our fluid is incompressible and that the height of any fluid element is free to vary.

We then place N fluid markers, $\underline{X}_j = (X_j, Y_j, Z_j)$, on the surface of the unit sphere.  These markers are used to partition the sphere into spherical polygons, so that each particle (fluid marker), $\underline{X}_j$, represents all the fluid that is closer to $\underline{X}_j$ than to any other marker particle (see figures 2.1 and 2.2).  That is

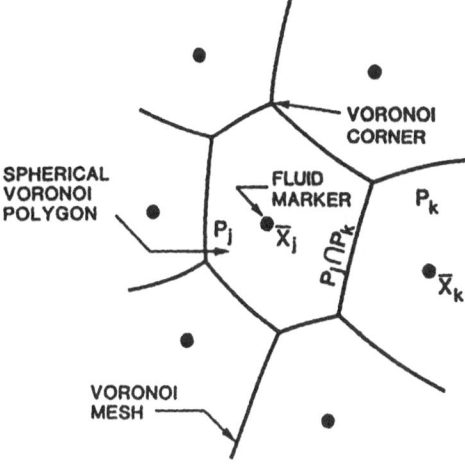

Figure 2.1 Portion of a Voronoi mesh

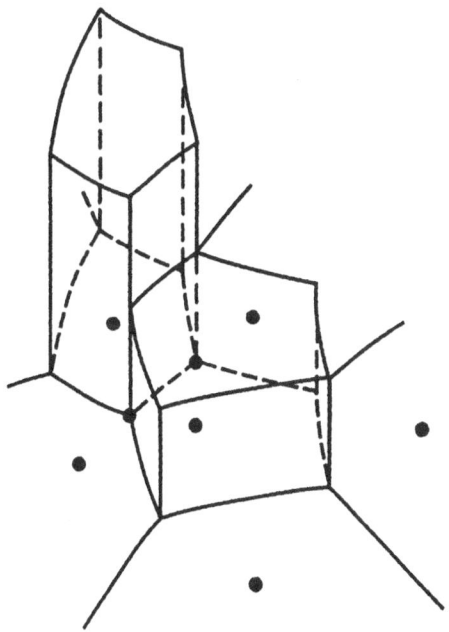

Figure 2.2 Portion of a Voronoi fluid cell

$$P_k = \bigcap_{\substack{j=1 \\ j \neq k}}^{N} \{\underline{x}: |\underline{x} - \underline{X}_k| \leq |\underline{x} - \underline{X}_j|, \text{ and } |\underline{x}|^2 = 1\} \qquad (2.2)$$

The set of spherical polygons make up the spherical voronoi mesh. (A complete description of the spherical voronoi mesh is given in Augenbaum and Peskin (1985). To each spherical polygon $P_k$, we associate its spherical surface area $A_k$ and a height $H_k$. The volume of each fluid cell, with base $P_k$ and height $H_k$ is given by

$$V_k = A_k H_k$$

where we have neglected the difference in surface area between the top and bottom of $P_k$ since $H_k$ is much smaller than the radius of the sphere.

The mass of such a cell is then

$$m_k = \rho(A_k H_k) \qquad (2.3)$$

where $\rho$ = density.

Note that since the particular polygonal structure depends on the distribution of points $\{\underline{X}_1, \ldots, \underline{X}_N\}$ the area is a function of these N points, i.e., $A_j = A_j(\underline{X}_1, \ldots, \underline{X}_N)$.

Equations of Motion

First, since we are concerned with the time evolution of the fluid particles (markers), all the quantites defined above are functions of time.

Incompressibility (Assume $\rho \equiv 1$)

Since the fluid is incompressible we have

$$m_j(t) = A_j(\underline{X}_1(t), \ldots, \underline{X}_N(t)) \cdot H_j(t) = m_j(0) = m_j^o \qquad (2.4)$$

which defines $H_j(t)$ as a function of the configuration, i.e.

$$H_j = H_j(\underline{X}_1(t), \ldots, \underline{X}_N(t)) = \frac{m_j^o}{A_j(\underline{X}_j, \ldots, \underline{X}_N)} \qquad (2.5)$$

Lagrangian Dynamics

Following the one-dimensional case (Augenbaum (1984)), we construct a Lagrangian for the system of particles $\underline{X}_j$ and derive their equations of motion as the minimization of the action integral subject to fixed initial and final values. This will yield a set of differential equations for the particle trajectories which can be solved (numerically) for arbitrary initial conditions.

## Potential Energy

The potential energy of the particles is given by

$$PE = \int_{A_s} \int_0^h gr \, dr \, dA_s$$

$$= \frac{1}{2} g \int_{A_s} h^2 \, dA_s$$

(2.6)

where $dA_s$ = surface area element on the sphere

$g$ = gravitation constant

$r$ = radial distance from surface of sphere

Equation (2.6) can be discretized as follows:

$$PE = \frac{1}{2} \sum_{k=1}^N gH_k (H_k A_k) = \sum_{k=1}^N \frac{g}{2} m_k H_k$$

(2.7)

## Kinetic Energy

To derive formulas for the kinetic energy, we need to use the shallow water assumptions. The shallow water theory assumes that the flow is essentially horizontal and that we can therefore ignore the velocity in the vertical direction (Haltiner and Williams (1980)). In the plane, this means that we can ignore $W^2$ in comparison to $U^2 + V^2$, i.e., $W^2 \ll U^2 + V^2$. On the sphere, this means that we can ignore the velocity normal to the sphere, and assume that the motion of the fluid markers are tangential to the sphere. In terms of our fluid cells, the relative postion of a fluid point in the fluid column is unchanged, and therefore the motion of fluid in a cell is determined by the motion of the fluid in the base on the sphere. Hence the motion is two-dimensional. Since we represent points on the sphere in terms of Cartesian coordinates, i.e., $\underline{X}_j = (X_j, Y_j, Z_j)$, the kinetic energy is given by the formula

$$KE = \sum_{k=1}^N \frac{1}{2} m_k |\underline{U}_k|^2 = \sum_{k=1}^N \frac{1}{2} m_k \left|\frac{d\underline{X}_k}{dt}\right|^2$$

(2.8a)

subject to the constraint that

$$|\underline{X}_k|^2 = 1$$

(2.8b)

where

$$\frac{dX_k}{dt} = U_k \qquad k=1,\ldots\ldots, N \qquad (2.9)$$

The Lagrangian is then given by

$$L = KE - PE$$

$$= \frac{1}{2} \sum_{k=1}^{N} m_k \left\{ \left|\frac{dX_k}{dt}\right|^2 -gH_k \right\} \qquad (2.10)$$

subject to $|X_k|^2 = 1$.

## Discrete Equations

Recall that equations of motion of a mechanical sysem are derived by minimizing $I = \int_0^T L\ dt$ subject to given initial and final values $X_j(0)$ and $X_j(T)$ and any additional constraints (Salmon (1983)).

Hence the discrete shallow water equations can be derived by minmizing

$$I = \int_0^T \sum_{j=1}^{N} \frac{1}{2} m_j \left\{ \left|\frac{dX_j}{dt}\right|^2 -gH_j \right\} dt \qquad (2.11a)$$

subject to

$$|X_j|^2 = 1 \qquad (2.11b)$$

where

$$H_j = \frac{m_j}{A_j\ (X_j,\ldots\ldots,X_N)} \qquad (2.11c)$$

A necessary condition for minimization is that $\delta I = 0$ subject to $|X_j|^2 = 1$ and that $\delta X_j(0) = \delta X_j(T) = 0$. We use the method of Lagrange multipliers.

Set

$$\tilde{I} = \int_0^T \sum_{j=1}^{N} \frac{1}{2} m_j \left|\frac{dX_j}{dt}\right|^2 dt - \frac{g}{2} \int_0^T \sum_{k=1}^{N} m_k H_k dt + \int_0^T \sum_{j=1}^{N} \mu_j (t)\ [|X_j|^2 - 1] dt \qquad (2.12)$$

where we have introduced the Lagrange multiplier $\mu_j(t)$. We have

$$\delta \tilde{I} = \int_0^T \left\{ \sum_{j=1}^{N} [-m_j \frac{d^2 X_j}{dt^2} - \frac{g}{2} \sum_{k=1}^{N} m_k \frac{\partial H_k}{\partial X_j} + 2\mu_j(t)X_j] \cdot \delta X_j \right\} dt = 0$$

Since $\delta \underline{X}_j$ is arbitrary, we have, for each j

$$-m_j \frac{d^2\underline{X}_j}{dt^2} - \frac{g}{2} \sum_{k=1}^{N} m_k \frac{\partial H_k}{\partial \underline{X}_j} + 2\mu_j(t)\underline{X}_j = 0 \qquad (2.13)$$

Using (2.11b) we also get

$$\underline{X}_j \cdot \frac{d\underline{X}_j}{dt} = 0 \qquad (2.14)$$

At this point we note that we use the notation $\frac{\partial A_j}{\partial \underline{X}_k}$ to mean the surface gradient of the function $A_j$ with respect to $\underline{X}_k$. That is, if $\underline{X}_k(t)$ is any curve on the surface of the sphere, then

$$\frac{dA_j}{dt} = \sum_{k=1}^{N} \frac{\partial A_j}{\partial \underline{X}_k} \cdot \frac{d\underline{X}_k}{dt}$$

and

$$\frac{\partial A_j}{\partial \underline{X}_k} \cdot \underline{X}_k = 0$$

This defines the tangential components of $\frac{\partial A_j}{\partial \underline{X}_k}$ because $\underline{X}_k(t)$ is an arbitrary curve in the surface. The normal component at $\underline{X}_k$ is arbitrary, but we set it equal to zero.

Equation (2.13) can now be rewritten in a form which is more convenient for computation. Since $H_j$ is a function of the $\underline{X}_j$'s through (2.5), we can rewrite $\frac{\partial H_k}{\partial \underline{X}_j}$ is another form.

$$0 = \frac{\partial A_k}{\partial \underline{X}_j} H_j + A_k \frac{\partial H_k}{\partial \underline{X}_j}$$

$$\Rightarrow \frac{\partial H_j}{\partial \underline{X}_k} = - \frac{H_j}{A_j} \frac{\partial A_j}{\partial \underline{X}_k} = - \frac{m_j}{A_j^2} \frac{\partial A_j}{\partial \underline{X}_k} \qquad (2.15)$$

Thus (2.13) can be written

$$m_j \frac{d^2\underline{X}_j}{dt^2} = \frac{g}{2} \sum_{k=1}^{N} m^2_k \frac{\partial A_k}{A_k^2 \, \partial \underline{X}_j} + 2\mu_j\underline{X}_j \qquad (2.16)$$

We can also use (2.14) to eliminate $\mu_j(t)$. Dot (2.16) with $\underline{X}_j$ to get

$$(m_j \frac{d\underline{U}_j}{dt}) \cdot \underline{X}_j = 2\mu_j\underline{X}_j \cdot \underline{X}_j = 2\mu_j$$

where we have used the fact that $\underline{X}_j \cdot \dfrac{\partial A_k}{\partial \underline{X}_j} = 0$ because $\dfrac{\partial A_k}{\partial \underline{X}_j}$ is tangential to the sphere

Now, since $\underline{X}_j(t) \cdot \underline{X}_j(t) = 1 \Rightarrow \underline{X}_j \cdot \dfrac{d\underline{X}_j}{dt} = 0$

$$\Rightarrow \underline{X}_j \cdot \dfrac{d^2\underline{X}_j}{dt^2} + \dfrac{d\underline{X}_j}{dt} \cdot \dfrac{d\underline{X}_j}{dt} = 0$$

or $\underline{X}_j \cdot \dfrac{d\underline{U}_j}{dt} = -|\underline{U}_j|^2.$ The result is that

$$\mu_j = \dfrac{-m_j}{2} |\underline{U}_j|^2 \qquad (2.17)$$

In summary, the spatially discrete shallow water equations on a non-rotating sphere can be written as

Conservation of Momentum

$$\dfrac{d\underline{U}_j}{dt} = \dfrac{g}{2m_j} \sum_{k=1}^{N} H_k^2 \dfrac{\partial A_k}{\partial \underline{X}_j} - |\underline{U}_j|^2 \underline{X}_j \qquad (2.18c)$$

Conservation of Mass

$$H_j = \dfrac{m_j^o}{A_j} \qquad (2.18b)$$

Move Points with Local Fluid Velocity

$$\dfrac{d\underline{X}_j}{dt} = \underline{U}_j \qquad (2.18c)$$

Note that the additional term $-|\underline{U}_j|^2\underline{X}_j$ which is not present in the one-dimensional case (Augenbaum (1984)) is a restoring force. This term keeps a particle from flying off the sphere, and it is directed toward the center of the sphere. The motion of each particle is therefore tangential to the sphere.

By a similar derivation, the continuous shallow water equations in Cartesian coordinates on the sphere are

$$\dfrac{DH}{Dt} = -H\nabla_s \cdot \underline{U} \qquad (2.19a)$$

$$\frac{DU}{Dt} = -g\nabla_s H - |\underline{U}|^2 \underline{X} \qquad (2.19b)$$

$$\frac{d\underline{X}}{dt} = \underline{U} \qquad (2.19c)$$

where $\nabla_s$ is the surface gradient on the sphere and $\nabla_s \cdot$ is the surface divergence.

## 3.  DISCRETE SHALLOW WATER EQUATION ON A ROTATING SPHERE

In meteorological applications we will need the equations of motion for flows on a rotating sphere.  Equations (2.18) can be transformed to rotating coordinates by considering a cartesian coordinate system rotating with angular velocity $\underline{\Omega}$ (i.e. let $\underline{U}_o = \underline{U}_r + \underline{\Omega} \times \underline{X}$ where $\underline{U}_o$ is the velocity relative to the stationary observer and $\underline{U}_r$ is the velocity relative to an observer rotating with angular velocity $\underline{\Omega}$).

The transformed equations, relative to the rotating frame are (for small $\underline{\Omega}$):

Conervation of Mass

$$m_j^o = A_j \, H_j \qquad (3.1a)$$

Consevation of Momentum

$$\frac{d\underline{U}_j}{dt} + 2(\underline{\Omega} \times \underline{U}_j)|| = \frac{g}{2m_j} \sum_{k=1}^{N} H_k^2 \frac{\partial A_k}{\partial \underline{X}_j} - |\underline{U}_j|^2 \underline{X}_j \qquad (3.1b)$$

Move Points with Local Fluid Velocity

$$\frac{d\underline{X}_j}{dt} = \underline{U}_j \qquad (3.1c)$$

where $(\ )||$ denotes the component of $(\ )$ parallel to the sphere at $\underline{X}_j$.

## 4.  DISCRETE DIFFERENTIAL OPERATORS ON SPHERE

By analogy between the discrete equations (2.18) and the continuous equations (2.19) we define discrete differential operators for the gradient $\underline{G}$ and the divergence $D$ on the sphere as follows:

$$A_j \, (G\psi)_j = - \sum_{k=1}^{N} \psi_k \frac{\partial A_k}{\partial \underline{X}_j}$$

$$A_j\ (D\underline{U}_j)\ =\ \sum_{k=1}^{N}\ U_k\ \cdot\ \frac{\partial A_j}{\partial \underline{X}_k}$$

Both differential opeators involve the derivatives $\partial A_k / \partial \underline{X}_j$. We now derive an explicit formula for them. In the one-dimensional case, Augenbaum (1982, 1984), there was a simple formula for $A_k$ in terms of its neighbors; on the sphere, however, formulas for $\partial A_k / \partial \underline{X}_j$ are not trival. We proceed by modifying an idea of Peskin from the plane to the sphere

We note first, that if points $\underline{X}_j$ and $\underline{X}_k$ are not neighbors then

$$\frac{\partial A_k}{\partial \underline{X}_j} = 0 \ , \quad \underline{X}_j, \ \underline{X}_k \ \text{not neighbors} \tag{4.1}$$

We now consider the case where points $\underline{X}_j$ and $\underline{X}_k$ are neighbors and $j \neq k$. Consider a spherical polygon $P_k$ and let $\underline{X}_k$ be fixed. Move the point $\underline{X}_j(t)$ and see how the edge $P_j \cap P_k$ moves and sweeps out a change in area (see figure 4.1), i.e. we move a point $\underline{X}_j$ while holding all others fixed and compute its effect on the area $A_k$.

Let $\underline{X}(t)$ be a point on the edge $P_j \cap P_k$ so that

$$|\underline{X}-\underline{X}_k|^2 = |\underline{X}-\underline{X}_j|^2 \tag{4.2a}$$

$$|\underline{X}|^2 = 1 \tag{4.2b}$$

$$|\underline{X}_j|^2 = 1 \tag{4.2c}$$

Differentiating (4.2) with respect to time, we get

$$2(\underline{X}-\underline{X}_k) \cdot \frac{d\underline{X}}{dt} = 2(\underline{X}-\underline{X}_j) \cdot (\frac{d\underline{X}}{dt} - \frac{d\underline{X}_j}{dt}) \tag{4.3a}$$

$$\underline{X} \cdot \frac{d\underline{X}}{dt} = 0 \tag{4.3b}$$

$$\underline{X}_j \cdot \frac{d\underline{X}_j}{dt} = 0 \tag{4.3c}$$

Using (4.3b) and (4.3c) in (4.3a) we have

$$\underline{X} \cdot \frac{d\underline{X}_j}{dt} = (\underline{X}_k-\underline{X}_j) \cdot \frac{d\underline{X}}{dt} \tag{4.4}$$

Divide (4.4) by $|\underline{X}_k - \underline{X}_j|$ and integrate over $P_j \cap P_k$ we get

$$- \int_{P_j \cap P_k} \frac{(\underline{X}_j - \underline{X}_k)}{|\underline{X}_j - \underline{X}_k|} \cdot \frac{d\underline{X}}{dt} \, ds = \int_{P_j \cap P_k} \frac{\underline{X}}{|\underline{X}_j - \underline{X}_k|} \cdot \frac{d\underline{X}_j}{dt} \, ds \qquad (4.5)$$

where ds is arch length.

Now, we notice that $(\underline{X}_j - \underline{X}_k)/|\underline{X}_j - \underline{X}_k| = \underline{n}_{jk}$ is normal to the arc $P_j \cap P_k$ and tangent to the sphere along this arc.

If we move the point $\underline{X}_j$, while holding all other points fixed, an compute its effect on the area $A_k$ we have

$$\frac{dA_j}{dt} = \int_{P_j \cap P_k} (\underline{n}_{jk} \cdot \frac{d\underline{X}}{dt}) \, ds \qquad (4.6)$$

So (4.5) becomes

$$\frac{dA_k}{dt} = - \int_{P_j \ P_k} \frac{\underline{X}}{|\underline{X}_j - \underline{X}_k|} \cdot \frac{d\underline{X}_j}{dt} \, ds$$

or

$$\frac{dA_k}{dt} = - \frac{1}{|\underline{X}_j - \underline{X}_k|} \left[ \int_{P_k \cap P_k} \underline{X} ds \right] \cdot \frac{d\underline{X}_j}{dt} \qquad (4.7)$$

To compute $\int_{P_j \cap P_k} \underline{X} ds$ we introduce perpendicular unit vectors $\underline{e}_j$ and $\underline{e}_k$ in the plane of the arc $P_j \cap P_k$. Let $\theta$ be the angle in the resulting $(e_j, e_k)$ plane, and let $\theta_j$, $\theta_k$ denote the end points of the arc $P_j \cap P_k$ corresponding to the points $\underline{XC}_j$ and $\underline{XC}_k$ (see figure 4.1).

Note that $d\theta$ is the same as arc length since we are working on the unit sphere. Using $\theta$ as a parameter we have

$$\underline{X} = \cos \theta \, \underline{e}_j + \sin \theta \, \underline{e}_k$$

and therefore

$$\int_{\theta_j}^{\theta_k} \underline{X} d\theta = \sin \theta \Big|_{\theta_j}^{\theta_k} \underline{e}_j - \cos \theta \Big|_{\theta_j}^{\theta_k} \underline{e}_k$$

Let $\Delta\theta = \theta_k - \theta_j$ and

$$\theta_k = \hat{\theta} + \frac{\Delta\theta}{2}$$

$$\hat{\theta}_j = \theta - \frac{\Delta\theta}{2}$$

So

$$\int_{\theta_j}^{\theta_k} \underline{X}(\theta)d\theta = 2 \sin \frac{\Delta\theta}{2} [\cos \theta \; \underline{e}_j + \sin \theta \; \underline{e}_k] \qquad (4.8)$$

Notice that $\cos \theta \; \underline{e}_j + \sin \theta \; \underline{e}_k$ is a unit vector through the midpoint of $\underline{XC}_j$ and $\underline{XC}_k$. In cartesian coordinates this becomes

$$\cos \theta \; \underline{e}_j + \sin \theta \; \underline{e}_k = \underline{X}_{jk} = \frac{1}{2} \frac{(\underline{X}_j + \underline{X}_k)}{|\underline{X}_j + \underline{X}_k|}$$

It is easily shown that $2 \sin \Delta\theta/2$ is equal to the Euclidean distance from $\underline{XC}_j$ to $\underline{XC}_k$. In general, we end up with

$$\int_{P_j \cap P_k} \underline{X}ds = \int_{\theta_j}^{\theta_k} \underline{X}(\theta)d\theta = \ell_{jk} \; \underline{X}_{jk}$$

where

$$\ell_{jk} = [(XC_j - XC_k)^2 + (YC_j - YC_k)^2 + (ZC_j - ZC_k)^2]^{1/2}$$

$$\underline{X}_{jk} = \frac{\underline{XC}_j + \underline{XC}_k}{|\underline{XC}_j + \underline{XC}_k|}$$

$\underline{XC}_j$ and $\underline{XC}_k$ are the endpoints of $P_j \cap P_k$. We can therefore write (4.7) as

$$\frac{dA_k}{dt} = - \frac{\ell_{jk}}{|X_j - X_k|} \; \underline{X}_{jk} \cdot \frac{d\underline{X}_j}{dt} \qquad (4.9)$$

If we consider $A_k(\underline{X}_1(t), \ldots\ldots, \underline{X}_N(t))$ as being extended off the sphere, we can differentiate $A_k$ to obtain

$$\frac{dA_k}{dt} = \sum_{k=1}^{N} \frac{\partial A_k}{\partial \underline{X}_j} \cdot \frac{d\underline{X}_j}{dt} \qquad (4.10)$$

The results, of course, are independent of the extension, since $d\underline{X}_j/dt$ is parallel to the sphere.

If all the $\underline{X}$'s are fixed, except for $\underline{X}_j$, we obtain

$$\frac{dA_k}{dt} = \frac{\partial A_k}{\partial \underline{X}_j} \cdot \frac{d\underline{X}_j}{dt} = - \frac{\ell_{jk}}{|\underline{X}_j - \underline{X}_k|} \; \underline{X}_{jk} \cdot \frac{d\underline{X}_j}{dt} \qquad j \neq k \qquad (4.11)$$

Note that this procedure determines only the tangential component of $\partial A_k/\partial \underline{X}_j$ at $\underline{X}_j$. The normal components have no significance and we set them equal to zero by projecting (4.11) onto the tangent plane to the sphere at $\underline{X}_j$.

The formula for projecting an arbitrary vector onto the sphere is $P\underline{v} = \underline{v} - \underline{N}(\underline{N} \cdot \underline{v})$ where $\underline{N}$ is the unit normal. The projection of $\partial A_k/\partial \underline{X}_j$ onto the sphere is given by

$$P_j \left( \frac{\partial A_k}{\partial \underline{X}_j} \right) = \sum_{q=1}^{3} [\delta_{pq} - (\underline{X}_j)_p (\underline{X}_k)_q] \left[ \frac{-\ell_{jk}(\underline{X}_{jk})_q}{|\underline{X}_j - \underline{X}_k|} \right] \qquad (4.12)$$

$$p = 1,2,3$$

From now on, by writing $\partial A_k/\partial \underline{X}_j$ we mean the projected derivative (4.12). We note that the formulas (4.12) are independent of the particular extension of $A_k(\underline{X}_1, \ldots, \underline{X}_N)$ off the sphere.

We also mention that $\underline{X}_{jk}$ is normal to the sphere at $\underline{X}_{jk}$ and therefore approximately normal at $\underline{X}_j$; nevertheless only the small tangential component at $\underline{X}_j$ has any significance. Because of the arbitrariness of the normal component we could use the equivalent formula

$$\frac{\partial A_k}{\partial \underline{X}_j} = - \ell_{jk} \frac{(\underline{X}_{jk} - \underline{X}_j)}{|\underline{X}_j - \underline{X}_k|}$$

which has the same form as in the planar case (Peskin (preprint)). In the case where $j = k$, formula (4.11) does no hold. However, we can get a formula for the diagonal derivatives $\partial A_j/\partial \underline{X}_j$ in terms of the off-diagonal derivatives.

We use the identify

$$\sum_{k=1}^{N} A_k(\underline{X}_1(t), \ldots, \underline{X}_N(t)) = \text{Area of sphere} = 4\pi \qquad (4.13)$$

apply $\partial/\partial \underline{X}_j$ to (4.13) we get

$$\sum_{k=1}^{N} \frac{\partial A_k}{\partial \underline{X}_j} = 0$$

and therefore

$$\frac{\partial A_j}{\partial \underline{X}_j} = - \sum_{\substack{k=1 \\ k \neq j}}^{N} \frac{\partial A_k}{\partial \underline{X}_j} \qquad (4.14)$$

NOTE: All terms in sum are associated with $\underline{X}_j$.

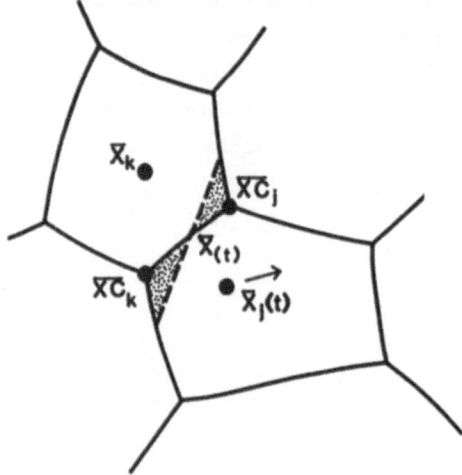

Figure 4.1   Change in area of $P_k$ swept out by edge $P_j \cap P_k$ as the point $\underline{X}_j(t)$ is moved, while holding all others fixed.

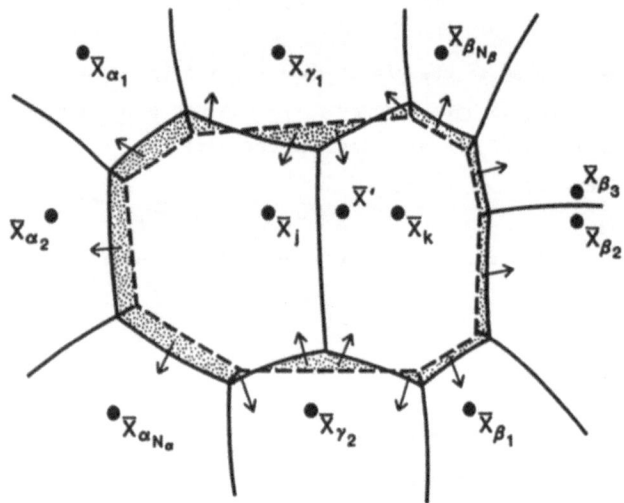

Figure 5.1   Transfer of fluid mass and momentum due to a collision. Arrows point in direction of fluid transfer.

## 5. DISSIPATION OF ENERGY

It seems attractive to solve the shallow water equation by computing the trajectories of the system (3.1) as geodesics along the manifold of allowed configuration. This approach, however, does not lead directly to a useful numerical method. The reason for this is that the derivatives $\partial A_k / \partial X_j$ becomes aribitrarily large as two generating points approach each other (eq. 4.11). Thus the manifold given by $m_j(X_1(t),$ $X_2,(t), \ldots X_N(t)) = m_j$ has singularities, near which the curvature will be arbitrarily large. It is not feasable to compute geodesic trajectories near these singular points.

This difficulty can be removed when dissipation is added to the equations. This is most often accomplished by adding fluid viscosity to the equations.

Our approach is to add dissipation only when it is needed by merging fluid points when particle distances decrease past some preassigned threshold. A practical distance threshold will be given in the next section.

We now present the particle merging procedure, first for the case of exact collision and then for approximate collisions. When one particle, say $X_j$, overtakes another particle, $X_k$, and collides with it, we replace them by one averaged particle, $X'$, with velocity and height chosen so that total mass and momentum are conserved.

After a collision has occurred, $X_j$ and $X_k$ are replaced by $X'$. Since the boundaries of $P_j$ and $P_k$ do not move during a colliison (except for $P_j \cap P_k$, which disappears) there is no transfer of mass and momentum to the non-colliding particle neighbors. The following set of equations are easily seen to hold. (Note: Primed quantities refer to after-collision values.)

$$X' = X_j = X_k \tag{5.1a}$$

$$A' = A_j + A_k \tag{5.1b}$$

Conservation of mass $$\qquad m' = m_j + m_k \tag{5.1c}$$

Conservation of momentum $$\qquad m'U' = m_jU_j + m_kU_k \tag{5.1d}$$

$$H' = m'/A' \tag{5.1e}$$

Note that $U'$ is normal to the sphere at $X'$, since $U_j$ and $U_k$ are.

After such a collision occurs the total number of fluid markers is decreased by one. This is consistent with the fact that energy is dissipated by a collision and the flow is now smoother in the immediate region of a collision. Thus we require less resolution in those regions where a collision has accured.

We also remark that one can show that the total energy decreases after such a collision has occurred. The proof is a straightforward generalization of the one-dimensional case (see appendix in [2]). Note that decreasing energy is the shallow-water equivalent of decreasing entropy in gas dynamics.

The above treatment does not work out well in our finite particle fluid systems since fluid particles may slide around other particles without actually colliding. Thus, collisions need not ever occur. Also, if two particles are allowed to get too close to each other, the Courant-Friedrichs-Lewy conditions is violated and instabilities are observed in the computational results.

We can solve both of the above difficulties by generalizing our model of a collision. As in the one-dimensional case, we consider a collision to occur when $\underline{X}_j$ and $\underline{X}_k$ come within a distance, d, of each other. When $|\underline{X}_j - \underline{X}_k| \leq d$, we replace $\underline{X}_j$ and $\underline{X}_k$ by one averaged point $\underline{X}'$, with mass, velocity and height chosen so that mass and momentum are conserved in the new cell $P_{X'}$ and its immediate neighbors. A natural choice for the collision radius, d, is given by the Courant-Fredrichs-Lewy condition. This is described in the next section.

Notation: We will need some convention to identify the points being affected by a collision. Let us assume that markers $\underline{X}_j$ and $\underline{X}_k$ collide and will be replaced by $\underline{X}'$ (which will eventually be given the designation $\underline{X}_j$, where $j < k$).

As Figure 5.1 makes clear, there are exactly two points which are neighbors of both $\underline{X}_j$ and $\underline{X}_k$. We denote them by $\underline{X}_{Y_1}$ and $\underline{X}_{Y_2}$ according to a counterclockwise convention. The set of neighboring points of $\underline{X}_j$, excluding $\underline{X}_{Y_{1,2}}$ will be denoted by $\underline{X}_{\alpha_1}$, $\alpha_1 = 1,2, \ldots, N_\alpha$. The set of neighboring points of $\underline{X}_k$, excluding $\underline{X}_{Y_{1,2}}$ will be denoted by $\underline{X}_{\beta_1}$, $\beta_1 = 1,2,\ldots,N_\beta$, where $N_\alpha(N_\beta)$ are the total number of neighboring points of $\underline{X}_j(\underline{X}_k)$ excluding $\underline{X}_{Y_{1,2}}$.

We first present the analysis for the plane case, then we make the necessary modifications on the sphere. We need to determine the position of the new point $\underline{X}'$ and its velocity $\underline{U}'$, height $\underline{H}'$, and mass m'.

Since the areas of the spherical polygons $P_{\alpha_1}$, $P_{\beta_1}$, $P_{Y_1}$ are changed by moving the points $\underline{X}_j$ and $\underline{X}_k$ to $\underline{X}'$, therefore, their masses $^m\alpha_1$, $^m\beta_1$ and $^mY_1$ change to $^{m'}\alpha_1$, $^{m'}\beta_1$, $^{m'}Y_1$. We choose to keep the heights $^H\alpha_1$, $^H\beta_1$, and $^HY_1$ the same after a collision, so we adjust $^{m'}\alpha_1$, $\underline{U}'\alpha_1$, $^m\beta_1$, $\underline{U}'\beta_1$ and $^{m'}Y_1$, $\underline{U}'Y_1$ in such a way the total mass and momentum as well as the heights of the non-colliding particles are conserved. The location of $\underline{X}'$ is arbitarily chosen to be weighted average (center of mass) of $\underline{X}_j$ and $\underline{X}_k$. We therefore have the following system of equations to solve.

## Conservation of Height of Non-Colliding Particles

$$H'_{\alpha_1} = H_{\alpha_1}$$

$$H'_{\beta_1} = H_{\beta_1} \qquad (5.2a)$$

$$H'_{Y_1} = H_{Y_1}$$

## Conservation of (Total) Mass

$$m_j + m_k + \sum_{i=1}^{N_\alpha} m_{\alpha_i} + \sum_{i=1}^{N_\beta} m_{\beta_i} + \sum_{i=1}^{2} m_{\gamma_i}$$

$$= \sum_{i=1}^{N_\alpha} m'_{\alpha_i} + \sum_{i=1}^{N_\beta} m'_{\beta_i} + \sum_{i=1}^{2} m'_{\gamma_i} + m' \tag{5.2b}$$

## Conservation of (Total) Momentum

$$m_j \underline{U}_j + m_k \underline{U}_k + \sum_{i=1}^{N_\alpha} m_{\alpha_i} \underline{U}_{\alpha_i} + \sum_{i=1}^{N_\beta} m_{\beta_i} \underline{U}_{\beta_i} + \sum_{i=1}^{2} m_{\gamma_i} \underline{U}_{\gamma_i}$$

$$= \sum_{i=1}^{N_\alpha} m'_{\alpha_i} \underline{U}'_{\alpha_i} + \sum_{i=1}^{N_\beta} m'_{\beta_i} \underline{U}'_{\beta_i} + \sum_{i=1}^{2} m'_{\gamma_i} \underline{U}'_{\gamma_i} + m' \underline{U}' \tag{5.2c}$$

## Create New Particle at Center of Mass of Colliding Particles

$$\underline{X}' = \frac{(m_j \underline{X}_j + m_k \underline{X}_k)}{(m_j + m_k)} \tag{5.2d}$$

## Mass and Momentum After Collision

The solution of equations (5.2) for the mass and momentum after a collision has occurred ($m'_\alpha$, $m'_\beta$, $m'_\gamma$, $m'$, $\underline{U}'_\alpha$, $\underline{U}'_\beta$, $\underline{U}'_\gamma$, $\underline{U}'$) is constructed according to the following principles:  When $\underline{X}_j$ and $\underline{X}_k$ are merged into a new point $\underline{X}'$, the boundaries of the neighboring spherical polygons change, thereby changing the areas of those polygons.  This has the affect of transferring fluid from $P_j$ and $P_k$ to $P_{\alpha_i}$ and $P_{\beta_i}$ and from $P_{\gamma_i}$ to $P_x$. since fluid is transferred between the polygons, we also transfer the corresponding mass and momentum.

We note that, since the only points that actually move are $\underline{X}_j$ and $\underline{X}_k$, the faces between any two adjacent neighboring polygons do not move.  Thus there is no transfer of fluid amongst neighbors of $\underline{X}_j$ or $\underline{X}_k$, only between neighboring polygons, to either $\underline{X}_j$ or $\underline{X}_k$.  (See Figure 5.1.)

When $\underline{X}_j$ or $\underline{X}_k$ move toward $\underline{X}'$ they are moving away from $\underline{X}_{\alpha_i}$ or $\underline{X}_{\beta_i}$.  The faces of the polygons between $\underline{X}_j$ and $\underline{X}_{\alpha_i}$ (or $\underline{X}_k$ and $\underline{X}_{\beta_i}$) move in, which has the same effect as transferring fluid out of $P_j$($P_k$) and into $P_{\alpha_i}$ ($P_{\beta_i}$).

Let $\Delta A_{\alpha_i}$ denote the change in area of polygon $P_{\alpha_i}$ due to a collision, i.e.,

$$\Delta A_{\alpha_i} = A'_{\alpha_i} - A_{\alpha_i}.$$

In order to keep the heights of $^H\alpha_i$ the same after a collision we look for a solution of (5.2) in the form

$$m'_{\alpha_i} = m_{\alpha_i} + (\Delta A_{\alpha_i})^H \alpha_i \qquad (5.3a)$$

$$m'_{\beta_i} = m_{\beta_i} + (\Delta A_{\beta_i})^H \beta_i \qquad (5.3b)$$

$$m'_{\gamma_i} = m_{\gamma_i} + (\Delta A_{\gamma_i})^H \gamma_i \qquad (5.3c)$$

$$m' = m_j + m_k - \sum_{i=1}^{N_\alpha} (\Delta A_{\alpha_i})^H \alpha_i + \sum_{i=1}^{N_\beta} (\Delta A_{\beta_i})^H \beta_i + \sum_{i-1}^{2} (\Delta A_{\gamma_i})^H \gamma_i \qquad (5.3d)$$

To solve equation (5.2b) for conservation of total momentum we note first that $^{\Delta A}\gamma_i$ can be positive or negative (see Figure 5.1). If $^{\Delta A}\gamma_i < 0$, the net flow of fluid particles from $^P\gamma_i$ is into the new polygon $P_{X'}$. If $^{\Delta A}\gamma_i > 0$, the net flow of fluid is out of the new polygon $P_{X'}$ into its neighbor $^P\gamma_i$. We therefore have the following solution to (5.2c)

$$\underline{U}'_{\alpha_i} = [{}^m\alpha_i \, \underline{U}_{\alpha_i} + (\Delta A_{\alpha_i})^H \alpha_i \, \underline{U}_j]/m'_{\alpha_i} \qquad (5.4a)$$

$$\underline{U}'_{\beta_i} = [{}^m\beta_i \, \underline{U}_{\beta_i} + (\Delta A_{\beta_i})^H \beta_i \, \underline{U}_k]/m'_{\beta_i} \qquad (5.4b)$$

Also, if $^{\Delta A}\alpha_i < 0$

$$\underline{U}_{\gamma_i} = [{}^m\gamma_i \, \underline{U}_{\gamma_i} + (\Delta A_{\gamma_i})^H \gamma_i \, \underline{U}_{\gamma_i}]/m'_{\gamma_i} \qquad (5.4c)$$

$$\underline{U}' = \{m_j \, \underline{U}_j + m_k \, \underline{U}_k - [\sum_{i=1}^{N_\alpha} (\Delta A_{\alpha_i})^H \alpha_i \, \underline{U}_j$$

$$+ \sum_{i=1}^{N_\beta} (\Delta A_{\beta_i})^H \beta_i \, \underline{U}_k + \sum_{i=1}^{2} (\Delta A_{\gamma_i})^H \gamma_i \, \underline{U}_{\gamma_i} \qquad (5.4d)$$

However, if $^{\Delta A}\gamma_i > 0$, define $\underline{U}_{jk} = (m_j \, \underline{U}_j + m_k \, \underline{U}_k)/(m_j + m_k)$, then

$$\underline{U}'_{\gamma_i} = [{}^m\gamma_i \, \underline{U}_{\gamma_i} + (\Delta A_{\gamma_i})^H \gamma_i \, \underline{U}_{jk}]/m'_{\gamma_i} \qquad (5.4e)$$

$$\underline{U}' = [m_j\underline{U}_j + m_k\underline{U}_k - (\sum_{i=1}^{N_\alpha} (\Delta A_{\alpha i})^H \alpha_i \underline{U}_j + \sum_{i=1}^{N_\beta} (\Delta A_{\beta_i})^H \beta_i \underline{U}_k$$

$$- \sum_{i=1}^{2} (\Delta A_{\gamma_i})^H \gamma_i \underline{U}_{jk})])/m' \qquad (5.4f)$$

It can easily be shown in the same manner as the one-dimensional case [2] that the solutions (5.3) and (5.4) solve the conservation equations (5.2).

We now describe the modifications that are necessary for a sphere. Equation (5.2d) yields a new point, $\underline{X}'$, which is not on the surface of the sphere. We therefore normalize $\underline{X}'$ to find the closest point on the sphere.

$$X' = \frac{m_j\underline{X}_j + m_k\underline{X}_k}{m_j + m_k} \Bigg/ \left| \frac{m_j\underline{X}_j + m_k\underline{X}_k}{m_j + m_k} \right| \qquad (5.5)$$

$$= \frac{m_j\underline{X}_j + m_k\underline{X}_k}{|m_j\underline{X}_j + m_k\underline{X}_k|}$$

Also, the new velocities $\underline{U}'_{\alpha_i}, \underline{U}'_{\beta_i}, \underline{U}'_{\gamma_i}, \underline{U}'$ have to be projected, tangent to the sphere, at their respective points, i.e., $\underline{U}'$ is projected on the sphere at $\underline{X}'$, $\underline{U}'_{\alpha_i}$ is projected at $\underline{X}_{\alpha_i}$, etc. Due to the projection, momentum is no longer conserved, exactly, however the effects of the change in total momentum is small since the projection is second order in space while the method appears, numerically, to be first order in space (see Section 7).

6.  Numerical Method

We now state a numerical method for the shallow-water equations on a rotating sphere.

The spatially discrete equations are (see Section 3)

$$m_j^o = A_j H_j \qquad (6.1a)$$

$$\frac{d\underline{U}_j}{dt} + 2(\underline{\Omega} \times \underline{U}_j)|| = \frac{g}{2m_j} \sum_{k=1}^{N} H_k^2 \frac{\partial A_k}{\partial \underline{X}_j} - |\underline{U}_j|^2 \underline{X}_j \qquad (6.1b)$$

$$\frac{d\underline{X}_j}{dt} = \underline{U}_j \qquad (6.1c)$$

with intial data

$$\underline{U}_j(0) = \text{given} \tag{6.2a}$$

$$H_j(0) = \text{given} \tag{6.2b}$$

$$\underline{X}_j(0) = \text{given} \tag{6.2c}$$

We discretize time by a fractional step method to arrive at the following explicit scheme:

$$m_j^o = A_j^n H_j^n \tag{6.3a}$$

$$\tilde{\underline{U}}_j^{n+1} = \underline{U}_j^n + (\Delta t)\ [-2(\underline{\Omega} \times \underline{U}_j^n)||\ +\ \frac{g}{2m_j}\ \sum_{k=1}^{N}\ (H_k^n)^2\ \frac{\partial A_k}{\partial \underline{X}_j}\bigg|_{\underline{X}_j = \underline{X}_j^n}\ -|\underline{U}_j^n|^2\underline{X}_j^n] \tag{6.3b}$$

$$\underline{U}_j^{n+1} = P_j \tilde{\underline{U}}_j^{n+1}$$

$$\underline{X}_j^{n+1} = \underline{X}_j^n + (\Delta t)\underline{U}_j^{n+1} \tag{6.3c}$$

$$\underline{X}_j^{n+1} = \frac{\underline{X}_j^{n+1}}{|\underline{X}_j^{n+1}|}$$

where $P_j\underline{\phi}$ = projection of $\underline{\phi}$ onto tangent plane to the sphere at $\underline{X}_j$.

and $\qquad\qquad H_j^n = H(\underline{X}_j^n, n\Delta t), \quad \underline{U}_j^n = \underline{U}(\underline{X}_j^n, n\Delta t)$

Initial conditions:

$$X_j^o = \text{given} \tag{6.3d}$$

$$U_j^o = \text{given} \tag{6.3e}$$

$$H_j^o = \text{given} \tag{6.3f}$$

We note that (6.3) is first order accurate in time. The order of accuracy of the spatial derivatives, however, is not known. Numerical experiments suggest that they are also first order accurate.

The algorithm is now as follows:

i) At time t=0, N points are chosen arbitrarily on the sphere (see Augenbaum and Peskin (1985)) and designated $(\underline{X}_1^o, \ldots \underline{X}_N^o) = \underline{X}^o$. These will be the generating points of the spherical polygons, and they will move with the fluid. Initial velocities and heights are assigned to these points according to

$$\underline{U}_k^o = \underline{U}(\underline{X}_k^o, 0) \qquad\qquad (6.4a)$$

$$H_k^o = H(\underline{X}_k^o, 0) \qquad\qquad (6.4b)$$

ii) Once the polygons are constructed we can calculate the area $A_j$ of each polygon. The area is then used, with the height $H_j^o$ to calculate the mass $m_j^o$ according to (6.3a).

iii) At each time step, the positions, velocities, and heights of the generating points are updated as follows:

a. Find $\underline{U}^{n+1}$ from known quantities $\underline{U}^n$, $H^n$, $\underline{X}^n$, using the explicit formula (6.3b).

b. Once $\underline{U}^{n+1}$ has been found, advect the generating points according to (6.3c). This yields the new generating points $\underline{X}^{n+1}$.

c. We can now compute the new polygonal structure at time $t = (n+1)\Delta t$ and therefore we can compute the new areas $A_j^{n+1}$.

d. Finally, we update the height by using (6.3a)

$$H_j^{n+1} = \frac{m_j^o}{A_j^{n+1}} \qquad\qquad (6.5)$$

This completes a time step.

The algorithm just presented must be slightly modified. Stability considerations require that, at all time, the points satisfy a Courant–Friedrichs–Lewy condition

$$\frac{\Delta t}{\Delta X} \le \frac{1}{c} \qquad\qquad (6.6)$$

or

$$\Delta X \ge c \, (\Delta t) \qquad\qquad (6.6a)$$

Since the points are not uniformly spaced from each other we have the condition

$$\min_j \, (\Delta X)_j \ge c \, (\Delta t) \qquad\qquad (6.7)$$

Now, recall from the previous section that when points get closer than some distance, 'd', we consider the particles to have collided and replace them by one averaged point. A natural choice for this distance, d, is is obtained from (6.7), i.e., d = c(Δt) = CFL.

The modification of the algorithm takes place after step iii.b). Once $\underline{X}_j^{n+1}$ is found we sweep through all the $\underline{X}_j$, j = 1, ...., N and see if the distance from $\underline{X}_j$ to its neighboring points is less then CFL or not. If it is, we follow the procedure outlined in Section 5, i.e., we remove the two points in favor of one averaged point, recompute the polygons and continue on.

## 7.  Results--Spherical Case

In this section we present the results of numerical computations on various test cases.

### Zonal Flow

The first test case tried was the so called zonal flow.  The zonal flow is an exact steady state solution of the shallow water equations.  In typical Eulerian calculations (fixed mesh and spherical coordinates) the zonal flow represents a trivial computation;  however, in our case, since the mesh it not uniform initially, and the points are free to move, this represents a non-trival test case.

The zonal flow is usually given in spherical coordinates (Gilliland (1981)). For our purposes we convert it to Cartesian coordinates.  We have on the unit sphere,

$$U = -QY$$

$$V = QX$$

$$W = 0 \tag{7.1}$$

$$H(X,Y,Z) = D - \frac{1}{g} (\Omega Q + \underline{Q}^2) Z^2$$

where Q = maximum velocity
Ω = rotation rate
D = hydrostatic height of fluid
g = gravitation constant

Thus the zonal flow is a rigid body rotation with constant angular velocity Q.

We use, in our calculations, data that is realistic for the earth, (Dutton (1976)) i.e.,

$$\text{Gravity} = g = 9.8 \text{ m/sec}^2$$
$$\text{Hydrostatic height of atmosphere} = D = 8.5 \text{ km}$$
$$\text{rotation rate} = \Omega = 7.3 \times 10^{-5} \text{ sec}^{-1}$$
$$\text{radius of earth} = r = 6371 \text{ km}$$
$$Q = 10 \text{ m/sec}$$

Since our code is set up for the unit sphere we scale all lengths. We introduce a new unit called an erad (earth radius) defined so that

$$1 \text{ erad} = 6.371 \times 10^8 \text{ cm (radius of earth)} \qquad (7.2)$$

With this scaling the other quantities become

$$r = 1 \text{ erad}$$
$$D = 1.33 \times 10^{-3} \text{ erads}$$
$$g = 1.53 \times 10^{-6} \text{ erads/sec}^2$$
$$c = \sqrt{gD} = 1.4302 \times 10^{-5} \text{ erads/sec (gravity wave speed)}$$
$$\Omega = 7.3 \times 10^{-5} \text{ sec}^{-1}$$
$$Q = 1.56 \times 10^{-6} \text{ erads/sec} \qquad (7.3)$$

To see how well out method works on the zonal flow, we computed the relative velocity error.

$$\text{Err} = \frac{|U_{exact} - U_{computed}|}{|U_{exact}|}$$

$$= \frac{\left\{ \sum\limits_{j=1}^{N} [(U_{ex}(J) - U_{comp}(J))^2 + (V_{ex}(J) - V_{comp}(J))^2 + (W_{ex}(J) - W_{comp}(J))^2]\text{Area}(J) \right\}^{1/2}}{\left\{ \sum\limits_{j-1}^{N} [(U_{ex}(J))^2 + ((V_{ex}(J))^2 + (W_{ex}(J))^2]\text{Area}(J) \right\}^{1/2}}$$

$$(7.4)$$

We now present the results of various test runs.

1. As $\Delta t(\Delta x)$ was decreased by a factor of 2, i.e., $\Delta t \rightarrow \Delta t/2$, the change in error also decreased by a factor of 2, i.e., $\Delta E \rightarrow \Delta E/2$. This indicates that the method is first order in time (space).

2. In the case of NPTS = 512 on a non-rotating sphere ($\Omega \equiv 0$) the error was less than .1% after 1 day with a time step of $\Delta t = 360$ sec.

3. In the case of NPTS = 326 pts on a rotating sphere ($\Omega$ = 7.3 x 10$^{-5}$/sec), the
   relative velocity error was 10.6% after 6 hours. After 12 hours the error
   was less than 15%. It should be noted that these meshes are extremely coarse
   by any standards.

4. A computer generated movie was made for NPTS = 488 with a time step $\Delta$t = 360
   sec and TMAX = 48 hours. The results clearly show a rigid body rotation,
   much better than the relative error would seem to indicate. Another observation is that each point also seems to move in a small circular orbit while
   also moving along the path of a rigid body rotation. The radius of these
   circular orbits is extremely small. In fact this feature is hardly noticeable
   to the eye, compared to the zonal flow. This probably explains why the flow
   looks so much like a rigid body rotation while the relative velocity error is
   as high as 20-30%.

5. When the distance between any two points was allowed to violate the Courant-
   Friedrich-Lewy condition, the flow became unstable. This instability mani-
   fested itself as a very fast gravity wave which shot through the fluid and
   completely destroyed the zonal flow. When we enforced the CFL condition
   by means of the merging procedure of Section 5, the flow was completely
   stable for test runs as high as 500 iterations. The CFL condition is $\Delta t / \Delta x$
   < 1/c, or

$$\Delta x > c(\Delta t), \quad \text{where c = gravity wave speed} = \sqrt{gD}.$$

For $\Delta t$ = 360 sec, we have $c\Delta t$ = .005 erads. Thus the CFL condition is that
any two points must be at least .005 erads from each other (for the case $\Delta t$ =
360 sec) for the flow calculations to be stable. This was borne out in
practice.

## Riemann Problem

The other test case tried on a sphere were a series of numerical experiments for
the analog of the Dam-Breaking problem that was treated in [2]. In this problem the
initial velocity of all fluid particles is zero. The height of the fluid is, initially, the hydrostatic height of the atmosphere, D, for all points except those in a
spherical cap around the north pole. In this spherical cap we set the height of the
fluid equal to 2D. We then let the flow evolve. The exact solution of this problem
is not known to this author, however there are certain qualitative features of the
flow which one would expect to find.

In the first experiment we tried this problem on a non-rotating sphere with the
same physical parameters as the zonal flow and with a time step $\Delta t$ = 360 sec and

TMAX = 48 hours. The results of this calculation are best displayed in a computer-generated movie for the changes in the height field. Several frames from the movie are shown in Figure 7.1.

We found the following:

1. There is a distinct shock wave that propagates downward past the equator and into the southern hemisphere. The shock converges on the south pole and is reflected back towards the north pole.

2. As the shock approaches the equator it seems to weaken and slow down; however, it speeds up and gets stronger as it approaches the south pole. This is consistent with the fact that as the shock gets closer to the equator it is more spread out and therefore to conserve energy, it must slow down and weaken. Whereas, when the shock is near the south pole, it is confined to a small area and each particle has a lot of energy and so the shock speeds up and strengthens.

As our last two experiments we tried the same problem on a rotating sphere with rotation rate $\Omega = 7.3 \times 10^{-5}$ sec $^{-1}$. In the first case the center of the cap of initially-raised fluid is at the north pole and in the second case, it is 45° away from the pole. Again, we displayed the results in a computer-generated movie, several frames of which are shown in Figure 7.2. We found the following:

1. The shock front that was present in the non-rotating case appears to be smoothed out after a short time. This is due to the fact that waves on a rotating sphere are dispersive. The wave speed is given by $c = \sqrt{gH + f^2/k^2}$, where $f = 2\Omega \sin \theta$, $k$ = wave number, and $\theta$ = latitude. On a non-rotating sphere the waves are non-dispersive i.e., $c = \sqrt{gh}$ and therefore the shock propagation is sharp. The dispersive nature of the wave in the rotating case explains why shocks are not seen in the atmosphere (personal communication by Michael Ghil).

2. There is, however, a reflection of the smoothed out wave in the southern hemisphere. In the polar rotation case (case 1), the reflection occurs at the south pole while in the off axis case (case 2), the reflection is deflected clockwise. In the southern hemisphere particles are deflected clockwise, while in the northern hemisphere they are deflected counter-clockwise. This effect is due to the coriolis term that is present in the equations on a rotating sphere.

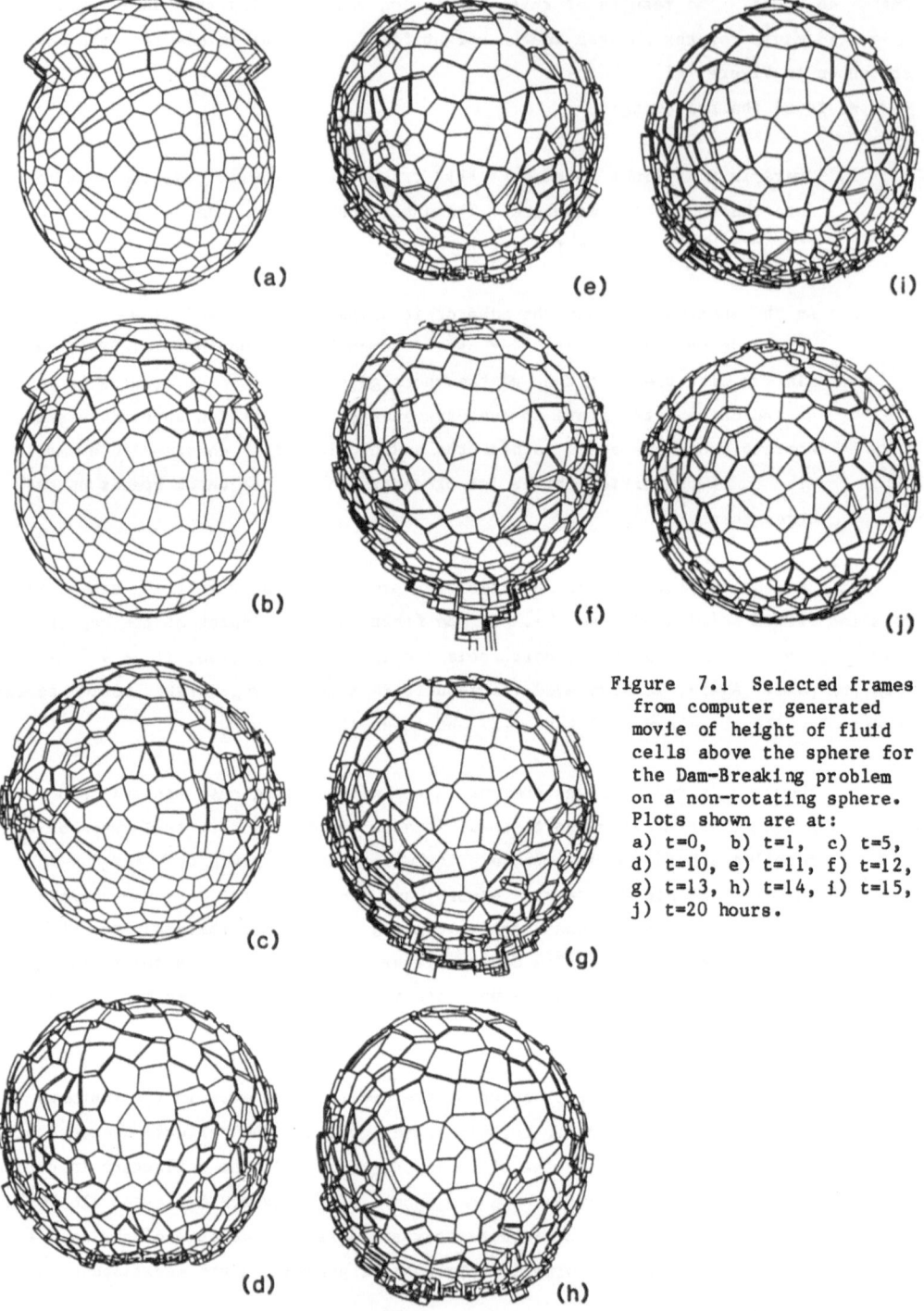

Figure 7.1 Selected frames
from computer generated
movie of height of fluid
cells above the sphere for
the Dam-Breaking problem
on a non-rotating sphere.
Plots shown are at:
a) t=0, b) t=1, c) t=5,
d) t=10, e) t=11, f) t=12,
g) t=13, h) t=14, i) t=15,
j) t=20 hours.

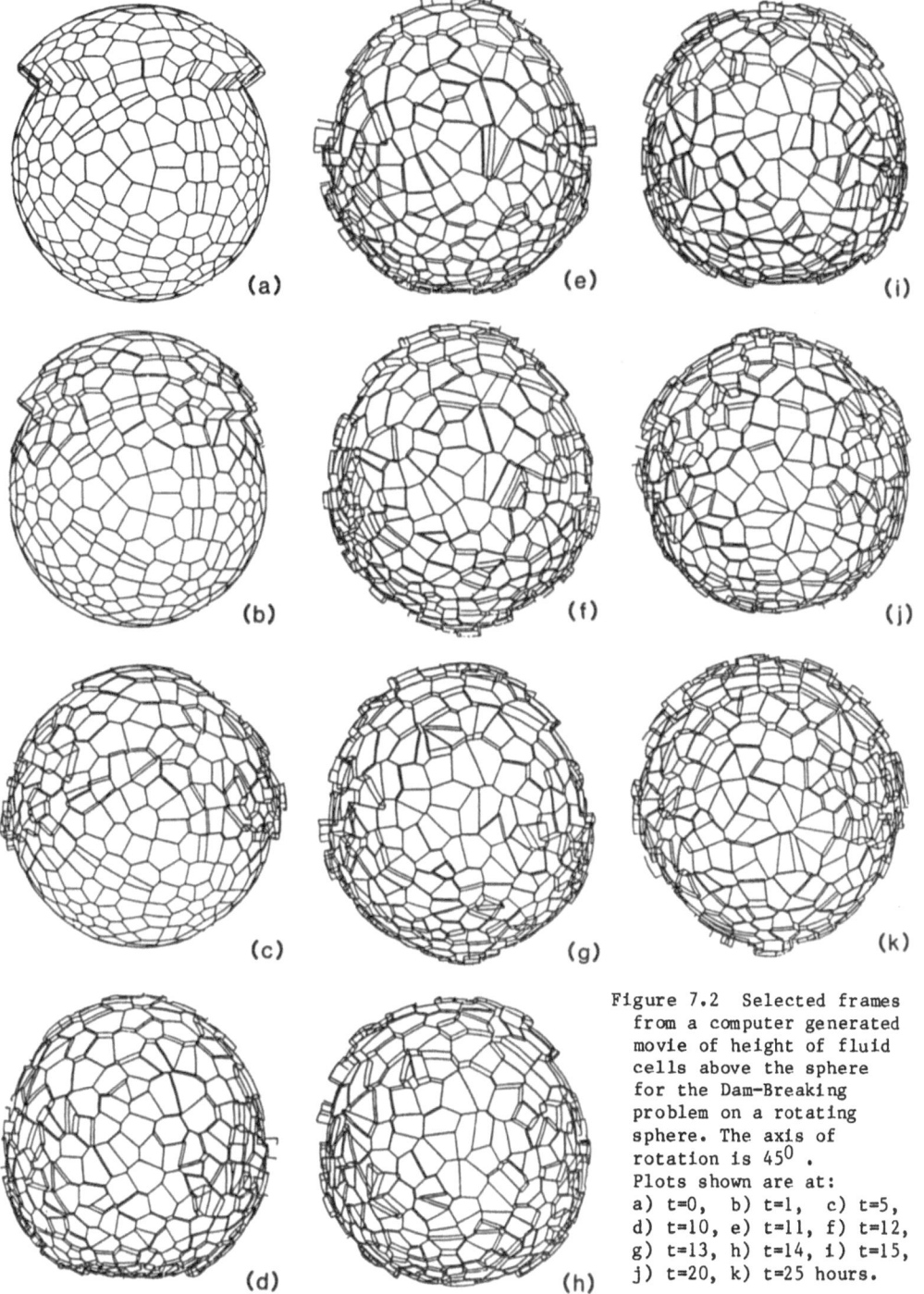

Figure 7.2 Selected frames from a computer generated movie of height of fluid cells above the sphere for the Dam-Breaking problem on a rotating sphere. The axis of rotation is $45^0$ .
Plots shown are at:
a) t=0,  b) t=1,  c) t=5, d) t=10, e) t=11, f) t=12, g) t=13, h) t=14, i) t=15, j) t=20, k) t=25 hours.

3. Finally, after the shock is smoothed out, the coriolis force appears to set up a smooth wave motion on the sphere. The exact nature of this wave is not known.

## 8. Conclusions

We have presented in this work and in a previous work [1, 2] a promising new Lagrangian method for the shallow water equations. The advantages of this method may be summarized as follows. First, there is no dependence on coordinate systems, and thus we can treat flows on a sphere without worrying about pole singularities. Second, unlike other Lagrangian schemes, there is no restriction that points have to retain their initial neighbors. On the contrary, at each time step the particles find their natural neighbors, and the derivatives are computed using these neighbors. Thus the method allows for large deformations. Third, a novel feature of the method is that it handles shocks in a very natural way. We consider a shock as one fluid particle overtaking another and colliding with it. This procedure not only makes it possible to handle shocks (which are unimportant in atmospheric flow calculations since the rotation of the earth causes waves to be dispersive) but it also guarantees the stability of the scheme by enforcing the Courant-Friedrichs-Lewy conditions in the neighborhood of each fluid marker. Fourth, since the markers can be placed anywhere, the fluid markers may be placed, initially, at points where satellite or other measured data are available. The main problem with Free Lagrangian methods however, is that the discrete operators seem to be of low order accuracy. Further research is needed to derive high order accurate schemes to compete with existing Eulerian schemes.

## References

1. J. Augenbaum, "A New Lagrangian Method for the Shallow Water Equations", Ph.D. Thesis, New York University, 1982.

2. _____, "A Lagrangian Method for the Shallow Water Equations Based on a Voronoi Mesh-One Dimensional Results", J. Comp. Phys., 53 (1984), pp. 240-265.

3. _____, and C. Peskin, "On the Construction of the Voronoi Mesh on a Sphere", J. Comp. Phys., in press (1985).

4. W. P. Crowley, "FLAG: A Free Lagrange Method for Numerically Simulating Hydrodynamic Flow into Two Dimensions", Proceedings of the Second International Conference on Numerical Methods in Fluid Dynamics, Lecture Notes in Physics, Vol. 8, pp. 37-43, Springer Verlag, New York (1971).

5. J. K. Dukowicz, "Lagrangian Fluid Dynamics Using the Voronoi Delaunay Mesh", in Numerical Methods for Coupled Problems. Pineridge Press, Swansea, U.K. (1981).

6. J. Dutton, The Ceaseless Wind. An Introduction to the Theory of Atmospheric Motion, McGraw Hill, New York (1976).

7. M. Fritts and J. Boris, "The Lagrangian Solution of Transient Problems in Hydrodynamics Using a Triangular Mesh", J. Comp. Phys., 31, pp 173-215 (1979).

8. R. Gilliland, "Solutions of the Shallow Water Equations on a Sphere", J. Comp. Phys., 43, pp 79-84, (1981).

9. G. Haltiner and R. T. Williams, Numerical Prediction and Dynamic Meteorology, 2nd edition, Wiley, New York (1980).

10. F. Mesinger, "Numerical Integration of the Primitive Equation with Floating Set of Computation Points: Experiments with a Barotropic Global Model", Mon. Wea. Rev., 99, pp 15-29 (1971).

11. C. Peskin, "A Lagrangian Method for the Navier Stokes Equation with Large Deformation", preprint.

12. R. Salmon, "Practical Use of Hamilton's Principle", J. Fluid. Mech., 132, pp 431-444 (1981).

13. H. E. Trease, "A Two Dimensional Free Lagrangian Hydrodynamics Model", Ph.D. Thesis, University of Illinois at Urbana-Champaign (1981).

14. G. Voronoi, J. Reine Angew Math., 134, p 198 (1908).

15. P. K. Wang,"A Brief Review of the Eulerian Variational Principle for Atmospheric Motion in Rotating Coordinates, Atmos. Oceans, 22, pp 387-392 (1984).

16. P. L. Williamson and G. L. Browning, "Comparison of Grids and Difference Approximation for Numerical Weather Prediction Over a Sphere", J. Appl. Meteor., 9, p 262 (1973).

## Appendix A:  Shallow Water Equations on a Rotating Sphere

In this appendix we transform equations (2.18) to rotating coordinates.
Consider a coordinate system rotating with angular velocity $\underline{\Omega}$.

Let

$$\underline{U}_o = \underline{U}_r + \underline{\Omega} \times \underline{X} \qquad\qquad (A.1)$$

where $\underline{U}_o$ is the velocity relative to a stationary observer and $\underline{U}_r$ is the velocity relative to an observer rotating with angular velocity $\underline{\Omega}$.

Differentiating (A.1) we get

$$\left(\frac{d\underline{U}_o}{dt}\right)_o = \frac{d(\underline{U}_r + \underline{\Omega} \times \underline{X})}{dt} + \underline{\Omega} \times (\underline{U}_r + \underline{\Omega} \times \underline{X}) = \frac{d\underline{U}_r}{dt} + 2(\underline{\Omega} \times \underline{U}_r) + \underline{\Omega} \times (\underline{\Omega} \times \underline{X}) \qquad (A.2)$$

### Conservation of Mass

The equation $m_j^o = A_j H_j$ is unchanged in rotating coordinates since no time derivatives occur.

### Conservation of Momentum

Using (A.2) in the momentum equation (2.18a) we get

$$\frac{d(\underline{U}_r^j)^o}{dt} + 2(\underline{\Omega} \times \underline{U}_r^j) + \underline{\Omega} \times (\underline{\Omega} \times \underline{X}_j) = \frac{g}{2m_j} \sum_{k=1}^{N} (H_k)^2 \frac{\partial A_k}{\partial \underline{X}_j} - |\underline{U}_r^j + \underline{\Omega} \times \underline{X}_j|^2 \underline{X}_j \qquad (A.3)$$

Let us now drop the subscript "r" with the understanding that $\underline{U} = \underline{U}_r$. We can now write each of the vectors $(\underline{\Omega} \times \underline{U}_j)$ and $\underline{\Omega} \times (\underline{\Omega} \times \underline{X}_j)$ as the sum of two vectors, one parallel to the sphere and one normal to the sphere.

### Centripetal Term

The normal component of the centripetal term $\underline{\Omega} \times (\underline{\Omega} \times \underline{X})$ is

$$[\underline{\Omega} \times (\underline{\Omega} \times \underline{X})]_\perp = \{[\underline{\Omega} \times (\underline{\Omega} \times \underline{X})] \cdot \underline{X}\}\underline{X} = [\underline{X} \cdot (\underline{A} \times \underline{B})]\underline{X}$$

where $\underline{A} = \underline{\Omega}$ and $\underline{B} = \underline{\Omega} \times \underline{X}$.

By the vector identity

$$\underline{X} \cdot (\underline{A} \times \underline{B}) = \underline{B} \cdot (\underline{X} \times \underline{A})$$

we arrive at

$$[\underline{\Omega} \times (\underline{\Omega} \times \underline{X})]_\perp = -|\underline{\Omega} \times \underline{X}|^2 \underline{X} \qquad\qquad (A.4)$$

The component of $\underline{\Omega} \times (\underline{\Omega}x\underline{X})$ parallel to the sphere will be denoted by

$$[\underline{\Omega} \times (\underline{\Omega}x\underline{X})]|| = \underline{\Omega} \times (\underline{\Omega}x\underline{X}) - [\underline{\Omega} \times (\underline{\Omega}x\underline{X})]_\perp .$$

## Coriolis Term

The normal component of the coriolis term $2(\underline{\Omega}x\underline{U})$ is

$$2(\underline{\Omega}x\underline{U})_\perp = 2[(\underline{\Omega}x\underline{U}) \cdot \underline{X}]\underline{X} = 2\underline{X}[\underline{X} \cdot (\underline{A}x\underline{B})]$$

where $\underline{A} = \underline{\Omega}$ and $\underline{B} = \underline{U}$.

By the previous vector identity we get

$$2(\underline{\Omega}x\underline{U})_\perp = -2\underline{X}[\underline{U} \cdot (\underline{\Omega}x\underline{X})] \qquad (A.5)$$

The component of $(\underline{\Omega}x\underline{U})$ parallel to the sphere will be denoted by

$$(\underline{\Omega}x\underline{U})|| = (\underline{\Omega}x\underline{U})-(\underline{\Omega}x\underline{U})_\perp .$$

We can also expand the term $-|\underline{U}_j+\underline{\Omega}x\underline{X}_j|^2\underline{X}_j$ as follows

$$-|\underline{U}_j+\underline{\Omega}x\underline{X}_j|^2 = - \{|\underline{U}_j|^2 + 2\underline{U}_j \cdot (\underline{\Omega}x\underline{X}_j) + |\underline{\Omega}x\underline{X}_j|^2 \} \qquad (A.6)$$

Therefore, using (A.4), (A.5) and (A.6) in the momentum equation (A.3) we get

$$\frac{d\underline{U}_j}{dt} + 2(\underline{\Omega}x\underline{U}_j)|| + [\underline{\Omega}x(\underline{\Omega}x\underline{X}_j)]|| = \frac{g}{2m_j} \sum_{k=1}^{N} H_k^2 \frac{\partial A_k}{\partial X_j} - |\underline{U}_j|^2\underline{X}_j \qquad (A.7)$$

In physical applications we will be concerned with the case where $\underline{\Omega}$ is small and therefore negelect the term $\underline{\Omega}x(\underline{\Omega}x\underline{X}_j)||$ since it is quadratic in $\underline{\Omega}$.

We note that all the terms in (A.7) are parallel to the sphere since $\partial A_k/\partial X_j$ is parallel to the sphere at $\underline{X}_j$ by definition.

The result of the above derivation is that, to study large scale flows on a rotating sphere, all we need to do is add a coriolis force term parallel to the sphere.

Finally, we remark that the equation of motion $d\underline{X}_j/dt = \underline{U}_j$ is unchanged in the rotating reference frame.

We can now summarize the discrete equations of motions in a rotating reference frame (relative to the rotating frame) for small $\underline{\Omega}$.

## Conservation of Mass

$$m_j^o = A_j H_j \qquad (A.8a)$$

Conservation of Momentum

$$\frac{d\underline{U}_j}{dt} + 2(\underline{\Omega} \times \underline{U}_j)|| = \frac{g}{2m_j} \sum_{k=1}^{N} H_k^2 \frac{\partial A_k}{\partial \underline{X}_j} - |\underline{U}_j|^2 \underline{X}_j \qquad (A.8b)$$

Move Points With Local Fluid Velocity

$$\frac{d\underline{X}_j}{dt} = \underline{U}_j \qquad (A.8c)$$

A Lagrangian Method Based on the Voronoi Diagram for the
Incompressible Navier Stokes Equations on a Periodic Domain

Christoph Börgers and Charles S. Peskin

1.  Introduction

We consider a fractional step method for the solution of

(1.1)     $\underline{u}_t + (\underline{u} \cdot \nabla)\underline{u} - \nu\Delta\underline{u} + \nabla p = \underline{f}$ ,   $\nabla \cdot \underline{u} = 0$

($\underline{u} = \underline{u}(\underline{x},t)$, $p = p(\underline{x},t)$, $\underline{x} \in \mathbb{R}^2$, $t \in \mathbb{R}$, $t > 0$) with the initial condition

(1.2)     $\underline{u}(\underline{x},0) = \underline{u}_0(\underline{x})$

and the periodicity conditions

(1.3)     $\underline{u}(\underline{x} + \underline{k},t) = \underline{u}(\underline{x},t)$, $p(\underline{x} + \underline{k},t) = p(\underline{x},t)$ , $\underline{k} \in \mathbb{Z}^2$.

We intend to use this method in problems involving immersed boundaries, modelling the boundaries by a chain of Lagrangian particles connected by springs [10]. In this context, it seems natural to use Lagrangian particles for the fluid as well.

In section 2, we describe a way of adapting the algorithm for the construction of Voronoi meshes suggested in [9] to the case of a periodic domain. We believe that this algorithm can be used in three dimensions with slight modifications. It can also be used as a triangulation algorithm which finds a triangulation with a prescribed set S of corners: The dual graph of the Voronoi mesh (associated with the set S) is such a triangulation, called the "Delaunay triangulation".

In [9], Peskin described discretizations of the Laplace, gradient and divergence operators on arbitrary sets of points in the plane. We review the derivation of these operators in section 3, and we prove that the discrete divergence and gradient operators are weakly consistent with the corresponding continuous operators.

In section 4, we describe a two-level iteration for the solution of discrete Helmholtz equations and present results of numerical experiments with this method which show that the efficiency of the method is similar to the efficiency of multigrid algorithms for the Helmholtz equation discretized on regular meshes.

In section 5, we discuss the problem of projecting a vector field onto its divergence-free component. We discuss an iterative algorithm which is applicable on unstructured meshes and appears to find quickly a good approximation; convergence to the exact solution of the discrete projection problem is guaranteed but slow.

Finally, we combine these tools to obtain a fractional step method for the Navier-Stokes equations (section 6). Preliminary numerical experiments indicate that we obtain a method which is "of first order" in the sense that the truncation error is roughly reduced by 1/2 if the total number N of fluid markers is multiplied by 4 and the time step is reduced by 1/2. (In two space dimensions, the effective mesh width is $O(N^{-1/2})$.) The work per time step required is $O(N\log N)$ but could easily be reduced to $O(N)$. The method is implicit: there is no stability condition on the size of the time step.

## 2. Construction of periodic Voronoi diagrams

Let S be a set of points in $\mathbb{R}^2$. For $x \in S$, we define

$$(2.1) \qquad P(\underline{x},S) : = \{\underline{y} \in \mathbb{R}^2 : \|\underline{y} - \underline{x}\| \leq \|\underline{y} - \underline{\tilde{x}}\| \text{ for all } \underline{\tilde{x}} \in S\} .$$

$P(\underline{x},S)$ is an intersection of half planes, i.e. a convex polygon. It is called the "Voronoi polygon" [11] or "Dirichlet region" of $\underline{x}$ with respect to S. The collection of all $P(\underline{x},S)$, $\underline{x} \in S$, is called the "Voronoi diagram" associated with S. We denote it by $V(S)$.

Interior edges in $V(S)$ are "generated" by two points, i.e. they belong to two polygons. Corners are "generated" by three or more points, i.e. belong to three or more polygons. We always assume that every corner is generated by three points. A degenerate case such as

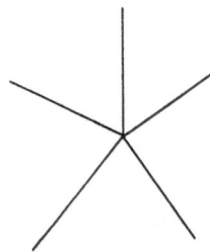

is "resolved" into

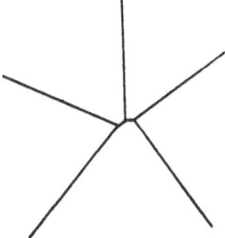

This "resolution" of "multiple" corners into sets of "simple" corners linked by edges of length zero is always possible but never unique.

In [7], Shamos and Hoey described a general algorithm for the construction of Voronoi diagrams requiring $O(N\log N)$ operations (N = number of points in S). This is known to be the best possible operation count for the general case.

If additional information on S is available, better operation counts can be achieved. The algorithm in [9] makes use of the assumption that the Voronoi diagram has already been constructed for a set $\tilde{S}$ which is "close to S", and that certain pieces of information about this construction have been saved. The algorithm used in [5] can be viewed as an algorithm which uses $V(\tilde{S})$ to compute $V(S)$. Experiments indicate that both algorithms construct $V(S)$ in linear time.

If S is random with a known distribution, the optimal expected value of operations may be less than $O(N\log N)$. This statement is also supported by numerical experience, see below.

We now consider the case

$$(2.2) \qquad S = S_N := \{\underline{X}_j + \underline{k}: 1 \leq j \leq N, \ \underline{k} \in Z^2\} \text{ with } \quad \underline{X}_j \in [0,1)^2 .$$

(In our fractional step method, the $\underline{X}_j$ will be fluid markers.) The corresponding Voronoi diagram can be viewed as a periodic Voronoi diagram in the plane or as a diagram on the torus

$$(2.3) \qquad T := \mathbb{R}^2/Z^2$$

with the metric

$$(2.4) \qquad d(\underline{x},\underline{y}) := \min_{\underline{k} \in Z^2} \| \underline{x} - \underline{y} + \underline{k} \| .$$

We describe a modified version of the algorithm in [9] suitable for this case.

It is convenient to use the following storage scheme for $V(S_N)$. We store information about the corners in $V(S_N)$ which lie in $[0;1)^2$. There are exactly 2N such corners.

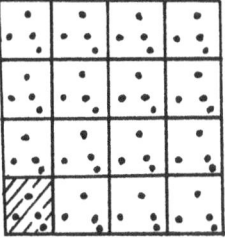

Proof: Consider m by m copies of the points $X_1, \cdots, X_N$:

the unit square

Consider the corresponding Voronoi diagram. Intersect it with $[0;m]^2$. The result is a polygonal net; we extend this net to one on the sphere $\mathbb{R}^2 \cup \{\infty\}$ by introducing edges that connect the four corners of $[0;m]^2$ with $\infty$. This introduces a degenerate corner at $\infty$ which we resolve into two non-degenerate corners connected by an edge of length 0. The net on the sphere has $m^2N + 4$ faces. Since every edge has 2 corners and every corner has 3 edges, the number of corners is $\frac{2}{3}$ times the number of edges. From Euler's formula we now conclude that the total number of corners, including those at $\infty$, is $2m^2N + 4$.

The $(m-2)$ by $(m-2)$ inner squares are identical up to translation. They contain, say, $(m-2)^2M$ corners. (We want to show M=2N.) For fixed N, the number of corners in the 4m-4 boundary squares is $O(m)$. Therefore

$$(m-2)^2M + O(m) = 2m^2N + 2 .$$

Letting $m \rightarrow \infty$, we obtain M = 2N.

We call the 2N corners in $[0;1)^2$ the "stored corners". We also call the points $\underline{X}_1, \cdots, \underline{X}_N$ "the stored points". For the j-th stored corner, we store its cartesian coordinates $(XC(j),YC(j))$ and the square RAD2(j) of its "radius". The "radius" of a corner is its distance from its generating points. Furthermore, we store integer indices

$$\text{IPT}(i,n,j) , \quad i=1,2,3 , \quad n=0,1,2 \quad \text{and}$$
$$\text{ICR}(i,n,j) , \quad i=1,2,3 , \quad n=0,1,2$$

which have the following meaning: the i-th neighbor corner of the j-th stored corner is the ICR(i,0,j)-th stored corner, shifted by ICR(i,1,j) in the x-direction and by ICR(i,2,j) in the y-direction. The i-th generating point of the j-th stored corner is the IPT(i,0,j)-th stored point, shifted by IPT(i,1,j) in the x-direction and by IPT(i,2,j) in the y-direction.

We now outline the algorithm for the construction of $V(S_N)$:

Step 1: Construct $V(S_1)$ .

Step 2: For $j=2, \cdots, N$:

    Step 2.1: Construct $V(S_{j-1} \cup \{\underline{X}_j\})$ from $V(S_{j-1})$ .

    Step 2.2: Construct $V(S_j)$ from $V(S_{j-1} \cup \{\underline{X}_j\})$ .

That is, in step 2.1 we add the stored point $\underline{X}_j$ without its periodic images to the otherwise periodic Voronoi diagram. Then in step 2.2 we add the periodic images of $\underline{X}_j$ (all at once).

Step 1 is trivial. The central part of the algorithm is a procedure for the construction of $V(S \cup \{\underline{z}\})$ if $\underline{z} \in S$ and $V(S)$ is known (step 2.1). This procedure was described in [2] and in [9] independently. We also present it here.

A corner $\underline{c}$ in V(S) is called "broken by $\underline{z}$" if it is not a corner in V(S $\cup$ {$\underline{z}$}). $\underline{z}$ breaks $\underline{c}$ if and only if $\underline{z}$ lies in the open disk around $\underline{c}$ with radius rad($\underline{c}$), where rad($\underline{c}$) is the radius of $\underline{c}$. $\underline{z}$ always breaks at least one corner. (This is, in general, guaranteed only if $\underline{z}$ lies in the convex hull of S. But the convex hull of a set of the form (2.2) is $\mathbb{R}^2$.)

Furthermore, the set of broken corners is connected in V(S).

Proof: For a fixed polygon in V(S), those of its corners which are broken by $\underline{z}$ clearly form a connected set.

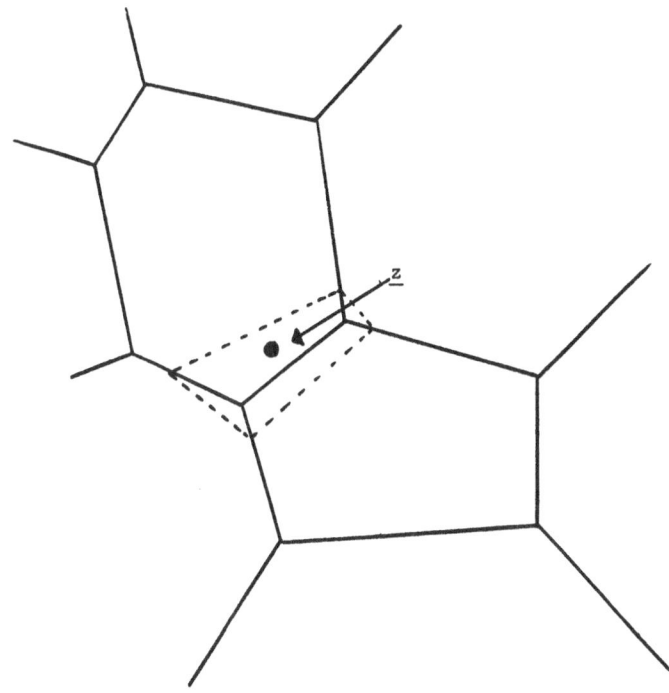

Walking along the boundary of P($\underline{z}$, S $\cup$ {$\underline{z}$}), one walks through all poygons which have corners broken by $\underline{z}$. [This walk was first pointed out to us by D. Goldfarb, unpublished.] Whenever $\partial$ P($\underline{z}$, S $\cup$ {$\underline{z}$}) crosses an edge between two polygons, this edge has exactly one endpoint which is broken by $\underline{z}$. This endpoint is a common corner of the two polygons. Therefore, following $\partial$ P($\underline{z}$, S $\cup$ {$\underline{z}$}) one constructs a walk through all broken corners.

We can therefore find all broken corners in $O(1)$ operations once we know one of them. We can find one broken corner in the following way. First choose some point $x \in S$ close to $z$. Good ways of choosing $x$ will be discussed below. The efficiency of the algorithm mainly depends on how this choice is made. Determine a corner $c$ of $P(x,S)$. (We use an array which contains for each polygon the basic and shift indices of one of its corners for this purpose.) Then search through the graph of corners of $V(S)$, starting at $c$, until a broken corner is found.

Once we know all corners broken by $z$, we can find all new corners (the corners of $P(z, S \cup \{z\})$) using the fact that each such corner lies on an edge between a broken and an unbroken corner in $V(S)$ (and on each such edge there is a new corner). It is useful (but not necessary) here to store the neighbor corners of a corner $c$ in counterclockwise order and to number the generating points such that the edge connecting $c$ with its i-th neighbor corner lies opposite to the i-th generating point:

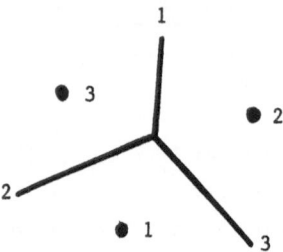

(This ordering seems to be the most important specifically two-dimensional feature of our code.)

It remains to describe step 2.2. To simplify the notation, we take $j=N$ and define

(2.5)  $\mathscr{P}_N := P(\underline{X}_N, S_{N-1} \cup \{\underline{X}_N\})$ and

(2.6)  $P_N := P(\underline{X}_N, S_N)$ .

It is clear then how to get $P_N$ from $\mathscr{P}_N$:

(2.7)  $P_N = \mathscr{P}_N \cap ([X_{N1} - 0.5; X_{N1} +0.5] \times [X_{N2} -0.5; X_{N2} + 0.5])$ .

($X_{N1}$ and $X_{N2}$ are the coordinates of $\underline{X}_N$ .)

We first assume that the decision which corners of $\mathscr{P}_N$ have to be "cut off" to obtain $P_N$ is simply made based on (2.7), checking coordinates. It seems easy then to find the radii of the new corners in $V(S_N)$, their neighbor corners and their neighbor points, as indicated in the figure below:

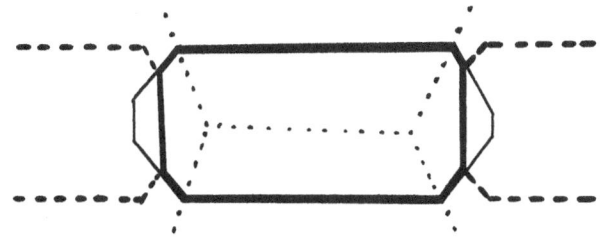

......... : edges in $V(S_{N-1})$

━━━━━━━━ : $\partial P_N$

- - - - - - : translations of edges of $P_N$

━━━━━━━━ : $\partial \mathcal{P}_N$

(To find the neighbors of the new corners, follow "••••••" and "------".)

Therefore the description of step 2.2 seems to be completed. This procedure works correctly in most cases; occasionally it breaks down, though. We show a typical situation which leads to a breakdown:

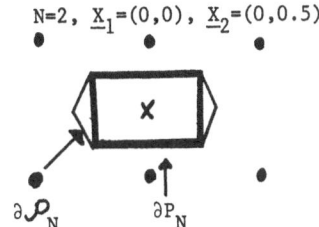

$P_N$ should have exactly 6 corners, two of which are identical up to translation with two others. In arbitrarily but finitely accurate arithmetic, the computed number of corners of $P_N$ may be 4, 5, 6, 7 or 8.

The situations which cause difficulties are those in which a polygon "wraps around" the torus; for $V(S_N)$, this means that a polygon is adjacent to its own translation. This is most likely to happen at early stages of the construction. It may even happen in $V(S_N)$ with arbitrarily large N.

We therefore have to "cut" $\mathcal{P}_N$ in a more careful way to obtain $P_N$. Let $\underline{c}$ be a corner of $\mathcal{P}_N$, generated by $\underline{\xi}$, $\underline{n}$, $\underline{X}_N$. Let $\underline{c}'$ be a corner of $\mathcal{P}_N$, generated by $\underline{\xi} + (1,0)$, $\underline{n} + (1,0)$, $\underline{X}_N$. The following figures show two (possible) such situations.

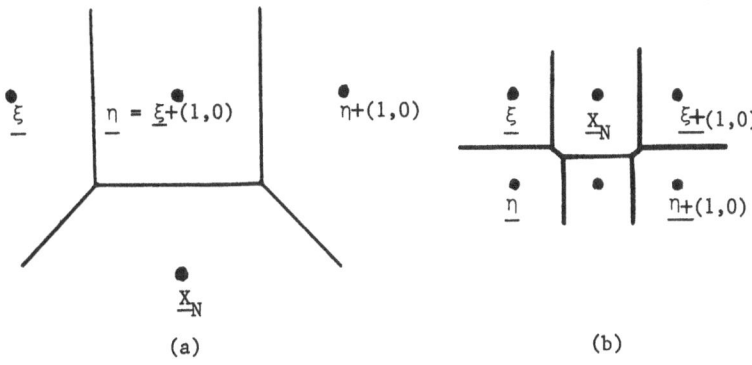

(a)                                    (b)

(Case (a) is the one discussed above as an example for a breakdown.) To decide whether $c$ is broken by $\underline{X}_N$ - (1,0) in situation (a), we consider the Voronoi diagram $V(\{\underline{\xi};\underline{\eta};\underline{X}_N;\underline{X}_N - (1,0)\})$.

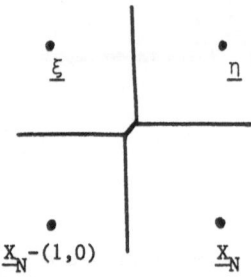

If we draw the diagram like this, we conclude that $c$ is not broken by $\underline{X}_N$ - (1,0) (since it exists in the four-point Voronoi diagram). Clearly, we can use the same diagram to decide whether $c'$ is broken by $\underline{X}_N$ + (1,0). We conclude that it is. If we had drawn the short edge of length 0 like this:

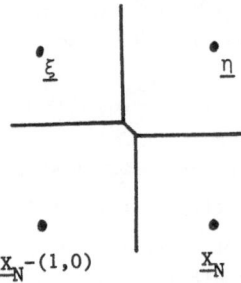

we would have concluded that $c$ is broken whereas $c'$ is not. In any case, we make opposite decisions for $c$ and $c'$, which is correct. We consider case (b) now. Neither $c$ nor $c'$ are broken here. Again, we can read this off from the Voronoi diagram $V(\{\underline{\xi};\underline{\eta};\underline{X}_N;\underline{X}_N - (1,0)\})$:

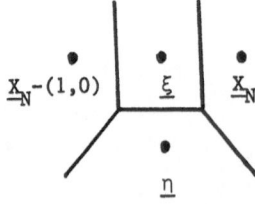

This motivates the following algorithm for "cutting" :

For each corner $\underline{c}$ of $\mathcal{P}_N$, decide whether or not it lies in $[X_{N1} - 0.5; \; X_{N1} + 0.5] \times [X_{N2} - 0.5; X_{N2} + 0.5]$, using the following procedure:

To see whether $\underline{c}$ is closer to $\underline{X}_N$ than to $\underline{X}_N + \underline{k}_0$, $\underline{k}_0 = \pm(1,0)$ or $\underline{k}_0 = \pm(0,1)$, compute

$$(2.8) \qquad V(\{\underline{\xi};\underline{n};\underline{X}_N;\underline{X}_N + \underline{k}_0\})$$

$(\underline{X}_N,\underline{\xi},\underline{n}$: generating points of $\underline{c}$). If $\underline{c}$ is a corner in this diagram, it is closer to $\underline{X}_N$ than to $\underline{X}_N + \underline{k}_0$, otherwise it is broken by $\underline{X}_N + \underline{k}_0$. If $\underline{c}'$ is a corner of $\mathcal{P}_N$ generated by $\underline{X}_N,\underline{\xi} - \underline{k}_0,\underline{n} - \underline{k}_0$, use the same diagram to decide whether or not $\underline{c}'$ is broken by $\underline{X}_N - \underline{k}_0$.

In our code, this procedure is simplified. The main simplification is a test whether cutting is necessary at all. (This can be done in a stable way by checking coordinates of corners of $\mathcal{P}_N$.) In the vast majority of cases, no cutting is necessary.

We finally describe two ways of finding a good starting point $x$ for the search for the first broken corner. If the points in S are known to be roughly uniformly distributed in $[0;1]^2$, we can divide $[0;1]^2$ into $n^2$ square cells of equal size with $n^2 \approx N$ and create pointers from those cells to the points in S which they contain. (A linked list is a good data structure to use for this purpose: In each cell, store a pointer to one point in the cell. Then let each point in the cell point to another with the last point containing a stop code.) We then introduce the points cell by cell in the order indicated in the following figure:

| n | n+1 | | |
|---|---|---|---|
| $\vdots$ | n+2 | | |
| 2 | $\vdots$ | $\vdots$ | $n^2-1$ |
| 1 | 2n | $\vdots$ | $n^2$ |

A good choice of $\underline{x}$ when introducing $\underline{X}_j$ is then $\underline{X}_{j-1}$. Using this method on sets of points which are random and uniformly distributed in $[0;1]^2$, we obtained the following CPU-times per point in msec on a VAX 11/780:

| N | worst case | best case | average |
|---|---|---|---|
| 400 | 19 | 17 | 18 |
| 800 | 19 | 18 | 18 |
| 1200 | . 19 | 18 | 18 |
| 1600 | 19 | 18 | 19 |

For each N, we tried 20 different random sets of points.

In our fractional step method, we use this method at time 0. At later times, we use the following method. Assume that $V(\tilde{S})$ has been constructed for a set $\tilde{S} = \{\tilde{X}_1; \cdots ; \tilde{X}_N\} + Z^2$, with $d(\underline{X}_j, \tilde{X}_j)$ small, $\tilde{X}_j \in [0;1)^2$. (d was defined in (2.4).) When inserting the k-th point $\underline{X}_k$, we start the search for the nearest neighbor with the suitable translation of the j(k)-th point $\underline{X}_{j(k)}$, where j(k) is such that a corner of $P(\tilde{X}_{j(k)}, \tilde{S})$ was broken when $\tilde{X}_k$ was inserted during the construction of $V(\tilde{S})$. (The indices j(k) $\in \{1; \cdots ; k\}$ must therefore be saved during the construction of $V(\tilde{S})$.)

Using this method, the construction becomes slightly faster at later times than at time 0.

We conclude this section with some remarks about other possible ways of constructing periodic Voronoi diagrams. It is, of course, possible to reduce the problem to nonperiodic Voronoi diagrams (in the plane): Consider

(2.9) $\qquad S^* := \{\underline{X}_j + \underline{k}: 1 \leq j \leq N, \underline{k} \in Z^2 , |k_1| \leq 1 \text{ and } |k_2| \leq 1\}$ .

It is easy to see that

(2.10) $\qquad P(\underline{X}_j, S^*) = P(\underline{X}_j, S)$ for all j .

Since $S^*$ contains 9N points, this procedure is quite inefficient. The definition of $S^*$ can be modified to reduce the number of auxiliary points, but the algorithm becomes unnecessarily complicated then.

An excellent alternative is the use of an algorithm similar to the one used in [5], which constructs the Delaunay triangulation rather than the Voronoi diagram and seems clearly more efficient than our method. It is also easily vectorizable, which doesn't seem to be the case for our method. Our algorithm may, however, be easier to generalize to Voronoi diagrams in spaces other than $\mathbb{R}^2$ because it uses the metric of the space but no angles. For a spherical version see [1]. From a non-practical point of view, our algorithm has the advantage that it will provably work for any case. Such a proof does not yet exist for the algorithm in [5], while the statement is obvious in the case of our algorithm – at least in the absence of rounding errors.

3.  Finite difference operators on arbitrary sets of points

We review the definitions of the discrete Laplace, divergence and gradient operators given in [9]. These definitions are dimension independent. We use the terminology appropriate for 2 dimensions for simplicity.

We introduce the following notations. For $\underline{X} \in S_N$, let $V[X]$ be the area of the Voronoi polygon $P(\underline{X}, S_N)$. For $\underline{Y} \in S_N$, $\underline{Y} \neq \underline{X}$, let $A[\underline{X}, \underline{Y}]$ be the length of the intersection of $P(\underline{X}, S_N)$ and $P(\underline{Y}, S_N)$. $A[\underline{X}, \underline{Y}] = 0$ if $P(\underline{X}, S_N)$ and $P(\underline{Y}, S_N)$ are not adjacent to each other. If $A[\underline{X}, \underline{Y}] \neq 0$, we say that $\underline{X}$ and $\underline{Y}$ are "neighbors" of each other. Our discrete operators couple two distinct points to each other only if they are neighbors. In the following, we let $Nb[\underline{X}]$ be the set of neighbors of $\underline{X}$. (Note that $\underline{X} \in Nb[\underline{X}]$).

We define a discrete Laplace operator L by

(3.1)    $V[\underline{X}](L\phi)(\underline{X}) := \sum_{\underline{Y} \neq \underline{X}} A[\underline{X}, \underline{Y}](\phi(\underline{Y}) - \phi(\underline{X}))/\|\underline{X} - \underline{Y}\|$    for $\underline{X} \in S_N$ ,

which is motivated by the formula

(3.2)    $\int_P \Delta\phi(\underline{x}) \, d\underline{x} = \int_{\partial P} \frac{\partial\phi}{\partial n}(\underline{x}) ds$

(for smooth functions $\phi$). The sum in (3.1) is formally infinite but contains only finitely many non-zero terms. As was pointed out by B. Mercier ([6]), the operator in (3.1), seen as a matrix, is the stiffness matrix obtained with piecewise linear finite elements on the Delaunay triangulation.

Proof: Both matrices have zero row sums. Therefore we need to consider non-diagonal entries only. Since both matrices are also symmetric, it makes sense to talk about the "coupling between $\underline{X}$ and $\underline{Y}$" ($\underline{X}, \underline{Y} \in S_N$, $\underline{X} \neq \underline{Y}$). We compare the couplings in the matrix in (3.1) and in the finite element stiffness matrix.

Without loss of generality, we assume that $\underline{X} = (0,0)$, $\underline{Y} = (0,1)$. Let $\underline{A}$, $\underline{B}$ be the points in $S_N$ such that $\underline{A}$, $\underline{X}$, $\underline{Y}$ and $\underline{B}$, $\underline{X}$, $\underline{Y}$ are triangles in the Delaunay triangulation:

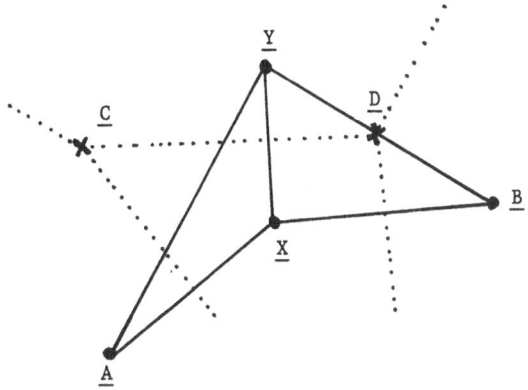

Let $\underline{C}$ be the center of the circle through $\underline{A}$, $\underline{X}$, $\underline{Y}$, and let $\underline{D}$ be the center of the circle through $\underline{B}$, $\underline{X}$, $\underline{Y}$. $\underline{C}$ and $\underline{D}$ are corners in $V(S_N)$. The claim is that the coupling between $\underline{X}$ and $\underline{Y}$ obtained with piecewise linear finite elements is $\|\underline{C}-\underline{D}\|$. This is shown by straightforward computation of this coupling and $\|\underline{C}-\underline{D}\|$ in terms of the coordinates of $\underline{A}$ and $\underline{B}$. Both turn out to be equal to

$$(3.3) \qquad -\frac{A_2^2 - A_2}{2\,A_1} - \frac{A_1}{2} + \frac{B_2^2 - B_2}{2\,B_1} + \frac{B_1}{2}$$

with $\underline{A} =: (A_1, A_2)$ and $\underline{B} =: (B_1, B_2)$.

Nevertheless, a discretization of the Poisson equation based on (3.1) is not a finite element discretization. The difference lies in the right-hand sides. Let $f$ be a given continuous right-hand side. The value in $\underline{X}$ of the discrete right-hand side obtained with the finite element method is a weighted sum of values of $f$ in neighbors of $\underline{X}$. The sum $M[\underline{X}]$ of the weights (elements of the mass matrix) is one third times the sum of the areas of the triangles to which $\underline{X}$ belongs. Using (3.1), one is lead to using $V[\underline{X}]f(\underline{X})$ as the value of the discrete right-hand side in $\underline{X}$. $V[\underline{X}]$ and $M[\underline{X}]$ are, in general, different from each other. They are, of course, equal "on the average". ($M[\underline{X}]$ is the area of a cell suggested in [4].)

We define a discrete divergence operator next.

$$(3.4) \qquad V[\underline{X}]D\underline{u}(\underline{X}) := \sum_{\underline{Y}} \underline{u}[\underline{Y}] \cdot \frac{\partial\,V[\underline{X}]}{\partial\,\underline{Y}} \quad \text{for } \underline{X} \in S_N .$$

Here $\dfrac{\partial\,V[\underline{X}]}{\partial\,\underline{Y}}$ is the gradient of $V[\underline{X}]$ with respect to $\underline{Y}$ (keeping all other points fixed). If $(\underline{X}_1, \cdots, \underline{X}_N) = (\underline{X}_1(t), \cdots, \underline{X}_N(t))$ are moving points, it follows that

$$(3.5) \qquad \frac{d}{dt}\,V[\underline{X}_j(t)] = V[\underline{X}_j(t)]D\underline{u}(\underline{X}_j(t))$$

for all $j$, in particular:

If $\underline{X}_1(t), \cdots, \underline{X}_N(t)$ are points in space which move in a discretely divergence-free velocity field, the Voronoi polygons maintain their areas exactly.

(However, the Voronoi polygons do change their areas, usually even quite drastically, if the points are moved in a continuously divergence-free field.)

An explicit formula for $\dfrac{\partial\,V[\underline{X}]}{\partial\,\underline{Y}}$ may be derived as follows. For a given $\underline{Y}$, there are three cases to consider:

    i)   $\underline{X} = \underline{Y}$

   ii)   $\underline{X} \in Nb[\underline{Y}]$ (so $\underline{X} \neq \underline{Y}$)

  iii)   $\underline{X} \neq \underline{Y}$ and $\underline{X} \notin Nb[\underline{Y}]$ .

It is obvious that

$$(3.6) \qquad \frac{\partial\,V[\underline{X}]}{\partial\,\underline{Y}} = 0 \text{ in case (iii) .}$$

Case (i) may be reduced to case (ii) as follows. From (3.6), it is easy to conclude that

(3.7)    $\dfrac{\partial}{\partial \underline{Y}} (V[\underline{Y}] + \sum\limits_{\underline{X} \in Nb(\underline{Y})} V[\underline{X}]) = 0$ ,

hence

(3.8)    $\dfrac{\partial}{\partial \underline{Y}} V[\underline{Y}] = - \sum\limits_{\underline{X} \in Nb(\underline{Y})} \dfrac{\partial V[\underline{X}]}{\partial \underline{Y}}$ .

One can also derive a formula dual to (3.8) by considering translation invariance of the Voronoi diagram on the torus. Let the points $\underline{X}_1 \cdots \underline{X}_N$ all move at the $\underline{same}$ constant velocity $\underline{u}$. Clearly, the diagram translates as a whole and the areas of the polygons do not change. Thus

(3.9)    $0 = \dfrac{d}{dt} V[X] = \sum\limits_{\underline{Y}} \dfrac{\partial V[X]}{\partial \underline{Y}} \cdot \underline{u}$

Since $\underline{u}$ was arbitrary

(3.10)    $0 = \sum\limits_{\underline{Y}} \dfrac{\partial V[X]}{\partial \underline{Y}}$

(3.11)    $\dfrac{\partial V[X]}{\partial \underline{X}} = - \sum\limits_{\underline{Y} \neq \underline{X}} \dfrac{\partial V[X]}{\partial \underline{Y}} = - \sum\limits_{\underline{Y} \in Nb[\underline{X}]} \dfrac{\partial V[X]}{\partial \underline{Y}}$

Note that (3.8) and (3.11) are $different$ formulae for the diagonal derivative. They are both written in terms of $\partial V[\underline{X}]/\partial \underline{Y}$, but (3.8) involves a sum over $\underline{X}$ and (3.11) a sum over $\underline{Y}$. Either of these formulae may be used to express the diagonal derivatives in terms of the off-diagonal derivatives.

We now consider case (ii) in which $\underline{Y}$ is a neighbor of $\underline{X}$.

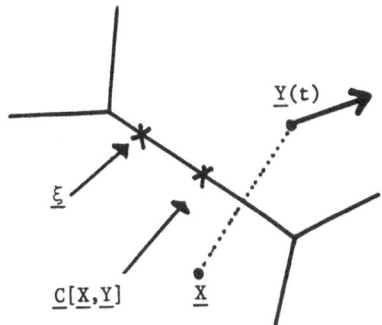

Suppose $\underline{Y}$ is moving and $\underline{X}$ is fixed. Also, all other points are fixed. Let $\xi$ be an arbitrary point on the edge common to $P[\underline{X}]$ and $P[\underline{Y}]$. Recall that this edge has length $A[\underline{X},\underline{Y}] \neq 0$. Let $\underline{C}[\underline{X},\underline{Y}]$ be the midpoint (center of mass) of this edge. Now the arbitrary point $\xi$ satisfies the equation

$$(3.12) \qquad \|\underline{\xi} - \underline{X}\|^2 = \|\underline{\xi} - \underline{Y}\|^2$$

Differentiating this with respect to t, we get

$$(3.13) \qquad \frac{d\underline{\xi}}{dt} \cdot (\underline{\xi} - \underline{X}) = (\frac{d\underline{\xi}}{dt} - \frac{d\underline{Y}}{dt}) \cdot (\underline{\xi} - \underline{Y})$$

$$(3.14) \qquad \frac{d\underline{\xi}}{dt} \cdot (\underline{Y} - \underline{X}) = \frac{d\underline{Y}}{dt} \cdot (\underline{Y} - \underline{\xi})$$

Dividing this expression by $\|\underline{Y} - \underline{X}\|$, integrating over the edge in question, and noticing that $(\underline{Y} - \underline{X})/\|\underline{Y} - \underline{X}\|$ is the unit normal to this edge, we obtain

$$(3.15) \qquad \frac{d}{dt} V[\underline{X}] = \int \frac{d\underline{\xi}}{dt} \cdot \frac{\underline{Y} - \underline{X}}{\|\underline{Y} - \underline{X}\|} ds = \frac{d\underline{Y}}{dt} \cdot \int \frac{\underline{Y} - \underline{\xi}}{\|\underline{Y} - \underline{X}\|} ds$$

where ds is the element of length along the edge. But $\int ds = A[\underline{X},\underline{Y}]$ and $\int \underline{\xi} ds = A[\underline{X},\underline{Y}]\underline{C}[\underline{X},\underline{Y}]$. Therefore

$$(3.16) \qquad \frac{d}{dt} V[\underline{X}] = \frac{d\underline{Y}}{dt} \cdot \frac{\underline{Y} - \underline{C}[\underline{X},\underline{Y}]}{\|\underline{Y} - \underline{X}\|} A[\underline{X},\underline{Y}]$$

Since $\partial \underline{Y}/\partial t$ was arbitrary, it follows that

$$(3.17) \qquad \frac{\partial V[\underline{X}]}{\partial \underline{Y}} = \frac{\underline{Y} - \underline{C}[\underline{X},\underline{Y}]}{\|\underline{Y} - \underline{X}\|} A[\underline{X},\underline{Y}] , \quad \underline{X} \in Nb[\underline{Y}]$$

For future reference we resolve this into normal and tangential components:

$$(3.18) \qquad \underline{Y} - \underline{C}[\underline{X},\underline{Y}] = (\underline{Y} - \frac{1}{2}(\underline{X} + \underline{Y})) + (\frac{1}{2}(\underline{X} + \underline{Y}) - \underline{C}[\underline{X},\underline{Y}])$$

$$= \frac{1}{2}(\underline{Y} - \underline{X}) + (\frac{1}{2}\underline{X} + \underline{Y}) - \underline{C}[\underline{X},\underline{Y}])$$

Therefore

$$(3.19) \qquad \frac{\partial V[\underline{X}]}{\partial \underline{Y}} = \frac{1}{2} A[\underline{X},\underline{Y}] \frac{\underline{Y} - \underline{X}}{\|\underline{Y} - \underline{X}\|} + \frac{\underline{T}[\underline{X},\underline{Y}]}{\|\underline{Y} - \underline{X}\|} A[\underline{X},\underline{Y}]$$

where

$$(3.20) \qquad \underline{T}[\underline{X},\underline{Y}] = \frac{1}{2}(\underline{X} + \underline{Y}) - \underline{C}[\underline{X},\underline{Y}]$$

is tangent to the edge in question.

We now make use of the foregoing results to prove that the operator D is <u>weakly</u> <u>consistent</u> (to first order) with the continuous divergence operator. That is, for arbitrary smooth, periodic (scalar and vector) functions $\phi$ and $\underline{u}$, we shall prove that

$$(3.21) \quad \sum_k V[\underline{X}_k]\phi(\underline{X}_k)D\underline{u}(\underline{X}_k) = \int_{[0,1]^2} \phi(\nabla\cdot\underline{u})dV + O(h)$$

where

$$(3.22) \qquad h = \underset{\underline{x}}{\text{Max}} \quad \underset{\underline{X}\in S}{\text{Min}} \|\underline{x} - \underline{X}\|$$

(That is, h is the maximum "radius" of the polygons in the diagram.) To show (3.21), we rewrite the left-hand side using the definition of D (Eq. 3.4) and the identity (3.8), which is used here with $\underline{X}$ and $\underline{Y}$ interchanged:

$$\sum_k V[\underline{X}_k]\phi(\underline{X}_k)D\underline{u}(\underline{X}_k)$$

$$= \sum_k \phi(\underline{X}_k) \sum_{\underline{Y}\in S_N} \frac{\partial V[\underline{X}_k]}{\partial\underline{Y}} \cdot \underline{u}(\underline{Y})$$

$$= \sum_k \phi(\underline{X}_k) \sum_{\underline{Y}\in Nb[\underline{X}_k]} \{\frac{\partial V[\underline{X}_k]}{\partial\underline{Y}} \cdot \underline{u}(\underline{Y}) - \frac{\partial V[\underline{Y}]}{\partial\underline{X}_k} \cdot \underline{u}(\underline{X}_k)\}$$

Next, we substitute into this the formula for $\partial V[\underline{X}]/\partial\underline{Y}$ , Eq. 3.19:

$$(3.24) \quad \sum_k V[\underline{X}_k]\phi(\underline{X}_k)D\underline{u}(\underline{X}_k)$$

$$= \sum_k \phi(\underline{X}_k) \sum_{\underline{Y}\in Nb[\underline{X}_k]} A[\underline{X}_k,\underline{Y}] \frac{\underline{Y} - \underline{X}_k}{\|\underline{Y} - \underline{X}_k\|} \cdot \frac{1}{2}(\underline{u}(\underline{Y}) + \underline{u}(\underline{X}_k))$$

$$+ \sum_k \phi(\underline{X}_k) \sum_{\underline{Y}\in Nb[\underline{X}_k]} A[\underline{X}_k,\underline{Y}]\underline{T}[\underline{X}_k,\underline{Y}] \cdot \frac{\underline{u}(\underline{Y}) - \underline{u}(\underline{X}_k)}{\|\underline{Y} - \underline{X}_k\|}$$

We analyze these two terms separately. In the last term, since $A[X,Y]$ and $\underline{T}[X,Y]$ are symmetric, we have the equivalent expression

$$(3.25) \quad \frac{1}{2}\sum_k \sum_{\underline{Y}\in Nb[\underline{X}_k]} (\phi(\underline{X}_k) - \phi(\underline{Y}))A[\underline{X}_k,\underline{Y}]\underline{T}[\underline{X}_k,\underline{Y}] \cdot \frac{\underline{u}(\underline{Y}) - \underline{u}(\underline{X}_k)}{\|\underline{Y} - \underline{X}_k\|}$$

Now $\phi(\underline{X}_k) - \phi(\underline{Y}) = O(h)$, $A[\underline{X}_k,\underline{Y}] = O(h)$, $\underline{T}[\underline{X}_k,\underline{Y}] = O(h)$, and $(\underline{u}(\underline{Y}) - \underline{u}(\underline{X}_k))/\|\underline{Y} - \underline{X}_k\|$ $= O(1)$. Therefore, each individual term in (3.25) is $O(h^3)$. Moreover, the number of terms is only $O(N) = O(h^{-2})$, since the number of elements in $Nb[\underline{X}_k]$ is $O(1)$. Therefore, the entire expression (3.25) is $O(h)$. In summary, we have shown that

(3.26) $\quad \sum_k V[\underline{X}_k]\phi(\underline{X}_k)D\underline{u}(\underline{X}_k)$

$$= \sum_k \phi(\underline{X}_k) \sum_{\underline{Y} \in Nb[\underline{X}_k]} A[\underline{X}_k,\underline{Y}] \frac{\underline{Y} - \underline{X}_k}{\|\underline{Y} - \underline{X}_k\|} \cdot \frac{1}{2}(\underline{u}(\underline{Y}) + \underline{u}(\underline{X}_k)) + 0(h)$$

The next step is to bring the integral $\int\phi(\nabla\cdot\underline{u})dV$ into a similar form. We have

(3.27) $\quad \displaystyle\int_{[0;1]^2} \phi(\nabla\cdot\underline{u})dV = \sum_k \phi(\underline{X}_k) \int_{P[\underline{X}_k]} (\nabla\cdot\underline{u})dV + 0(h)$

$$= \sum_k \phi(\underline{X}_k) \sum_{\underline{Y} \in Nb[\underline{X}_k]} A[\underline{X}_k,\underline{Y}] \frac{\underline{Y} - \underline{X}_k}{\|\underline{Y} - \underline{X}_k\|} \cdot \underline{u}^*[\underline{X}_k,\underline{Y}] + 0(h)$$

where $\underline{u}^*[\underline{X}_k,\underline{Y}]$ is the average of $\underline{u}$ over the edge common to $P[\underline{X}_k]$ and $P[\underline{Y}]$. Subtracting Eq. (3.27) from (3.26) we get

(3.28)

$$\sum_k V[\underline{X}_k]\phi(\underline{X}_k)D\underline{u}(\underline{X}_k) - \int\phi(\nabla\cdot\underline{u})dV + 0(h)$$

$$= \sum_k \phi(\underline{X}_k) \sum_{\underline{Y} \in Nb[\underline{X}_k]} A[\underline{X}_k,\underline{Y}] \frac{\underline{Y} - \underline{X}_k}{\|\underline{Y} - \underline{X}_k\|} \cdot \{\frac{1}{2}(\underline{u}(\underline{Y}) + \underline{u}(\underline{X}_k)) - \underline{u}^*[\underline{X}_k,\underline{Y}]\}$$

$$= \frac{1}{2} \sum_k \sum_{\underline{Y} \in Nb[\underline{X}_k]} (\phi(\underline{X}_k) - \phi(\underline{Y}))A[\underline{X}_k,\underline{Y}] \frac{\underline{Y} - \underline{X}_k}{\|\underline{Y} - \underline{X}_k\|} \cdot \{\frac{1}{2}(\underline{u}(\underline{Y}) + \underline{u}(\underline{X}_k)) - \underline{u}^*[\underline{X}_k,\underline{Y}]\}$$

As above, we have $\phi(\underline{X}_k) - \phi(\underline{Y}) = 0(h)$, $A[\underline{X}_k,\underline{Y}] = 0(h)$, $\{\frac{1}{2}(\underline{u}(\underline{Y}) + u(\underline{X}_k)) - \underline{u}^*[\underline{X}_k,\underline{Y}]\}$ $= 0(h)$, and $(\underline{Y} - \underline{X}_k)/\|\underline{Y} - \underline{X}_k\| = 0(1)$. Thus, the individual terms are $0(h^3)$ and there are (as before) $0(N) = 0(h^{-2})$ terms in the sum. It follows that the entire expression is $0(h)$ and hence that

(3.29) $\quad \displaystyle\sum_k V[\underline{X}_k]\phi(\underline{X}_k)D\underline{u}(\underline{X}_k) - \int_{[0;1]^2} \phi(\nabla\cdot\underline{u})dV = 0(h)$

as claimed.

This completes the proof that $D$ and $\nabla\cdot$ are weakly consistent. In fact, they are _not_ pointwise consistent. This was first noticed by M. McCracken, who constructed counterexamples (unpublished).

Note that the foregoing argument is exactly the same in three dimensions. We just have to define $\underline{C}[\underline{X}_k,\underline{Y}]$ as the "center of mass" of the face common to $P[\underline{X}_k]$ and $P[\underline{Y}]$, and $A[\underline{X}_k,\underline{Y}]$ as the area of that face. Then $A[\underline{X}_k,\underline{Y}] = 0(h^2)$ and $N = 0(h^{-3})$, so the extra factors of $h$ cancel.

Next, we introduce an operator $\underline{G}$ which is the negative adjoint of $D$ with respect to the discrete $L^2$ inner product

(3.30) $\quad (\phi,\psi) := \sum_k \phi(\underline{X}_k)\psi(\underline{X}_k)V[\underline{X}_k]$

(3.31) $\quad (\underline{u},\underline{v}) = \sum_k \underline{u}(\underline{X}_k)\cdot\underline{v}(\underline{X}_k)V[\underline{X}_k]$

The definition of $\underline{G}$ is

(3.32) $\qquad V[\underline{X}]\underline{G}\phi(\underline{X}) = -\sum_{\underline{Y}} \phi(\underline{Y}) \frac{\partial V[\underline{Y}]}{\partial \underline{X}}$

It follows directly from the definitions of D and $\underline{G}$ that

(3.33) $\qquad (\underline{u},\underline{G}\phi) + (D\underline{u},\phi) = 0$

which is the analog of the continuous identity

(3.34) $\qquad \int (\underline{u}\cdot\nabla\phi)dV + \int (\nabla\cdot\underline{u})\phi dV = 0$

The weak consistency of $\underline{G}$ with the continous gradient operator follows directly from these identities and the weak consistency of D. That is, combining Eqs. (3.29), (3.33), and (3.34), we obtain

$$\sum_k V[\underline{X}_k]\underline{G}\phi(\underline{X}_k)\cdot\underline{u}(\underline{X}_k) = \int_{[0;1]^2} \underline{u}\cdot\nabla\phi \ dV + O(h)$$

4. Solution of discrete Helmholtz equations on the Voronoi mesh

In this section, we consider the fast numerical solution of

(4.1) $\qquad -L\phi + c\phi = f \qquad (c \geq 0)$ .

We use a two-level iteration which is identical with the well-known multigrid correction cycle (see [8]), replacing coarsening of the grid by "regularization". We give a motivating derivation of the method here.

Consider a regular grid of the form

(4.2) $\qquad G^h = \{((i + 0.5)h,(j + 0.5)h) : i,j \in Z\}$

with $h = 1/n$ and $n \approx \sqrt{N}$. Let R be a linear interpolation operator which maps functions defined on $S_N$ onto functions defined on $G^h$. let P be a linear interpolation operator in the opposite direction.

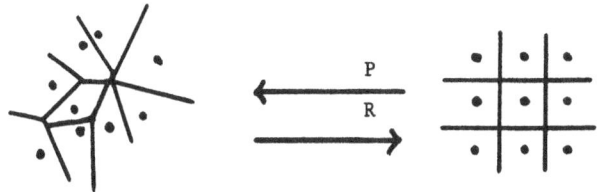

We introduce the short notation

(4.3)          $A := -L + cI$ ,

(4.4)          $B :=$ discretization of $-\Delta + cI$ on the grid, using
               the standard 5-point operator.

We consider the following simple iterative method for (4.1).

(4.5)          $\phi^{n+1} := \phi^n + PB^{-1}R(f - A\phi^n)$ .

(If $c=0$, $B^{-1}$ does not exist. "$B^{-1}r$" is obtained as follows in that case: Subtract the constant component from $r$, find some solution $v$ of $Bv = r$ and subtract the constant component from $v$.)

   If P and R are chosen in a simple, straightforward way (see below), the iteration (4.5) converges very poorly if at all. The reason is that P and R introduce averaging, which prevents highly oscillatory erros from converging fast. Therefore (4.5) has to be supplemented by a method which is efficient on highly oscillatory errors; we choose a suitable relaxation method for this. We obtain then the two-level correction cycle

                  Step 1: m relaxation sweeps
(4.6)             Step 2: (4.5)
                  Step 3: l relaxation sweeps

m and l are small integers, typically 1 or 2. To solve problems with B (as required in (4.5)), we use the Fast Fourier Transform. This is the only piece in our fractional step method which costs $O(N\log N)$ operations rather than $O(N)$ operations per time step. Instead of Fast Fourier Transform, an approximate solver (for example a multigrid solver) requiring only $O(N)$ operations could be used without significant deterioration of the convergence rates, see [8].

   We have to choose P, R and the relaxation scheme. We discuss the choice of P and R first. We list some "desirable properties":

CP : P preserves constants:

$$P(1)(\underline{X}_j) = 1 \text{ for all } j .$$

CR : R preserves constants:

$$R(1)(\underline{x}) = 1 \text{ for all } \underline{x} \in G^h .$$

IOP : P preserves the property of having discrete integral 0:

$$\text{If } \sum_{\underline{x} \in G^h} \phi(\underline{x})h^2 = 0, \text{ then } \sum_k (P\phi)(\underline{X}_k)V[\underline{X}_k] = 0 .$$

IOR : R preserves the property of having discrete integral 0:

$$\text{If } \sum_k \Phi(\underline{X}_k)V[\underline{X}_k] = 0 , \text{ then } \sum_{\underline{x}} (R\Phi)(\underline{x})h^2 = 0 .$$

IOP and IOR state the preservation of the compatibility condition if c=0; for our multigrid algorithm, it is desirable to have IOR at least, to ensure solvability of the problems on $G^h$ if c=0.

We focus attention on operators defined in the form

$$(4.7) \qquad (P\phi)(\underline{X}_k) := \sum_{\underline{x} \in G^h} \phi(\underline{x})\delta_h(\underline{X}_k - \underline{x})h^2 ,$$

$$(4.8) \qquad (R\Phi)(\underline{x}) := \sum_k \Phi(\underline{X}_k)\delta_h(\underline{x} - \underline{X}_k)V[\underline{X}_k] ,$$

where $\delta_h$ is a spread-out model of the delta distribution with support of (linear) size O(h). ((4.7), (4.8) imply P=R$^*$, where R$^*$ is the adjoint of R with respect to the discrete $L^2$-products on $S_N$ and on $G^h$.) It is clear then that CR cannot be satisfied in general. (It is possible that no point in $S_N$ lies close to $\underline{x} \in G^h$, in which case $R\Phi(\underline{x})$ is 0 no matter what $\Phi$ is.) CP, on the other hand, is easy to satisfy, for example with

$$(4.9) \qquad \delta_h(\underline{x}) = \delta_h(x_1)\delta_h(x_2) , \text{ with}$$

$$(4.10) \qquad \delta_h(x) = \begin{cases} (1 - \dfrac{|x|}{h}) \text{ if } |x| \leq h , \\ \\ 0 \text{ otherwise} \end{cases}$$

(piecewise bilinear interpolation). It is easy to conclude from this that IOR is satisfied while IOP is not. In fact, R preserves discrete integrals in general. These conclusions depend on nothing but the fact that P = R$^*$.

Eqs. (4.7) - (4.10) specify our choice of P and R.

It remains to choose a relaxation scheme. Using Gauss-Seidel relaxation, we obtain a performance which is disappointing in comparison with many multigrid methods on regular meshes. The following table contains the reduction of the discrete $L^2$-norm of the residual with Gauss-Seidel relaxation, m=2 and n=1, after

1,3 and 5 two-level iterations. For each N, we performed experiments on 10 different random sets $S_N$ (uniformly distributed in the unit square). We show the worst and the best results as well as the geometric means over all 10 experiments. In all cases, the continuous problem being solved has the solution

(4.11)        $\phi(x) = \sin(2\pi(x - 2y))$ .

The Helmholtz constant is c = 10.0.

| N | worst case | best case | geometric mean |
|---|---|---|---|
| 100 | 0.22E+00 | 0.69E-01 | 0.13E+00 |
|  | 0.34E-02 | 0.37E-04 | 0.23E-02 |
|  | 0.56E-03 | 0.28E-06 | 0.60E-04 |
| 400 | 0.26E+00 | 0.16E+00 | 0.22E+00 |
|  | 0.17E-02 | 0.29E-03 | 0.78E-02 |
|  | 0.26E-03 | 0.84E-05 | 0.57E-03 |
| 1600 | 0.29E+00 | 0.21E+00 | 0.25E+00 |
|  | 0.18E-02 | 0.39E-03 | 0.86E-02 |
|  | 0.30E-03 | 0.13E-04 | 0.76E-03 |

The results can be improved by a modification of the relaxation method. The modified relaxation scheme is defined as follows. Introduce the notation

(4.12)        $(L\phi)(\underline{X}_k) =: \sum_{\underline{Y}} L[\underline{X}_k,\underline{Y}]\phi(\underline{Y})$ .

When "relaxing $\underline{X}_k$", determine the set of all neighbors $\underline{Y}$ of $\underline{X}_k$ for which

(4.13)        $L[\underline{X}_k,\underline{Y}] \geq -\delta\, L[\underline{X}_k,\underline{X}_k]$ ,

where $\delta \in [0;1]$ remains to be chosen. Change values in all these neighbors and in $\underline{X}_k$ simultaneously such that their equations become satisfied. If $\delta$ is at least 0.5, it follows that the blocks which are relaxed simultaneously are at most of size 2. (To see this, recall that the diagonal elements of L are negative, that the off-diagonal elements are positive or zero with at least 3 positive elements in each row, and that the row sums of L are zero. Thus there is at most one neighbor $\underline{Y}$ of any point $\underline{X}_k$ for which (4.13) is satisfied when $\delta \geq 0.5$.) Therefore we choose $\delta = 0.5$. We repeat the above experiments with this modified relaxation scheme. The results, for c = 10.0, are as follows:

| N | worst case | best case | geometric mean |
|---|---|---|---|
| 100 | 0.90E-01<br>0.55E-04<br>0.70E-06 | 0.30E-01<br>0.15E-05<br>0.33E-07 | 0.52E-01<br>0.81E-04<br>0.77E-06 |
| 400 | 0.15E+00<br>0.11E-03<br>0.18E-05 | 0.96E-01<br>0.17E-04<br>0.18E-06 | 0.12E+00<br>0.48E-03<br>0.40E-05 |
| 1600 | 0.16E+00<br>0.12E-03<br>0.33E-05 | 0.13E+00<br>0.28E-04<br>0.60E-06 | 0.14E+00<br>0.63E-03<br>0.12E-04 |

The results for c = 0.0 look hardly different:

| N | worst case | best case | geometric mean |
|---|---|---|---|
| 100 | 0.83E-01<br>0.70E-04<br>0.98E-06 | 0.33E-01<br>0.20E-05<br>0.27E-07 | 0.54E-01<br>0.10E-03<br>0.71E-06 |
| 400 | 0.16E+00<br>0.12E-03<br>0.19E-05 | 0.10E+00<br>0.19E-04<br>0.19E-06 | 0.12E+00<br>0.53E-03<br>0.45E-05 |
| 1600 | 0.16E+00<br>0.13E-03<br>0.34E-05 | 0.13E+00<br>0.29E-04<br>0.72E-06 | 0.15E+00<br>0.65E-03<br>0.12E-04 |

(Here we have projected the discrete right-hand side onto its constant-free part to ensure that there is a solution.)

We still obtain convergence rates which seem to increase with growing N. Good multigrid methods have convergence rates which are bounded independent of the number of unknowns, the bound being far below 1; see [8]. We do not know whether the method presented here has this property. It is, in any case, satisfactory when used within our fractional step method, see section 6.

5. Numerical projection of a vector field onto its divergence free part

The last tool needed for our fractional step method is an algorithm which orthogonally projects a given vector field $\underline{u}$ on $S_N$ onto the kernel of the discrete divergence operator D.

We want to find $\underline{Pu}$ and q such that

(5.1) $\qquad \underline{u} = \underline{Pu} + \underline{Gq}$

and

(5.2) $\qquad D\underline{P}\underline{u} = 0$ .

To do this exactly, we have to solve

(5.3) $\qquad DG\underline{q} = D\underline{u}$ .

No difficulty arises from the fact that $DG$ is singular. $D\underline{u}$ satisfies the compatiblity condition and $G\underline{q}$ is unique.

On a regular square grid with mesh width h, $DG$ is the standard 5-point discretization of the Laplace operator with mesh width 2h. Simple relaxation schemes have no smoothing effect for this discretization, as can easily be verified by Fourier analysis. (We ignore the fact that the system can be decoupled into 4 smaller systems which can be solved easily, since this will not be the case on irregular meshes.) The difficulty is related to lack of "discrete ellipticity", see [3].

We therefore consider replacing the discrete projection operator

(5.4) $\qquad I - \underline{G}(D\underline{G})^{-1}D$

by

(5.5) $\qquad I - \underline{G}L^{-1}D$ .

($DG$ and $L$ are singular; nevertheless "$G(DG)^{-1}D$" and "$GL^{-1}D$" have obvious well-defined meanings.)

(5.5) is not a projection operator. In fact, its discrete $L^2$-norm

(5.6) $\qquad \|I - \underline{G}L^{-1}D\| = \rho(I - \underline{G}L^{-1}D)$

is often larger than 1. (The eigenvalues with largest absolute value must be smaller than -1 if that is the case.) If this norm is smaller than 1, then

(5.7) $\qquad \lim_{j \to \infty} (I - \underline{G}L^{-1}D)^j = I - \underline{G}(D\underline{G})^{-1}D$ .

Since (5.5) is a discretization of the continuous projection operator, it is to be expected that eigenvalues corresponding to smooth eigenfunctions are larger than -1. This motivates the following modification of (5.5):

(5.8) $\qquad I - \underline{G}(I + \omega L)^k L^{-1} D$ ,

with $\omega \approx 1/\rho(L)$ but $\omega < 1/\rho(L)$ , k integer. For a fixed k, $(I + \omega L)^k$ is consistent with the identity operator and (5.8) can therefore still be considered a discretization of the continuous projection operator. For sufficiently large k, the powers of (5.8) converge to $I - \underline{G}(D\underline{G})^{-1}D$.

This leads to the following algorithm.

> Given $\underline{u}$, generate $\underline{u}^0, \underline{u}^1, \underline{u}^2, \cdots$ with $\underline{u}^j \to P\underline{u}$
> as follows:
> $\underline{u}^0 := \underline{u}$ .
> Given $\underline{u}^j$, define
> $k := 0$; $\underline{u}^{j,0} := \underline{u}^j$, $q^{j,1'} := L^{-1}D\underline{u}^{j,0}$; $\underline{u}^{j,1} := \underline{u}^{j,0} - Gq^{j,1}$ .

(5.9)   While $\| \underline{u}^{j,k+1} \| > \| \underline{u}^j \|$ :

> $k := k+1$; $q^{j,k+1} := \omega L q^{j,k}$; $\underline{u}^{j,k+1} := \underline{u}^{j,k} - Gq^{j,k+1}$.
> $\underline{u}^{j+1} := \underline{u}^{j,k+1}$ .

In our experiments, this algorithm usually chooses $k = 0$ when computing $\underline{u}^1$ from $\underline{u}^0$, $k > 0$ for $j > 1$, sometimes even $k \gg 0$ .

We give a numerical example. Consider

(5.10)        $\underline{u}(\underline{x}) = (\sin(2\pi x_1)\cos(2\pi x_2) , 0)$ .

The divergence-free part of $\underline{u}$ is

(5.11)        $0.5(\sin(2\pi x_1)\cos(2\pi x_2) , -\cos(2\pi x_1)\sin(2\pi x_2))$ .

Computing the divergence-free part of $\underline{u}$ numerically, we obtain truncation errors with the following discrete $L^2$-norms:

| N | iteration | worst case | best case |
|---|---|---|---|
| | 1 | 0.35 | 0.21 |
| | 2 | 0.34 | 0.19 |
| | 3 | 0.36 | 0.18 |
| 25 | 4 | 0.34 | 0.17 |
| | 5 | 0.37 | 0.17 |
| | 10 | 0.36 | 0.16 |
| | 20 | 0.37 | 0.16 |
| | 1 | 0.17 | 0.12 |
| | 2 | 0.16 | 0.11 |
| | 3 | 0.16 | 0.11 |
| 100 | 4 | 0.16 | 0.11 |
| | 5 | 0.16 | 0.11 |
| | 10 | 0.17 | 0.11 |
| | 20 | 0.17 | 0.11 |
| | 1 | 0.079 | 0.067 |
| | 2 | 0.076 | 0.060 |
| | 3 | 0.077 | 0.062 |
| 400 | 4 | 0.077 | 0.064 |
| | 5 | 0.078 | 0.064 |
| | 10 | 0.079 | 0.063 |
| | 20 | 0.078 | 0.065 |
| | 1 | 0.040 | 0.035 |
| | 2 | 0.040 | 0.033 |
| | 3 | 0.043 | 0.034 |
| 1600 | 4 | * | 0.035 |
| | 5 | * | 0.036 |
| | 10 | * | 0.037 |
| | 20 | * | 0.037 |

For each N, 10 sets of points were tested (random, uniformly distributed). "*" means: The iteration was aborted because k became larger than 50.

One iteration generates an approximation to the continuous solution which is as good or even better than the approximation obtained after more iterations. What we have gained by replacing (5.5) with (5.9) is guaranteed stability, not higher accuracy.

It is a surprising fact that the truncation error does decrease with growing N here, even though D is not pointwise consistent with the divergence operator. We also observe that the variance of the truncation error seems to decrease with growing N.

## 6. A fractional step method for the Navier-Stokes equations

We consider the following fractional step method for the Navier-Stokes equations:

(6.1)
$$\frac{\tilde{u}^{n+1} - u^n}{\Delta t} - \nu L^n \tilde{u}^{n+1} = f^{n+1}$$
$$\underline{u}^{n+1} = P^n \tilde{\underline{u}}^{n+1}$$
$$\underline{x}^{n+1} = \underline{x}^n + \Delta t \underline{u}^{n+1} .$$

Here $L^n$ denotes the discrete Laplacian on "$S_N^n$", the set $S_N$ at time $n\Delta t$; the points in that set are the components of the infinite vector $\underline{x}^n$. $P^n$ denotes the orthogonal projection onto the kernel of $D^n$, where $D^n$ is the discrete divergence operator on $S_N^n$.

Writing $P^n \tilde{\underline{u}}^{n+1} =: \tilde{\underline{u}}^{n+1} - G^n q^{n+1}$, we obtain

$$\frac{\underline{u}^{n+1} - u^n}{\Delta t} - \nu L^n \tilde{\underline{u}}^{n+1} + \frac{G^n q^{n+1}}{\Delta t} = \underline{f}^{n+1} \quad \text{and}$$

(6.2)
$$D^n \underline{u}^{n+1} = 0 .$$

We therefore call $p^{n+1} := q^{n+1}/\Delta t$ the "approximation for the pressure".

We notice that (6.1) is stable in the sense that

(6.3)
$$\| \underline{u}^{n+1} \|_{(n)} \leq \| \underline{u}^n \|_{(n)}$$

where "$\| \bullet \bullet \|_{(n)}$" denotes the discrete $L^2$-norm on $S_N^n$. This follows because $L^n$ is symmetric and negative semidefinite in the corresponding inner product. Using the projection method described in section 5, stability is guaranteed even though we do not apply the operator $P^n$ exactly.

As a test example, we use

(6.4)
$$u_1(\underline{x},t) = \sin(\tfrac{\pi}{2}t)\cos(2\pi x_1)\sin(2\pi x_2)$$
$$u_2(\underline{x},t) = -\sin(\tfrac{\pi}{2}t)\sin(2\pi x_1)\cos(2\pi x_2)$$
$$p(\underline{x},t) = \sin(\tfrac{\pi}{2}t)\cos(2\pi x_1)\cos(2\pi x_2)$$

with $\nu = 0.1$. We use

(6.5)
$$\Delta t = 1/\sqrt{N}$$

and measure discrete $L^2$-norms of the errors in the velocity at time 1. In every time step, we use only one step of the iteration (5.9). Initially, the $\underline{X}_j$ are at the positions $((k_1 + 0.5)h, (k_2 + 0.5)h)$, $0 \leq k_1 \leq \sqrt{N}-1$, $0 \leq k_2 \leq \sqrt{N}-1$. If all Helmholtz and Poisson problems are solved up to rounding error accuracy, we obtain:

| N | error in $u_1$ | error in $u_2$ |
|---|---|---|
| 100 | 0.35 | 0.35 |
| 400 | 0.12 | 0.12 |
| 1600 | 0.067 | 0.067 |

It is reasonable to expect that iteration up to rounding error accuracy is necessary only at time 0. At later time, the values from the previous time step should provide an excellent initial guess, and little work should be required to improve this guess such that the truncation error level is reached. We repeat our experiments with only one two-level cycle with m=2, l=1, confirming these considerations:

| N | error in $u_1$ | error in $u_2$ |
|---|---|---|
| 100 | 0.35 | 0.35 |
| 400 | 0.12 | 0.12 |
| 1600 | 0.065 | 0.065 |

## Acknowledgement

This work was supported by the Department of Energy at the Courant Mathematics and Computing Laboratory of New York University under contract DE-ACO2-76ERO3077. Börgers was also supported by a Dean's Dissertation Fellowship at New York University.

Bibliography

[1]   J.M. Augenbaum and C.S.  Peskin: "On the Construction of the Voronoi Mesh on  a
        Sphere"; in press.  J. Comput.  Phys.

[2]   A.Bowyer: "Computing Dirichlet Tessellations"; Computer J. 24, 162, 1981.

[3]   A. Brandt:  "Guide  to  Multigrid  Development";  in Multigrid Methods, Lecture
        Notes in Mathematics, vol.  961, Springer-Verlag, 1982.

[4]   J. Dukowicz: "Lagrangian Fluid Dynamics Using the  Voronoi-Delaunay  Mesh";  in
        Numerical  Methods  for  Coupled  Problems, Pineridge Press, Swansea, U.K.,
        1981.

[5]   M.J. Fritts:  "Two-Dimensional  Lagrangian  Fluid  Dynamics  Using   Triangular
        Grids";  in  Finite-Difference  Techniques  for  Vectorized Fluid Dynamics
        Calculations, David L. Book, ed., Springer-Verlag, 1981.

[6]   B. Mercier: personal communication.

[7]   M.I. Shamos and D. Hoey: in Proc.  16th Annual IEEE Symposium on Foundations of
        Computer Science, Berkeley, 1975.

[8]   K. Stueben  and  U.  Trottenberg:  "Multigrid  Methods: Fundamental Algorithms,
        Model Problem Analysis and Applications"; in  Multigrid  Methods,  Lecture
        Notes in Mathematics, vol.  961, Springer-Verlag, 1982.

[9]   C.S. Peskin:  "A  Lagrangian  Method for the Navier-Stokes Equations with Large
        Deformations"; preprint.

[10] C.S. Peskin: "Numerical Analysis of Blood Flow in the Heart"; J. Comput.  Phys.
        25, 220-252, 1977.

[11]  G. Voronoi: in Z. Reine Angew.  Math.  134, 198, 1908.

# Free-Lagrangian Methods, Independent Time Steps, and Parallel Processing[†]

*Peter G. Eltgroth*

**Parallel Processing Project**
**Lawrence Livermore National Laboratory**
**Livermore, Ca. 94550**

## 1. Introduction

The use of Free-Lagrangian methods for hydrodynamic problems has many advantages, including that of freeing the calculational process from the constraint of inappropriate spatial coordinate choices. A new method for performing hydrodynamic computations has been developed which permits different spatial regions of a problem to be advanced with time steps appropriate to local conditions (Reference 1). This new method, called the Independent Time Step Method (ITSM), frees the calculational process from the constraint of inefficient time coordinate choices. It would seem that a marriage of the two approaches should lead to physics codes which can efficiently deal with complex hydrodynamic computations. In addition, the resulting codes can be used efficiently on parallel processors, since they need no synchronization points or bottlenecks at which searches over all points must be carried out.

A one-dimensional Free-Lagrange code has been developed which runs on any computer supporting FORTRAN 77. Further, the code has been written (with macros) so that it executes different portions of the problem domain in parallel on a parallel processor. The code has been tested on a variety of Unix machines and the Denelcor HEP multiprocessor. It has been found that approximately a factor of 10 fewer zone cycles are needed to compute results for a colliding blast wave problem when compared to a classic "lockstep" lagrangian calculation on a uniprocessor. The number of time step calculations is less by a factor of approximately 4. The overhead of the hydrodynamics and time step computations (needed to support parallelism) is sufficiently increased, however, that the ITSM code is no faster than the lockstep code for one-dimensional problems on uniprocessors. On parallel processors, speedups of order 20 have been demonstated. It appears that as the complexity of a computation goes up (as for multi-dimensional problems), the advantages will swing toward the favor of the ITSM Free-Lagrangian approach, even for uniprocessors.

---

[†]Work performed under the auspices of the U. S. Department of Energy by the Lawrence Livermore National Laboratory under contract No. W-7405-ENG-48.

## 2. Time Step Choice

The essential physical concept behind the Independent Time Step Method is the restriction of the advance of a node so that information from its neighbors can only affect that node's future (Courant condition). This is coupled with the statement that a node should have its physical attributes advanced only when needed because of some accuracy requirement or because that node restricts the advance of another. Communication of information is local so that many different regions can be worked on at once. The order of selection of regions is arbitrary and might be tuned to suit specific problems.

We will now give some of the details of the time step choice for the one-dimensional code. We define the position of the j-th node to be $x_j$ and assume it to be evaluated at time $t_j$. When a node is to have its time step computed, the first quantity of interest is the local Courant time step $\Delta t_{lcj}$. A local sound speed $c_{l\pm}$ is known at the right and left interfaces of the j-th node with its neighbors. From neighbor node positions, velocities, and times we can construct interface positions $x_{i\pm}$ and velocities $u_{i\pm}$ at the average time between neighboring nodes. The local Courant time step is constructed by demanding that information from a neighboring node reach the interface no earlier than the time at which nodal quantities are to be evaluated.

$$\Delta t_{lcj} = \frac{(x_{i-}-x_{j-1})-(c_{l-}-u_{i-}/2)(t_j-t_{j-1})}{(c_{l-}-u_{i-})} \tag{1}$$

The expression above gives the local Courant condition as determined from the left. A similar expression with the sign of the interface velocity reversed gives the same condition determined from the right. The minimum of the two values is assigned to $\Delta t_{lcj}$. A simple expression for the local Courant condition such as $\Delta t = \Delta x/c$ does not adequately protect a node when its neighbors are undergoing substantial changes.

A local Courant condition is not sufficient to guarantee that information from anywhere in a problem will not violate causality for a given node. In the usual formulation of hydrodynamics, a search is made over all local conditions and a global value for the Courant condition is set to the minimum. In parallel systems such a global search can be a serious bottleneck. In ITSM a global restriction on the advance of each node is constructed by local means using one piece of information known at all nodes. That piece of information is the fastest speed of information propagation which can occur anywhere in the problem. We call it the global speed of sound $c_g$. A Courant time $t_{Cj}$ is constructed as the envelope of allowable propagation times from all nodes in the problem. The global Courant time step $\Delta t_{gcj}$ is given by

$$\Delta t_{gcj} = \frac{(x_{i-}-x_{j-1})-(c_g-u_{i-}/2)(t_j-t_{Cj-1})}{(c_g-u_{i-})} \tag{2}$$

The expression above gives the global Courant condition as determined from the left. It is exactly analogous to Eqn. 1 except for the use of the Courant time as a reference time and the use of the global speed of sound. Again the minimum of the left and right values is assigned to $\Delta t_{gcj}$. As information propagates through the problem via $t_C$, the global Courant time step can be larger than the local Courant time step. The final Courant time step $\Delta t_{cj}$ is given by

$$\Delta t_{cj} = minimum(\Delta t_{lcj}, \Delta t_{gcj}) \tag{3}$$

but is never allowed to be less than zero. The Courant time is given by

$$t_{Cj} = t_j + \Delta t_{cj} \tag{4}$$

In ITSM the Courant time step is not ordinarily used when it is desired to advance the physical quantities associated with a node. Instead it is used to inhibit the advance of a node and to determine when to signal other nodes to advance.

An accuracy time step $\Delta t_{aj}$ is computed for each node by limiting the allowable change of physical quantities during an update. For the one-dimensional code, it was found sufficient to restrict the time step so that volume changes were limited to 5 or 10% of the original volume associated with a node. The expression used for the accuracy time step was

$$\Delta t_{aj} = 0.05 \frac{(\tilde{x}_j - \tilde{x}_{j-1})}{|\tilde{u}_j - \tilde{u}_{j-1}| + \epsilon} \tag{5}$$

where quantities such as $\tilde{x}$ and $\tilde{u}$ are estimated at the time $t_{Cj}$ and $\epsilon$ is a suitable small quantity.

Given the two time steps as outlined above, the decision to update a node can now be made. First compare $\Delta t_{cj}$ and $\Delta t_{aj}$. If the Courant time step is smaller, then the update of the node is not carried out. Instead a flag is set for the node which has caused the restriction of the Courant time step. This flag is used to force that restricting node to update its variables when it is next scheduled for examination. If the accuracy time step is smaller, then the node is scheduled for update with $\Delta t_{aj}$.

If the flag is on for the node being considered, then the node is scheduled for update with time step equal to minimum $(\Delta t_{aj}, \Delta t_{cj})$. The flag forcing the update of a node is set whenever that node restricts a neighbor through the local Courant condition or whenever that node restricts all of its neighbors through the global Courant condition. This last condition is not necessary for correct code execution but enhances efficiency.

The time step logic remains essentially the same for multi-dimensional schemes or for other choices of hydrodynamics analysis (eg., Lagrangian or Eulerian).

## 3. Physical Variables Update

The Free-Lagrangian code was implemented by tracking the position and velocity of each node, while attributing internal energy to each region between nodes. At each node, information had to be kept about the position and velocity of that node at the time of each of its neighbors. This information was needed for estimation of volume, specific internal energy, and pressure.

We will now give some of the details of the procedures used to update physical variables in the one-dimensional Free-Lagrange code. Quantities which needed to be evaluated between nodes (ie., at an interface) were estimated at the average time and position between nodes. Many alternative schemes were tried, but the average placement gave the most accurate results.

Upon entering the update procedure in which node j is to be advanced by $\Delta t_j$ , the volume and energy density evaluated between nodes are used to determine the pressure acting between each node pair. This pressure includes artificial viscosity using the usual von Neumann formulation. For the one-dimensional problem denote pressure at the left side by $p_{lj}$ and at the right side by $p_{rj}$. The update of node velocity is then accomplished by careful accounting of the momentum transfer from each interface. If $t_{j-1} \leq t_j$ then

$$I_l = p_{lj} \Delta t_j \tag{6}$$

and

$$Z_j = Z_j + I_l \tag{7}$$

where $I_l$ is the impulse imparted to the left interface of the node and $Z_j$ is the sum of all impulses already applied to whichever of nodes j-1 or j is at the later time.

If $t_{j-1} > t_j$ then

$$I_l = Z_j \ minimum[1, \Delta t_j/(t_{j-1} - t_j)] + p_{lj} \ maximum(0, t_j + \Delta t_j - t_{j-1}) \tag{8}$$

and

$$Z_j = Z_j \ \{1 - minimum[1, \Delta t_j/(t_{j-1} - t_j)]\} + p_{lj} \ maximum(0, t_j + \Delta t_j - t_{j-1}) \tag{9}$$

An exactly analogous calculation is performed for the right interface.

Assuming a node mass $M_j$ , the node velocity $u_j$ is then given by

$$u_j = \frac{I_l - I_r}{M_j} \tag{10}$$

The node position is given by

$$x_j = x_j + u_j \ \Delta t_j \tag{11}$$

The node acceleration is kept in order to facilitate velocity estimation at future times. The specific internal energy is then advanced for each node-neighbor pair to the average time between the pair. In the one-dimensional problem, the specific internal energy at the left side of node j is found by

$$e_j = e_j - \frac{p_{lj} \ (u_j - u_{j-1}) \ \Delta t_j}{M_j + M_{j-1}} \tag{12}$$

Note that Eqn. 12 effectively uses one half of the time step by which node j is advanced. An exactly analogous equation is used to advance the specific internal energy at the right side, $e_{j+1}$. A local sound speed at each interface region can be constructed from the internal energy and the artificial viscosity value. At the end of the physical variables update all flags are cleared and the node and its neighbors are scheduled for re-examination.

## 4. Derivation of the Energy Equation

We will here include a short justification for the form of the energy equation given as Eqn. 12. The quantity $e_j$ is not strictly identical to the specific internal energy between nodes since the nodes are not necessarily even known at a given time. Instead the quantity $e_j$ as defined in Eqn. 12 represents a space-time average of the specific internal energy which reduces to a standard definition when adjacent nodes are evaluated at the same time. All normal edits of the code are taken with all nodes at one time in order to facilitate checks of physical quantities.

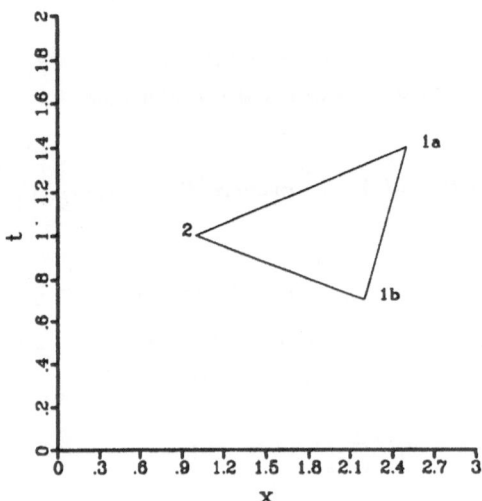

Fig. 1 : Schematic diagram of space-time region covered when a node is updated from position 1b (before) to position 1a (after).

Consider the update of node 1 in Figure 1. Call the node to its left node 2. The initial information about $e_j$ is known between space-time points 1b (for before) and 2. We wish to evaluate $e_j$ between points 1a (for after) and 2. Denote the actual microscopic value for the specific internal energy between 1b and 2 by $e_0$; between 1a and 2 by $e$. Then approximately

$$e = e_0 + \frac{de}{dt}\delta t \tag{13}$$

throughout the region enclosed by 1b, 1a, 2. $\delta t$ is the time difference between an original point on line 2,1b and the point at which we evaluate $e$. The total rate of change of $e$ is given by

$$\frac{de}{dt} = -\frac{p}{\rho}\frac{\partial u}{\partial x} \tag{14}$$

We choose to parametrise our path integrals using the spatial variable $x$. Call $E$ the total internal energy average along the path 2,1a.

$$E = \int_{x_2}^{x_{1a}} \rho\, e\, dx = \int_{x_2}^{x_{1a}} \left( \rho e_0 - p\frac{\partial u}{\partial x}\delta t\right) dx \tag{15}$$

$$E = E_0 - p\,\delta t\,(u_{1a} - u_2) \tag{16}$$

where $E_0$ is the original internal energy between the nodes and we have assumed that mass between nodes is constant (true for one-dimension). Since, by inspection $\delta t = \Delta t/2$ when averaged over the space-time region just covered, this last result yields Eqn. 12.

## 5. Results

The one-dimensional code has been tested on a problem involving two blast waves launched from opposite ends of a shock tube (Reference 2). This problem has been found to be a convenient basis for comparison of different numerical methods. At the beginning of the computation, only a few nodes are involved in violent action. At finish time, the two blast waves have collided and all nodes have reached high internal energies.

A short summary of execution times has been given in the Introduction. On the two blast wave test problem, momentum conservation was perfect (having been constructed into the code) while total energy in the problem increased slowly, changing by about 0.2% at problem termination.

Since a primary motivation behind the development of the ITSM was the desire to develop efficient code for parallel processors, much effort has been expended on executing the code on a variety of machines. The code has been run on uni-processors, multi-processors, and multi-processor simulators. The high level description of the code for one process is the same for all incarnations:

    Node chosen from queue for examination
    Identify neighbors
    Lock neighbors (no other process can update)
    Determine time step
    Update physics if time step acceptable
    Put appropriate nodes in queue(s)
    Unlock neighbors
    Return to top

A variety of different schemes have been attempted for the selection of a node for examination. Even though the physical algorithm has no inherent synchronization penalty, the problem of finding work to do can be a bottleneck.

The original approach used low level synchronization constructs to parcel out jobs. When a fixed partitioning of work to a process was employed, not all processes could be kept busy for arbitrary problems. When any process could access any part of the work space, a critical section of code was created (only one process active at one time) in order to protect against simultaneous updates. This overhead caused a significant slowdown of the code. In addition, the low level of the implementation caused debugging and modification to be difficult. In one dimension, the low level synchronization mechanisms were tricky. In two and three dimensions, the implementation details seemed overwhelming.

At this point, it was decided to attempt a high level job selection approach using macros developed at Argonne National Laboratory (Reference 3). The high level approach allows more experimentation with different schemes. Some of the approaches tried include several queues containing nodes ready to examine (ASKFOR monitor), a master queue pointing to the individual queues above (Queue of queues), and a distributed set of subroutines accessing the queues containing node numbers (distributed ASKFOR). In all cases the code performance was directly related to the size of the critical section employed. Every effort must be made to keep critical section size small if large (>16) numbers of processes are to be kept busy. A table of run times on the Denelcor HEP (4 Process Execution Modules) is given below.

| Independent Time Step Run Times | | | | | | | | | | | | |
|---|---|---|---|---|---|---|---|---|---|---|---|---|
| | P | | | | | | | | | | | |
| Method | 1 | 4 | 8 | 12 | 16 | 20 | 24 | 28 | 32 | 36 | 40 | 48 |
| Self-Scheduled | 4795 | 1686 | 1076 | 987 | 1018 | 1123 | 1268 | | 1568 | | 1897 | 2214 |
| ASKFOR Monitor | | 1188 | | 512 | 384 | 388 | 431 | | 510 | | | |
| Queue of queues | | 1211 | | 471 | 433 | 437 | 481 | 528 | 582 | 631 | | |
| Distributed ASKFOR | | 1479 | | | 413 | | 340 | 324 | 326 | 343 | 360 | 396 |

Times in seconds
P = number of processes
The test problem was two colliding blast waves with 480 zones.
Four PEMs - processes distributed evenly.

Table 1 : Run times for a variety of work distribution methods.

An advantage of methods which have queues of queues is that they provide a natural mechanism for vectorizing the procedures within the multiprocessor framework. The real usefulness of the vector code will be tested by the two dimensional Free Lagrange approach which is now being developed.

The two dimensional Free Lagrange Independent Timestep (FLIT) code uses a straightforward extension of ideas from the one dimensional approach. The primary difficulty is in maintaining a consistent accounting of the momentum exchange between nodes which are, were, or will be neighbors within a given time window. In addition, the procedure for determining neighbors at a given instant of the calculation must use only local information in order to have an efficient multiprocessor implementation. The first attempt at this code scans only old neighbor nodes and their neighbors in order to determine new neighbors.

## References

1. Eltgroth, P. G. and Porter, A. P., "The Independent Time Step Method for Hydrodynamics" UCRL-89853, Rev.1, October 1984

2. Woodward, P. R. and Colella, P., Journal of Computational Physics 54 (1984), 115-173

3. Lusk, E. L. and Overbeek, R. A., "Implementation of Monitors with Macros: A Programming Aid for the HEP and Other Parallel Processors." ANL-83-97, December 1983.

THREE-DIMENSIONAL ALGORITHMS FOR GRID RESTRUCTURING
IN FREE-LAGRANGIAN CALCULATIONS

Martin Fritts
Science Applications International Corporation
134 Holiday Court, Suite 318
Annapolis, Maryland 21401

INTRODUCTION

The Free-Lagrange method is unique among adaptive grid techniques
in that it may incorporate the full range of mesh mapping, nesting,
blending and movement algorithms in conjunction with local grid restructuring. For example, the method offers the capability of automatically increasing the resolution in the neighborhood of different
physical flow features, such as shocks, while simultaneously using
local grid adjustments to track material interfaces which are frag-
menting or merging. The price that is paid for this flexibility is
increased storage overhead, the complex numerical templates associated
with general-connectivity grids, and decreased efficiency due to the
lack of global indexing. This paper focuses on techniques which lower
that price for three-dimensional calculations. Since many of the
global adaptive grid schemes are both efficient and readily vectorized, the discussion will be restricted to local grid restructuring
algorithms.

Two-dimensional triangular grid simulations have shown that the
three basic grid restructuring algorithms of line reconnection, vertex
addition and vertex deletion are a sufficient basis for building
higher order, automated local grid restructuring algorithms. Figure 1
is a sequence of frames taken from a two-dimensional calculation of
drop shattering in an incompressible flow.[1] As seen in the figure,
these algorithms allow an automatic transition from a singly-connected
fluid region, a single drop, to a multiply-connected fluid region, a
series of elongated droplets. A triangular grid is well suited to
this application since triangle sides can remain aligned along the
interface throughout its development. The placement of densities on
triangles leads to proper vorticity generation at the interface while
strict vorticity conservation is maintained during grid advancement
and grid restructuring.

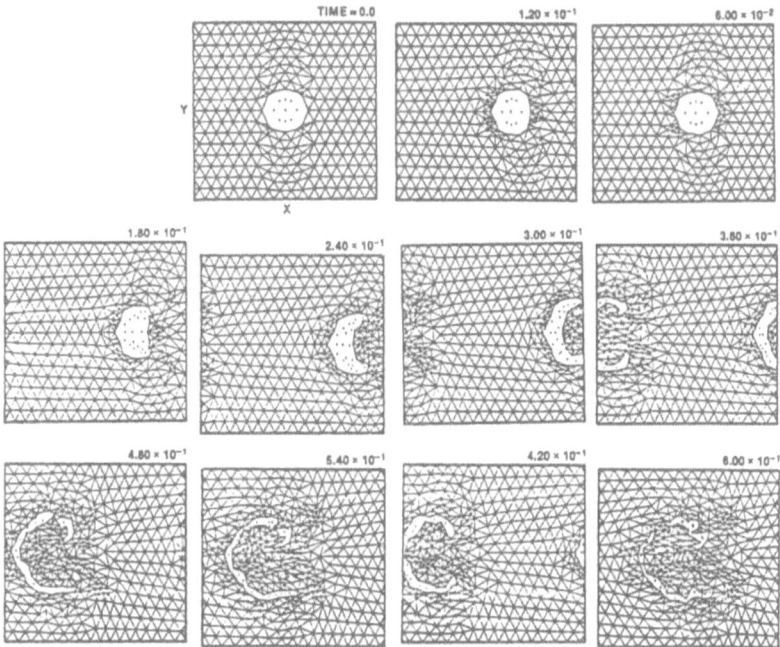

Figure 1. Droplet deformation and shattering due to an external flow, from a free-Lagrangian simulation without viscosity or surface tension. The grid lines have been omitted within the drop for clarity.


There are many reasons for wanting to extend this technique to three dimensions. First, there has been considerable progress in two-dimensions in quantifying the meaning of the "best" approximating mesh for an arbitrary distribution of mesh vertices. Much of that progress has been linked to the properties of the Voronoi mesh. Every vertex is associated with its Voronoi cell, the region of space surrounding that vertex which contains all points in space closer to that vertex than to any other. Figure 2 illustrates a Voronoi cell, bounded by bisectors of line segments joining neighboring vertices. As shown in the figure, the vertex connections form a triangular mesh which is the dual of the Voronoi mesh, namely, the Delaunay mesh. The bisectors which form the Voronoi cell determine which vertices are neighbors in the Delaunay mesh. Differencing over a Voronoi mesh is particularly desirable for Lagrangian calculations because a change in neighboring vertices, as in Figure 3, is signaled by smooth changes in cell side lengths rather than the abrupt grid connection changes in the Delaunay mesh. Reconnection algorithms have been derived for triangular meshes which ensure that the mesh is identically a Delaunay mesh[2], saving the

large calculational overhead of tracking Voronoi cells. These algorithms preserve diagonal dominance of solution matrices of Laplacian
operators differenced over a general triangular mesh. The "best" mesh
for other operators may then be viewed as a generalized "Voronoi" mesh
dictated by grid connection algorithms more appropriate to those operators.

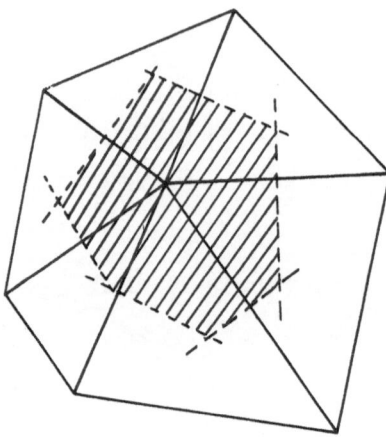

Figure 2. The construction of a Voronoi cell in two dimensions from
bisectors of lines connecting neighboring vertices. The outline of
the Voronoi cell is given by the dashed lines. The solid lines constitute the Delaunay triangular mesh.

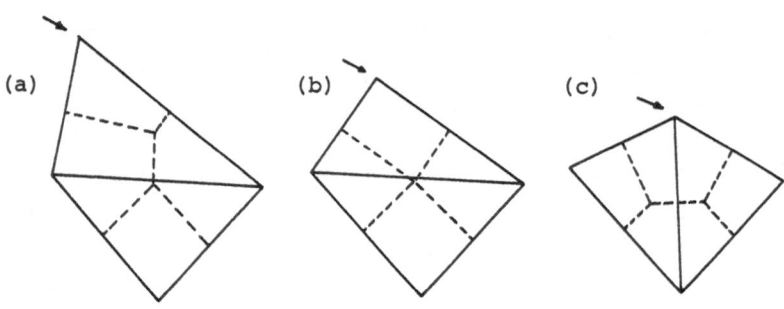

Figure 3. Changes in a Voronoi mesh. Movement of the uppermost vertex causes the separation of two neighboring vertices and the establishment of two new neighbors. In the Voronoi mesh this change is
evidenced by the gradual disappearance of the nearly vertical bisector
segment and the emergence of the nearly horizontal segment. In the
Delaunay mesh there is an abrupt change in the mesh connectivity at
inset (b), when either connection is equally valid.

A second reason for extending the technique to three dimensions is the success of local grid restructuring strategies in two dimensions. The tracking and resolution of physical features and interfaces is vastly simplified when the local connectivity between vertices is stored. Grid adaption becomes a straightforward question of what local vertex distributions are desired for the finer mesh. The placement and storage of that fine mesh is accomplished trivially. Restructuring algorithms ensure a regular mesh structure despite arbitrary movement of physical features embedded in the mesh and the local changes in resolution about them.

A third reason follows from advances in conservative differencing techniques that were realized in two dimensions. As shown in Figure 4, conservation for Lagrangian schemes involves additional work. Conservation must be assured not only for the updating of physical variables, as in Eulerian methods, but for grid advancement and grid restructuring as well. Knowledge of intercell fluxes is central to the formulation of exact conservative schemes, and requires access to grid connectivity information.

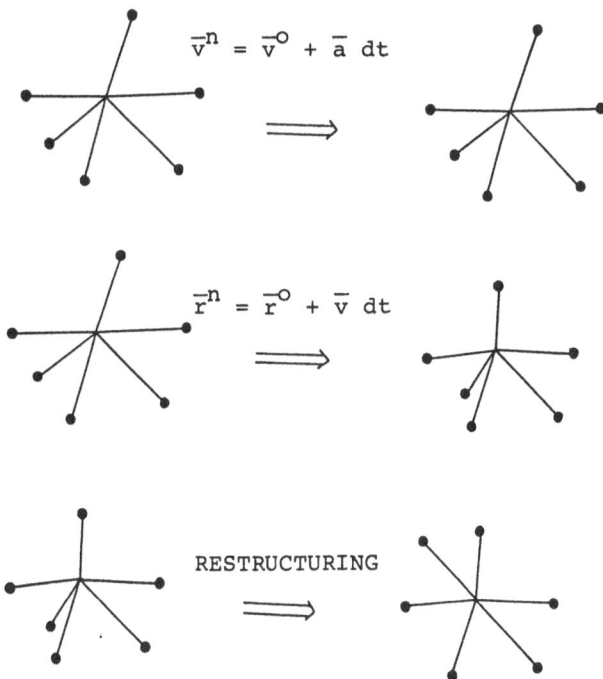

$$\vec{v}^n = \vec{v}^0 + \vec{a}\ dt$$

$$\vec{r}^n = \vec{r}^0 + \vec{v}\ dt$$

RESTRUCTURING

Figure 4.   The three steps in updating a Lagrangian calculation. Physical variables must be conserved throughout each of these steps.

Finally, two-dimensional free-Lagrangian schemes have generally allowed complete freedom for the proper placement of physical variables. Three-dimensional methods should also permit placement of physical variables at either vertices or within cells. However, the trade-offs between flexibility and efficiency are very real. Totally vertex-centered schemes are much more compact in storage and require far simpler grid restructuring algorithms.

Unfortunately, the simplest extensions of triangular mesh data structures to tetrahedral elements are not practical. Two-dimensional storage is normally redundant, typically storing information about vertices, edges and triangles. Storage for lists linking tetrahedral quantities to all these arrays is both expensive and difficult to implement efficiently. The grid reconnection algorithms are much more cumbersome in three dimensions, as shown in Figure 5. Grid reconnection in two dimensions entails redefinition of just two triangles and one line: in three dimensions tetrahedra may be added or deleted in the course of reconnection. In the face of such complexity, it is clearly advantageous to reduce the amount of grid bookkeeping and storage to a minimum.

**FIVE-VERTEX CONFIGURATIONS**

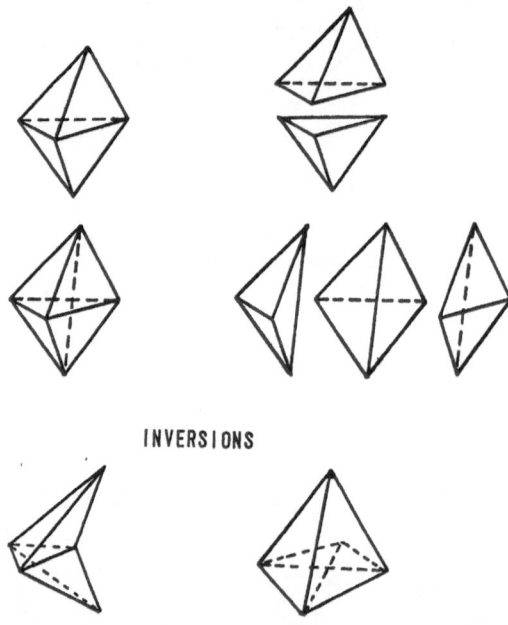

**INVERSIONS**

Figure 5. Three-dimensional reconnection. Choosing an alternate diagonal may result in a change in the number of tetrahedra. A good reconnection algorithm must also automatically eliminate inversions of tetrahedra, which are signs of the presence of gross inaccuracies.

A large reduction in the complexity of algorithms and data structures can be achieved by eliminating redundant information and by
carefully evaluating what should be stored and what should be regenerated. This paper presents algorithms for grid generation in the third
dimension starting from embedded two-dimensional surfaces. In two dimensions, this is analogous to storing only a series of lines which
connect all the vertices, and regenerating the triangular grid whenever triangle information is needed. Such a scheme stores a minimum
of information (the vertex positions, the vertex ordering and interface information) and yet is capable of quickly yielding Voronoi
neighbors. If the interior of a region contains a single fluid, the
scheme may even eliminate the need to reconnect vertices, and addition
and deletion of vertices on surfaces and in the interior is simpler.
However, the grid must be regenerated at every timestep. For two-
dimensional problems the storage overhead is low, so that local grid
restructuring is faster than the regeneration of the grid. In three
dimensions the additional complexity of reconnection and the additional storage of tetrahedral information make partial grid regeneration
the more attractive alternative.

This paper presents the grid restructuring algorithms in three
stages. First the algorithms are presented for the case of planar
triangulated surfaces embedded in and spanning a three-dimensional
region. Although this case is restrictive, it permits an examination
of the basic techniques needed for fast regeneration of grid connections in the third dimension. The resulting algorithms are simple and
efficient, and require storage only for the triangular meshes defining
the surfaces. Furthermore, the technique can be used to improve resolution in the third dimension by inserting additional planar surfaces where needed. It is shown that the tetrahedra generated by this
technique form a Delaunay mesh if the interplane spacing is comparable
to the resolution within the planes. The algorithm can therefore be
used for efficient determinations of Voronoi connections for initial
grids. Modifications to the algorithm for the case of closely spaced
surfaces appear later in the paper in the context of restructuring
algorithms which can accommodate colliding surfaces.

In the second stage, the restriction to planar surfaces is removed and regular surfaces are examined. The basic algorithms remain
the same with an additional operation to project the vertices of one
surface onto another. In the final stage, vertices on the surfaces

are allowed to migrate anywhere in space. This step can be accomplished in two ways. First a general reconnection algorithm is given which
extends the simple two-dimensional algorithms. If the cost of computer
storage is not a factor, this algorithm can be used in exactly the same
manner: a vectorizable search for all reconnections, followed by the
scalar operations themselves. A more efficient implementation requires
an additional algorithm to group points onto arbitrary smooth surfaces
for gridding purposes, as well as an additional "splitting" operation.
The splitting operation accounts for actual vertex positions, since
four vertices which are grouped onto the same surface can form a
tetrahedron within that surface. Algorithms will also be presented
for the resolution of gridding problems associated with interface
merging, intersection and fragmentation.

This paper will not discuss the physical aspects of numerical
modeling using these techniques. Although conservation of physical
quantities and the placement of physical variables on a mesh are important topics, this paper will center on the algorithms for grid
generation and not the use of the grids in calculations. The relationship of this scheme to other numerical schemes will similarly be left
unexplored. For the purposes of this paper it is not necessary to consider whether finite-difference, finite-element or other methods will
be used. The operators used in the partial differential equations are
assumed unknown, except for the illustrative use of the Laplacian.
Finally, none of the unfruitful methods developed and discarded will be
covered, even though these are in many cases particularly instructive.

TESSELLATION BASICS

There are many simple extensions of differencing schemes for two-
dimensional, triangular meshes to tetrahedral meshes in three dimensions. Control volumes may be constructed by similar algorithms in
both cases, whether using equi-division of tetrahedra, as in Figure 6,
Voronoi cells or other subdivisions. Numerical integration of the
divergence and curl of constant cell-based quantities is trivially performed, and a substantial literature is available on higher-order
approximations. As mentioned above, data structures to support general
meshes may be formed by adding appropriate tetrahedral lists linked to
the triangular lists.

As was shown in Figure 5, the primary problem with the extension to three dimensions is one of geometrical complexity. An assembly of tetrahedra which undergoes reconnection may be subdivided into a different number of tetrahedra. The problem of inversions of tetrahedra is similar to that of triangles, but in three dimensions additional tests must be performed to identify the inverted cells. Figure 7 shows all possible reconnections for a six-vertex configuration of four tetrahedra, and Figure 8 shows one possible reconnection for a seven-vertex configuration. As is evident, in many cases the number of tetrahedra before and after reconnection will change, and the logic of the reconnection algorithms becomes fairly complex for even the small number of vertex configurations shown.

## CONTROL VOLUMES

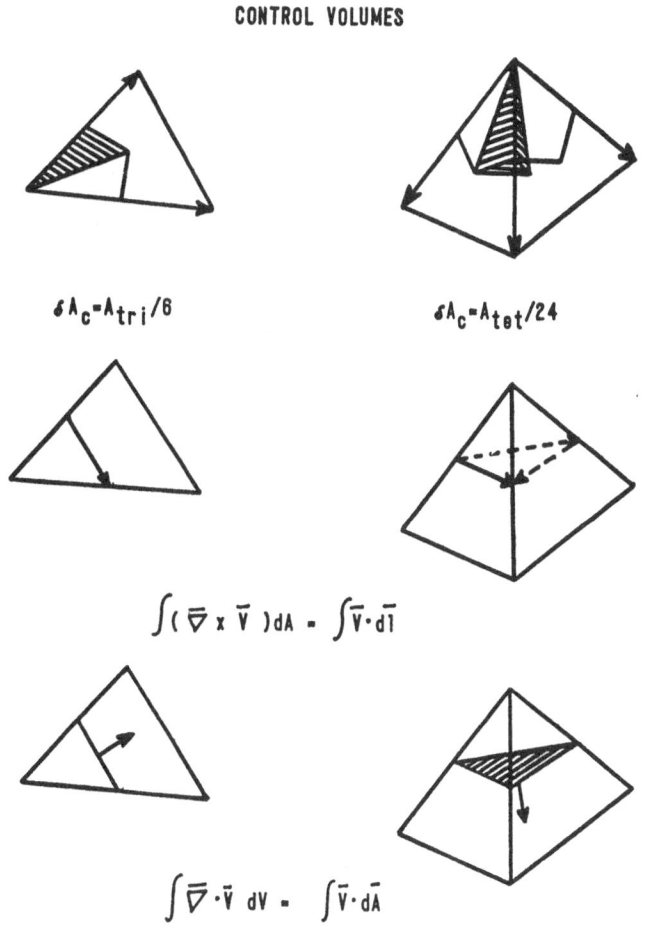

$$\delta A_c = A_{tri}/6 \qquad\qquad \delta A_c = A_{tet}/24$$

$$\int ( \bar{\nabla} \times \bar{V} ) dA = \int \bar{V} \cdot d\bar{l}$$

$$\int \bar{\nabla} \cdot \bar{V} \ dV = \int \bar{V} \cdot d\bar{A}$$

Figure 6. A comparison of cell construction and integration schemes for two and three dimensions.

The approach taken in this paper is to simplify both the restructuring algorithms and the data storage by taking advantage of fast and efficient gridding algorithms in the third dimension. If the meshes are used in Eulerian calculations, these procedures are merely efficient adaptive grid initializers which may be utilized at different points in the calculation to inject regions of finely zoned mesh or to delete unwanted vertices. If the mesh is moving, then the algorithms are the backbone of the restructuring scheme and are used at every timestep.

The simplest case will be covered first. The computational region is assumed to be spanned by a series of surfaces, some of which are associated with material boundaries or embedded features of the flow. For the initial discussion, those surfaces are coparallel planes.

### SIX-VERTEX CONFIGURATIONS

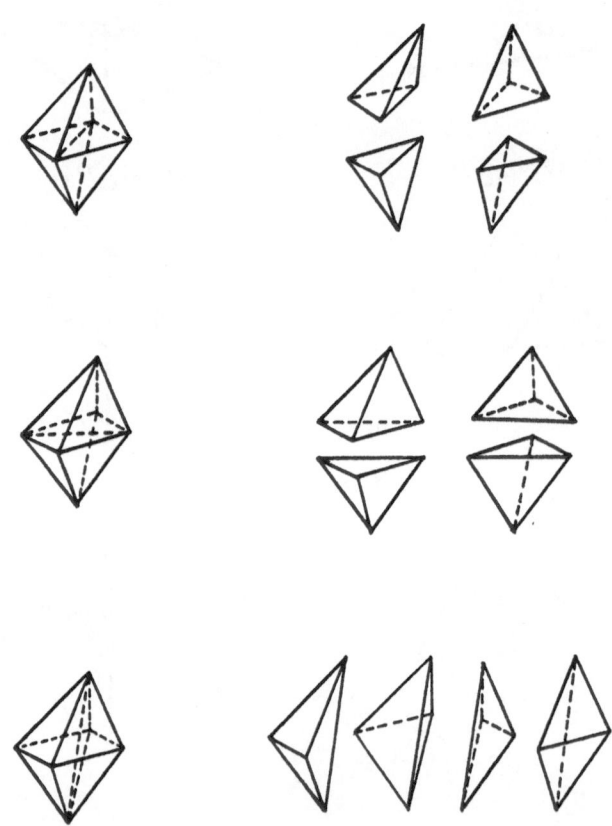

Figure 7. Reconnections for an assembly of tetrahedra sharing six common vertices.

All the vertices are found within these planes, which may degenerate to a single point or a single line. Each plane has been triangulated using two-dimensional restructuring and adaptive grid techniques so that they are Delaunay triangulations. The principle focus of this study is on how the region between the surfaces is filled with tetrahedra.

Two further assumptions are made. First, the spacing between the planes is assumed to be comparable to the intraplane grid spacings. Although the second assumption is not needed at this juncture, it is also assumed that the numerical approximation involves the use of the Laplacian. The reconnection algorithms are based on the difference operators used in the solution of the relevant partial differential equations. If the operator is the Laplacian, then the reconnection algorithm yields a mesh with the same connectivity as the Voronoi

## SEVEN-VERTEX CONFIGURATIONS

Figure 8. One possible reconnection for an assembly of tetrahedra sharing seven common vertices. The number of tetrahedra in the assembly varies.

mesh.  Further, the coefficients in the matrix representation of that
operator are the basis of the restructuring algorithm.  No additional
calculations are required: the computational work must be done for the
matrix inversion in any case.

The crux of the procedure outlined above is then an efficient algorithm which fills the interplane region.  Figure 9 illustrates two
triangulated planes, one having six triangles, the other two.  The
lower section of the figure shows the two planes superposed.  The
volume between the two planes must be filled by tetrahedra, and, by
the prescription above, must yield a Delaunay tessellation.  In
essence, the procedure to be used is very simple.  The first step is
to construct all tetrahedra  with three vertices on one of the planar
surfaces and one vertex on the other; that is, all "3x1" tetrahedra.

TRIANGULATED   PLANES

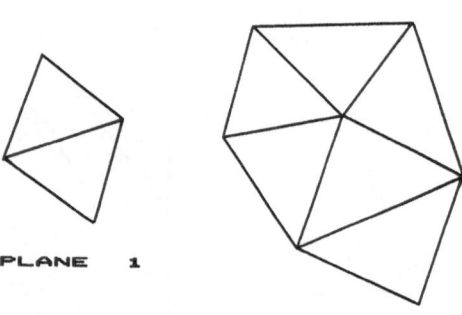

PLANE   1

PLANE   2

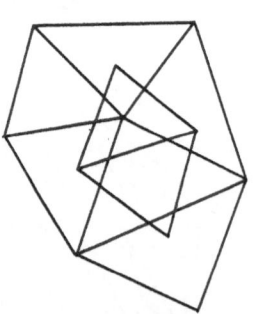

SUPERPOSITIONED

Figure 9.  Two simple triangulated planar segments.  In the lower
section of the figure, the two planes are superimposed to illustrate
their relative positioning.

The regions between the "3x1" tetrahedra are then filled with tetrahedra having two vertices on each of the surfaces, the "2x2" tetrahedra. If all the vertices lie on a series of planes, the interplane region is then completely filled.

Figure 10 illustrates the interconnections for the different types of tetrahedra for the planar segments shown in Figure 9. In the figure, "3x1" tetrahedra have three vertices in plane 2, whereas "1x3" tetrahedra are those having three vertices on plane 1. Figure 10 highlights the tetrahedra themselves for clarity. The heavy lines are those associated with plane 1, the lighter lines are on plane 2, and the interplane lines are transitionally shaded. All interplane lines not associated with a given type of tetrahedra are omitted.

**PLANE   INTERCONNECTIONS**

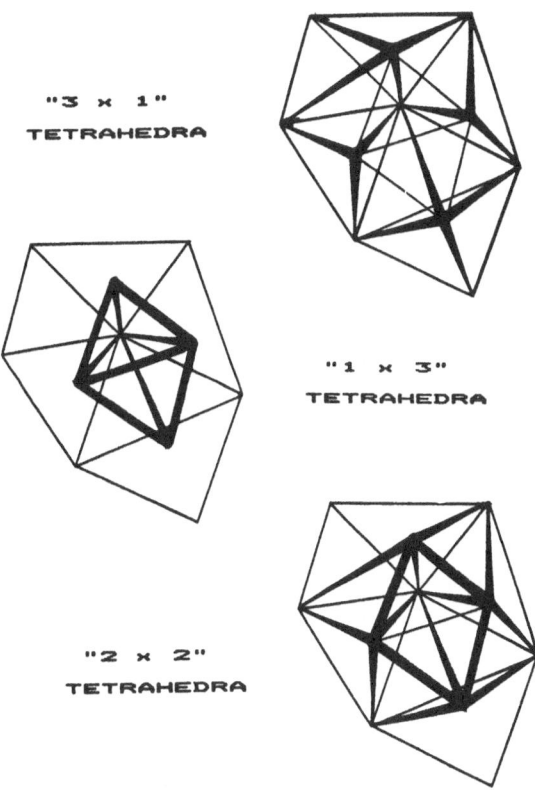

"3 x 1"
TETRAHEDRA

"1 x 3"
TETRAHEDRA

"2 x 2"
TETRAHEDRA

Figure 10.  Interconnections for the planes shown in Figure 9 according to the type of tetrahedron. Note that the tetrahedra completely fill the region between the planes.

The procedure for constructing the tetrahedra shown in Figures 9 and 10 guarantees a Delaunay tessellation. Every triangle has associated with it the center of its circumscribing circle, found at the intersection of its perpendicular side bisectors. The circumcenters of triangles are the corner points of Voronoi meshes in two dimensions. In three dimensions the circumcenter of a sphere around a tetrahedron must fall on the line perpendicular to a triangle and passing through the triangle's circumcenter. A Delaunay tetrahedra is then one which has the smallest radius circumsphere possible; that is, no other neighboring vertex could form a tetrahedron with the other three vertices which would have a smaller circumradius.[3] Since all planes are already triangulated by Delaunay triangles, choosing the vertex in the neighboring plane which is closest to the circumcircle center for the triangle ensures that all the "3x1" and "1x3" tetrahedra are Delaunay tetrahedra. The only case for which other connections are possible are those in which a line passes through an intermediate plane, forming a configuration similar to that in the middle section of Figure 5. The restriction that the interplane distance is comparable to the intraplane spacing removes that possibility.[4]

The construction of "2x2" tetrahedra follows from similar arguments about the minimum circumradii. Figure 11 illustrates a slightly more complicated configuration for which the vertices in one plane are all clustered above a triangle in another plane. The one possible "3x1" tetrahedron is formed as above, as are the five "1x3" tetrahedra. The "2x2" tetrahedra fill the remaining space. The procedure for constructing these tetrahedra must ensure that all space is filled by Delaunay tetrahedra without having to resort to extensive checks of neighboring vertices.

Figure 12 gives one possible procedure for selecting "2x2" tetrahedra. It is called the "nibble" method since it entails removing all vertices which are completely enclosed. Figure 12 continues the configuration presented in Figures 9 and 10. The procedure specifies that all "2x2" tetrahedra are found first at the boundaries. Once the boundary tetrahedra are specified, the boundary vertices may be eliminated from consideration, since, for the two planes under consideration, they will be completely surrounded. Deletion of these vertices results in new boundaries, for which all "2x2" tetrahedra are found, starting the cycle anew. Once the entire interplane region is completed, the procedure is repeated for the next interplane region.

Boundary tetrahedra are used because it is simpler to visualize that all connections are properly made at the boundaries. The "2x2" tetrahedra cannot, in general, cluster about a line. Lines within a plane which emanate from a single boundary vertex must be included within triangles in that plane unless the lines are themselves the surface; that is, unless there is a hole in the surface with the lines forming a ridge. In all other cases there must be intervening "1x3" or "3x1" tetrahedra. The "nibble" method merely makes it evident which vertices are to be combined into "2x2" tetrahedra. Checking the

## MULTIPLE VERTICES PER TRIANGLE

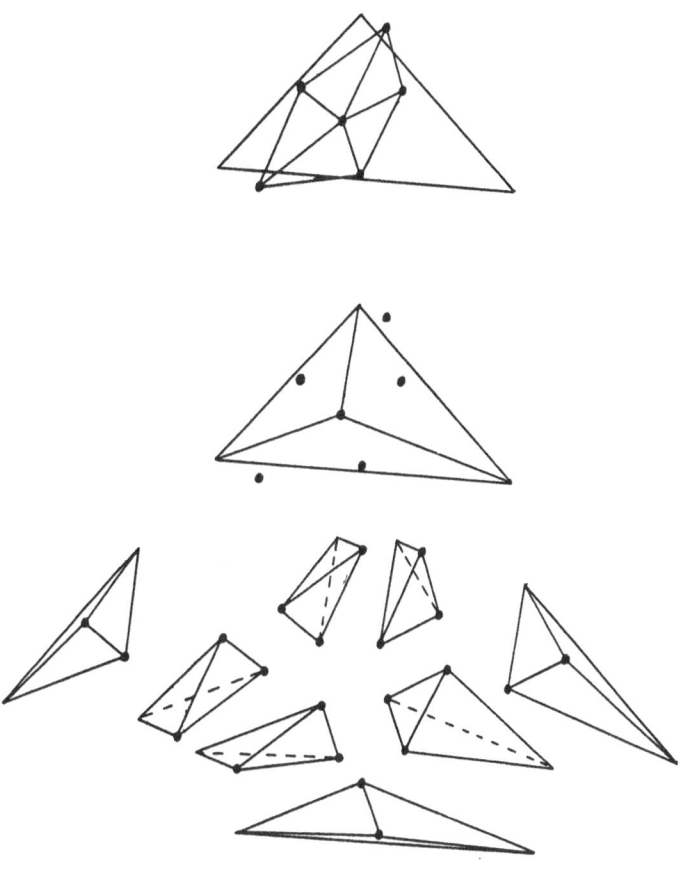

Figure 11. Construction of tetrahedra for the case in which several points in one plane lie within a triangle projected from the neighboring plane. The upper figure shows the two planes superimposed. Directly below is the one "3x1" tetrahedron. The five "1x3" tetrahedra are clustered at the bottom center, framed by the three "2x2" tetrahedra. All tetrahedra mirror their proper orientations in the assembly.

circumradius of the vertices ensures that they form Delaunay tetrahedra. Since the "3x1" and "1x3" tetrahedra were Delaunay tetrahedra,
the check is generally not required if boundary vertices are removed
one at a time.


RECONNECTION ALGORITHMS

For grids which are composed of assemblies of planar surfaces the
gridding procedures described above are sufficient. If vertices are
allowed to migrate out of these planes, however, more general reconnection procedures must be used. The reconnection formulas will be
derived by examining the coupling coefficients between two neighboring
vertices. The basic grid entities are therefore a line and the shell
defined by exterior faces of tetrahedra having that line in common.

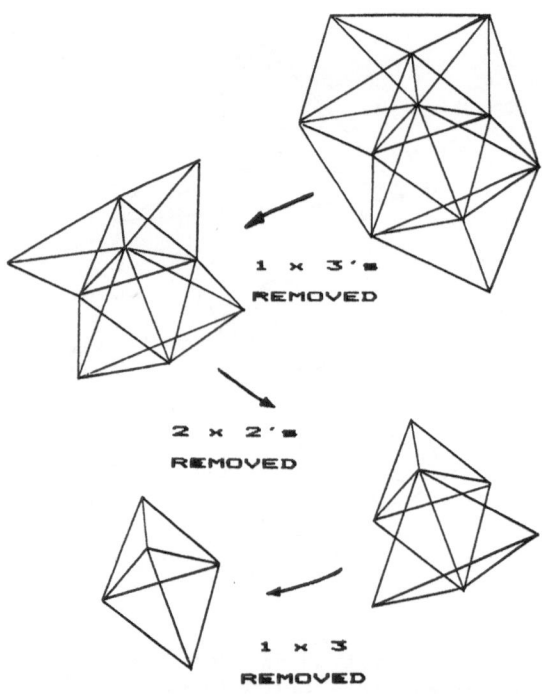

Figure 12. A procedure for selecting which vertices form "2x2" tetrahedra. Bookkeeping is simplified by proceeding inward from the boundaries in forming, then eliminating, the tetrahedra.

It will be instructive to first derive reconnection algorithms in two dimensions. If a variable f is defined at vertices on a triangular mesh, then the gradient of f is most simply approximated as a constant over triangles. The divergence of the gardient is found by summing all contributions from triangles about a central vertex. The Laplacian is therefore a vertex-centered quantity. Higher-order schemes may also be derived, but this simple scheme is second-order for constant mesh spacing and yields the standard second-order templates. For general triangular meshes, the approximation is first-order in the distance between the actual vertex position and the cell centroid, a much smaller distance than the mesh spacing for all but the most distorted meshes. For restructuring meshes, large distortions are not permitted, and the first-order error term is small.

An approximation to the Laplacian for two-dimensional, general triangular meshes is therefore

$$A_c <\nabla \cdot \nabla f> = \sum_{i;c} \frac{1}{2A_{i+\frac{1}{2}}} \{ f_i \hat{z} \times (\bar{r}_c - \bar{r}_{i+1}) + f_{i+1} \hat{z} \times (\bar{r}_i - \bar{r}_c) + f_c \hat{z} \times (\bar{r}_{i+1} - \bar{r}_i) \}$$

$$\times \frac{(\bar{r}_{i+1} - \bar{r}_i)}{2} \cdot \hat{z} \qquad (1)$$

where the vertices surrounding the central vertex, vertex c, are numbered in a counter-clockwise order, $\bar{r}_i$ is the position of the ith vertex, $A_{i+\frac{1}{2}}$ is the area of the triangle with vertices i, i+1 and c, $A_c$ is the cell area defined as one-third the sum of all the surrounding triangle areas and $\hat{z}$ is the unit vector in the third direction. The symbol $\sum_{i,c}$ denotes a summation over all triangles about the vertex c. The coefficient of the $f_c$ term is

$$a_c \simeq -\sum_{i;c} |\bar{r}_{i+1} - \bar{r}_i|^2 / 4A_{i+\frac{1}{2}} \qquad (2)$$

and is always negative. The coefficient of the $f_i$ term is

$$a_i \simeq -\frac{(\bar{r}_c - \bar{r}_{i+1}) \cdot (\bar{r}_{i+1} - \bar{r}_i)}{4A_{i+\frac{1}{2}}} - \frac{(\bar{r}_{i-1} - \bar{r}_c) \cdot (\bar{r}_i - \bar{r}_{i-1})}{4A_{i-\frac{1}{2}}} \qquad (3)$$

and should be positive if the matrix of coefficients is to be diagonally dominant.  As shown in Figure 13, this coefficient may be written
in the form

$$a_i \simeq \tfrac{1}{2} \ (\cot\Theta_1 + \cot\Theta_2) \tag{4}$$

where $\Theta_1$ and $\Theta_2$ are as shown.  Therefore, if the sum of $\Theta_1$ and $\Theta_2$ is
less than 180 degrees for each i, the matrix is diagonally dominant.
If the sum is greater than 180 degrees for any i, then the line from
i to c must be reconnected.  This reconnection algorithm yields exactly a Delaunay mesh.[2]

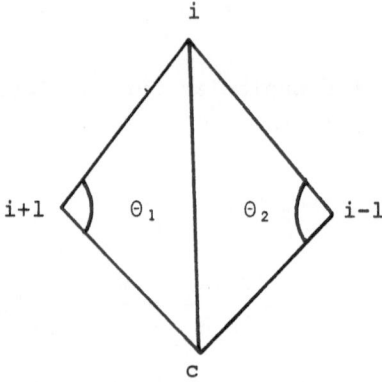

Figure 13.  The angles subtended by a line segment in a triangular
mesh.  The sum of the angles is used to determine line reconnection.


It should be noted that although this procedure gives exactly the same
mesh connections as a Voronoi mesh, it is very efficient.  The evaluation of coefficients proceeds through Equation 3, and involves triangle areas and dot products only, not angle equations.  Furthermore,
there is no extra calculation performed for the reconnection.  The
coefficients would have to be evaluated in any case if the original
set of partial differential equations included the Laplacian.

The fundamental difference in the three-dimensional problem is
that the grid line may be replaced by either another line or by a
tetrahedral face.  The converse is also true: a plane may be pene-
trated by a grid line.  In the context of the previous section, this
situation arises when two planes approach closely enough that the more
appropriate connection passes through one plane and into the plane

above or below it.   That is, a planar surface develops a hole.

A recasting of Equation 4 will prove convenient:

$$a_i \simeq -\sum_{j=1}^{2} \frac{|r_{cj}||r_{ij}|}{4A_j} \, \hat{n}_{cj} \cdot \hat{n}_{ij} \qquad (5)$$

As shown in Figure 14, the $\hat{n}$'s are the outward-directed normals to the appropriate line segments.   The analogous formula for three dimensions is derived in exactly the same manner

$$a_i \simeq -\sum_{j=1}^{n} \frac{A_{cj} A_{ij}}{12V_j} \, \hat{n}_{cj} \cdot \hat{n}_{ij} \qquad (6)$$

Figure 14b illustrates that the normals $\hat{n}_{ij}$ and $\hat{n}_{cj}$ are now directed outward from tetrahedral faces of area $A_{ij}$ and $A_{cj}$.   $V_j$ is the volume of the jth tetrahedron, and the sum runs over all n tetrahedra having the line segment in common.   The sum is not a simple sum over the angles between faces.   Its evaluation remains simple, however, and involves areas, volumes and dot products.   The coefficients are those arising from the differenced form of the Laplacian, and therefore must be evaluated anyway.      A proof that this algorithm yields a Delaunay mesh is not appropriate for this paper, and will be reported elsewhere.[4]

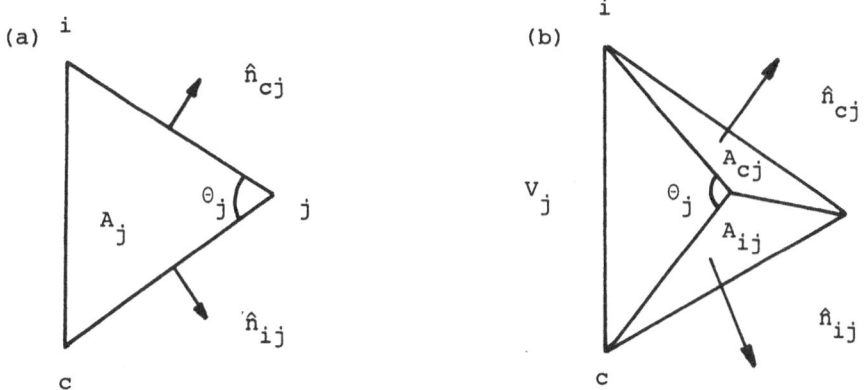

Figure 14.   The reconnection formula is derived from the coupling co-efficients for the line from vertex c to vertex i. In two dimensions the angles subtended by the line determine the reconnection, as shown in section a.   The three-dimensional analogue is shown in part b.

The statement that there is a minimum of extra computational work
to be performed with this algorithm needs clarification. For a configuration such as shown at the top of Figure 5, it is not possible to
immediately evaluate whether to make the alternate connection, choosing the line instead of the face, since there is no line connection
at the time. What is known is that some of the five vertices in the
configuration have coefficients which preclude diagonal dominance.
The local site of the problem is known; the cure is not. What remains
to be determined is which vertices form each of the final Delaunay
tetrahedra. In practice this process is made simple by an examination
of the magnitude and sign of individual terms in the sum. In general
the largest terms arise from couplings for the vertex to which a new
connection must be made. An efficient check is available through a
search for the minimum circumscribing radii for candidate tetrahedra.
Since the number of candidates is small, the search is minimal.

With this reconnection procedure, it is possible to now proceed
in two different directions. If all the relevant tetrahedral information can be stored, depending on the size of the problem, the reconnection algorithm can be used directly to retructure the mesh for arbitrary migrations of vertices. The algorithms for a general restructuring mesh are then complete except for the special treatment required
for lines or faces which cannot be reconnected, such as interface
lines.

If storage is costly, then the approach of saving primarily surface information must be taken. The modifications to the planar algorithms are fairly straightforward. An initial grid is tessellated
as before. If points lie on regular surfaces, then the vertices on one
surface must be projected onto its neighbors. The "3x1" tetrahedra
are then found as before. At this point the existing triangular
arrays may be expanded to store a minimal amount of tetrahedral information. For example, the triangle arrays may store the three triangle
vertices, the vertex in the "3x1" tetrahedra lying above it, the
vertex in the "3x1" tetrahedra lying below it and the "types" of the
tetrahedra. Note that only the "3x1" tetrahedra are associated with
a surface. The "1x3" tetrahedra are stored in the "3x1" arrays for
the neighboring surface. The tetrahedra types allow different physical models to be used for each of several fluid types. No further
vertex information is reqired, so that the cost is a doubling in the
size of the triangle arrays. Since the "2x2" tetrahedra fill the

space between the "1x3" and "3x1" arrays, they can be regenerated using the same algorithms as used to generate the initial grid. The "2x2" tetrahedra may also be stored compactly. The four vertex numbers and the type of the tetrahedron are all that is required.

If the surfaces approach each other, allowances must be made for reconnections through a surface. If all the tetrahedral information is being stored, holes in the surface need no special treatment. If the grid is being regenerated from the surface information, then the inclusion of a special triangle type, a hole, accommodates these re-connections with minimal changes.

Thus far little mention has been made of vertex addition or dele-tion. If the vertices all lie on regular surfaces, extra surfaces may be easily added or portions of surfaces deleted. Additions and dele-tions within surfaces are accomplished in exactly the same manner as in the two-dimensional case. In either case, the reconnection algo-rithm is used after the addition or deletion to ensure that the neighbors appropriate to the new configuration are chosen. Figure 15 illustrates one adaptive scheme for doubling resolution. It may be implemented either as a global restructuring technique or as a sequence of local grid adjustments.

The collision or fragmentation of surfaces requires special algo-rithms, as in two dimensions. Whenever a physical feature is being tracked by lines or planes, reconnection is not permitted across those lines or planes. For example, reconnection at a material interface would mix two otherwise distinct fluids. Diagonal dominance can still be maintained in such cases by increasing resolution at those lines or surfaces which cannot be reconnected. The basic test for where and when to increase resolution is identical to that for reconnection. The response of the gridding routines is to alter the resolution by adding a vertex at the line or plane whenever reconnection is forbid-den.

There remains one major consideration. Although vertices migrate from regular surfaces they may still be grouped by those surfaces. If the surfaces remain logically ordered and do not interpenetrate every triangle on a surface may still be associated with a vertex on a sur-face both above and below it. If the surfaces collide or mix, then the surfaces must be redefined. For Lagrangian calculations this is only rarely a problem, since the vertices are carried with the flow.

ADAPTIVE GRIDDING

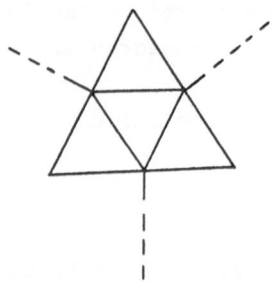

TWO DIMENSIONS

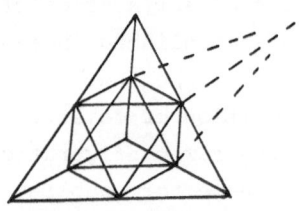

THREE DIMENSIONS

Figure 15. An adaptive gridding technique for doubling resolution. Each triangle is subdivided into four triangles in the finer mesh. Triangles which are transition elements between the coarse and fine meshes are halved. In three dimensions every tetrahedron is sub-divided into eight tetrahedra. One diagonal out of three possible diagonals must be chosen.

If the surfaces do collide, then a procedure is needed which regroups the vertices onto pseudo-surfaces which do not intersect. Since partial surfaces are accommodated as easily as surfaces which span the computational region, the grouping can be made a function of the local resolution. All vertices lying within a distance $\Delta/2$ from the pseudo-surface are include on that surface, as shown in Figure 16. Here $\Delta$ is the local resolution which is determined by the vertex density.

The remaining required algorithm is one which finds "4-0" tetra-hedra, tetrahedra which lie completely within a pseudo-surface. The solution is simple. If such a tetrahedra is dictated by the reconnec-tion algorithm, then the vertex lying furthest from the pseudoplane

is removed from the plane and assigned either to a neighboring plane
or a plane by itself. The decision is arrived at by considering if
other "4x0" tetrahedra would be formed by the switch to a neighboring
surface.

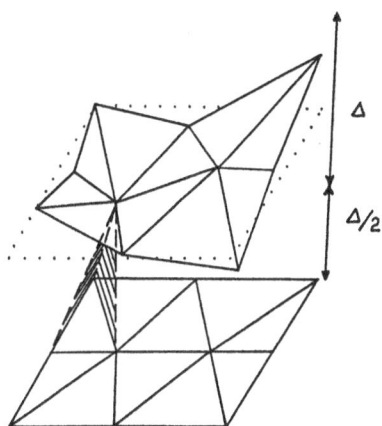

Figure 16. Grouping of vertices on pseudo-surfaces. All vertices
within a distance Δ from the pseudoplane are associated with it.

In summary, alternative, efficient data structures for three-
dimensional calculations are feasible. The only storage that must be
permanent is that which is associated with surfaces embedded within
the computational domain. Reconnection algorithms can ensure diagonal dominance for matrices of coupling coefficients and generate
mesh connections identical to those derived from Voronoi meshes.

ACKNOWLEDGMENTS

The author is pleased to acknowledge years of collaboration on
the Free-Lagrange Method with Dr. Jay Boris and the staff of the
Laboratory for Computational Physics at the Naval Research Laboratory.
The vital and challenging atmosphere created by Dr. Boris within the
Plasma Physics Division and later in the Laboratory for Computational
Physics provided both impetus and a wealth of applications for the
method. This work was supported by the Office of Naval Research, the
Air Force Office of Scientific Research, NASA-Lewis and the Naval
Research Laboratory. Current work on three-dimensional algorithms has
been supported by SAIC.

REFERENCES


1.  Fritts, M.J., D.E. Fyfe and E.S. Oran, "Numerical Simulations of
    Fuel Droplet Flows Using a Lagrangian Triangular Mesh," NRL Memo-
    randum Report 5408, September 1984.

2.  Fritts, M.J., "Numerical Approximation on Distorted Lagrangian
    Grids," In Advances in Computer Methods for Partial Differential
    Equations-III, (Proceedings of the Third IMACS International
    Symposium of Computer Methods for Partial Differential Equations,
    Bethlehem, PA, 20-22 June 1979).

3.  Tanemura, M., T. Ogawa and N. Ogita, "A New Algorithm for Three-
    Dimensional Voronoi Tessellation," J. Comp. Phys. 51, (1983), 191.

4.  Fritts, M.J., "Grid Reconnection, Numerical Accuracy and General-
    ized Voronoi Meshes," in preparation.

# THREE-DIMENSIONAL FREE LAGRANGIAN HYDRODYNAMICS

Dr. Harold E. Trease

Computational Physics Group (X-7)

Los Alamos National Laboratory

Los Alamos, New Mexico

## INTRODUCTION:

The purpose of the following discussion is to describe the development of a 3-D free Lagrangian hydrodynamics algorithm. The 3-D algorithm is an outgrowth of the 2-D free Lagrange model that is fully described in Ref. 1. Only the more pertinent issues of the free Lagrange algorithm will be presented, the details of the rest of the code development project are interesting but not appropriate in the context of the free Langrange conference. Let it suffice to say that a complete production code is being developed to support the free Lagrange algorithm to be described. A graphic description that outlines this code development project is presented in Figure 1.

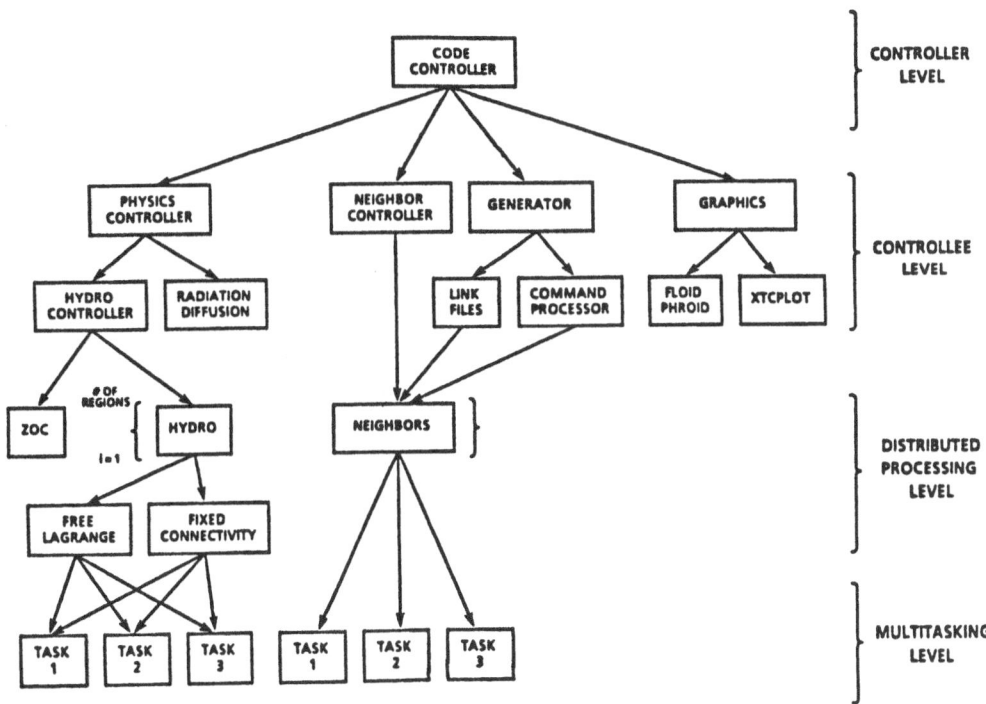

Figure 1. Block structure of the 3-D code, illustrating the four levels that make up the code.

The main objective of this project is to develop a computer model that can be used to simulate fluid flow in three dimensions. The inspiration for using free Lagrange as a basis for a hydrodynamics code was gained from the work of Crowley (Ref. 2), Fritts and Boris (Ref. 3), and Kirkpatrick (Ref. 4). The 2-D code, described in Ref. 1, was based on this previous work. This two dimensional model showed several attractive features about free Lagrange. First, it showed that free Lagrange can be used to handle fluid flow problems that exhibit strong shearing forces which classically could only be handled by Eulerian type algorithms.

Second, the accuracy of a free Lagrange algorithm was shown to be sufficient, with the irregular mesh, to produce credible solutions. Third, due to the arbitrary connectivity of the free Lagrange logic, meshes can be variably zoned. This allows the user to put the resolution where it is needed.

All three of these features of free Lagrange have been exploited to extend the 2-D model to include the third dimension. In doing so many aspects of large scale code development have been investigated. Several modern software tools have been used to make the manipulation of the arbitrary connectivity matrix, that is associated with free Lagrange, easier. Just for reference these tools include; a dynamic (heap) memory manager, a storage block manager and also a relational data base manager.

## DESCRIPTION OF THE FREE LAGRANGE ALGORITHM:

The main features of the free Lagrange algorithm that identifies it from a standard Lagrange algorithm is the connectivity matrix that both defines the nearest neighbors for each point and the shape of the computational cells over which the fluid equations are integrated. A construction technique, that creates a VORONOI mesh, is used to identify nearest neighbors and define the mesh cells. The Voronoi mesh that is constructed has several properties that make it an excellent choice for maintaining the connectivity matrix for the 3-D code, these are:

A) The set of resulting polyhedra map the space defined by the mass points and bounding surfaces uniquely, i.e., none of the polyhedra overlap and nearest neighbors are guaranteed to be reciprocal.

B) Each polyhedron remains convex. This is accomplished by changing the area and the number of faces, i.e., neighbor swapping.

C) The volume and surface area of each polyhedron changes continuously. These and other aspects of the Voronoi mesh will be discussed more thoroughly in later sections.

One of the more important goals of the 3-D code is to be able to couple various hydrodynamic algorithms together. This means that free Lagrangian hydro will be used in regions where the flow field is most distorted. Then, in the (more) well behaved regions we will use an adaptive rezoning technique, with a mesh composed of mass points that have fixed connectivity. These two algorithms will then be coupled through a third algorithm called a ZOC. The free Lagrange and the ZOC algorithms will now be described in detail. The detailed hydrodynamic equations, that are solved by these algorithms, along with their finite difference representations are discussed in detail in Appendix A.

The basic features that describe the free Lagrange algorithm are listed below:

A) All mesh quantities are cell centered.

B) The computational domain is described by an arbitrary distribution of mass points.

C) The code automatically constructs its connectivity matrix.

D) Mass points can be merged and/or added to the mesh.

The code determines its connectivity matrix by constructing a unique polyhedron about each mass point. The resulting polyhedral mesh is known as a Voronoi mesh. The faces of the polyhedron determine the set of "nearest" neighbors with which a mass point interacts. The faces of the polyhedron are represented by intersecting perpendicular, bisecting planes between a given mass point and each of its neighbors. The details of this construction process are described more fully in Appendix B. Figure 2 shows several examples of Voronoi cells. The set of polyhedra that describe the mesh completely and uniquely span the space over which the mass points are distributed. Figure 3 shows a 2-dimensional projection of a 3-dimensional mesh, where the arbitrary polyhedra reduce to arbitrary polygons.

Due to the fact that all physical quantities are carried at cell centered mass points, each point can change the set of "nearest" neighbors that it associates with by changing the shape of the polyhedron surrounding it while still retaining its Lagrangian definition. The neighbor changing process is smooth and continuous because of the integral nature of the algorithm. Two points become neighbors when a face with "epsilon" surface area appears between them. These points will drop each other as "nearest" neighbors when (and if) this face area shrinks below "epsilon".

There are several advantages and disadvantages associated with free Lagrange hydro. The advantages are obvious to anyone doing hydrodynamic calculations. Due to the arbitrary connectivity of the mesh and the ability to change this connectivity, highly distorted flows can be modeled by using a Lagrangian algorithm. Also, since the mesh maintains itself, no manual rezoning is needed (this is extremely important in a 3-dimensional code). Complex geometries that require variable zoning can be setup relatively easily since the code figures out the connectivity matrix from an arbitrary distribution of mass points. The main disadvantage of this method is the overhead associated with maintaining and processing the connectivity lists, but since the neighbor lists are unique, they are very amenable to a calculation using a multitasking algorithm (i.e., the neighbor searches can be done in any order, but the resulting global connectivity matrix is the same). Also, future machines that support hardware gather-scatter operations

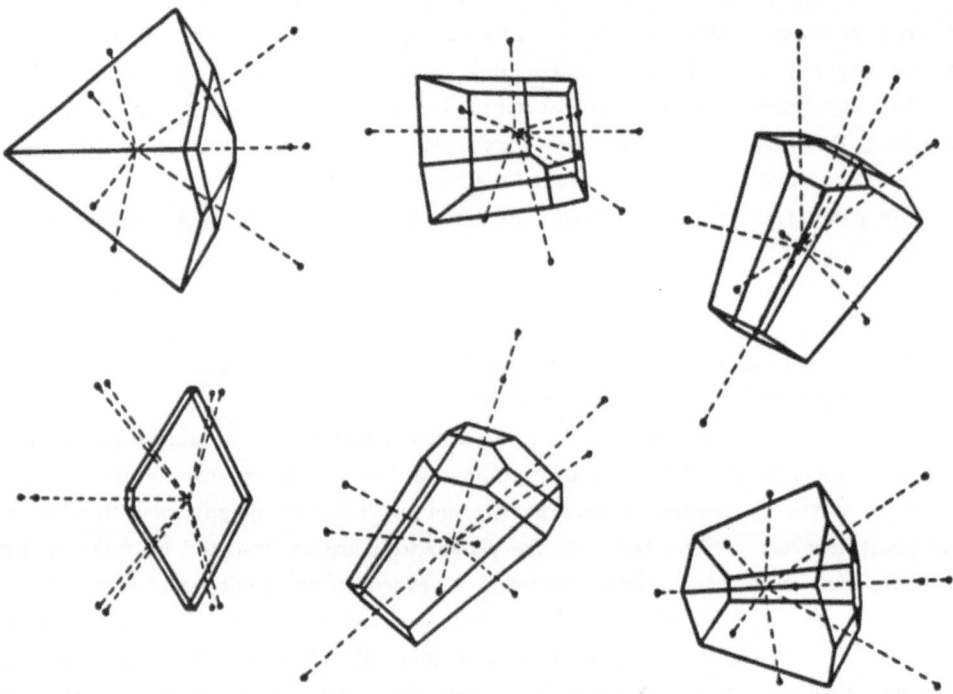

Figure 2.  Examples of several Voronoi cells.  The Voronoi cells are the polyhedral
    shaped objects.  The straight lines that end at a point represent the
    connections between the central mass point and its "nearest" neighbors.

will improve the efficiency of this algorithm.

One of the weakest aspects of the free Lagrange method just described is the treatment of continuous interfaces.  This results from the fact that the edges of the computational volumes are arbitrarily defined to be midway between two "nearest" neighbors.  This definition, while consistent with the Lagrangian equations, leads to a poor definition of a continuous interface.  The realization of this fact suggested that an  interface tracking  algorithm was needed to follow the motion of interfaces.  We will now describe the algorithm that is being used and how we intend to develop it into a full blown interface tracking algorithm, along with some of the  positive side effects, in  relation to  distributed  processing  and slip-line treatment.

A little reflection on two key properties of interfaces will help the reader's understanding of the algorithm to be described.  First, a continuous interface separates what could be considered immiscible fluids, i.e., fluid "A" remains distinct from fluid "B" even though interpenetration may occur.  Second, an interface can be described in a space that is one dimension less than the rest of the problem.  In one dimension an interface is a point, in two dimensions it is a line, and  in  three  dimensions it is a surface.  Generalizing  this  idea  we  can

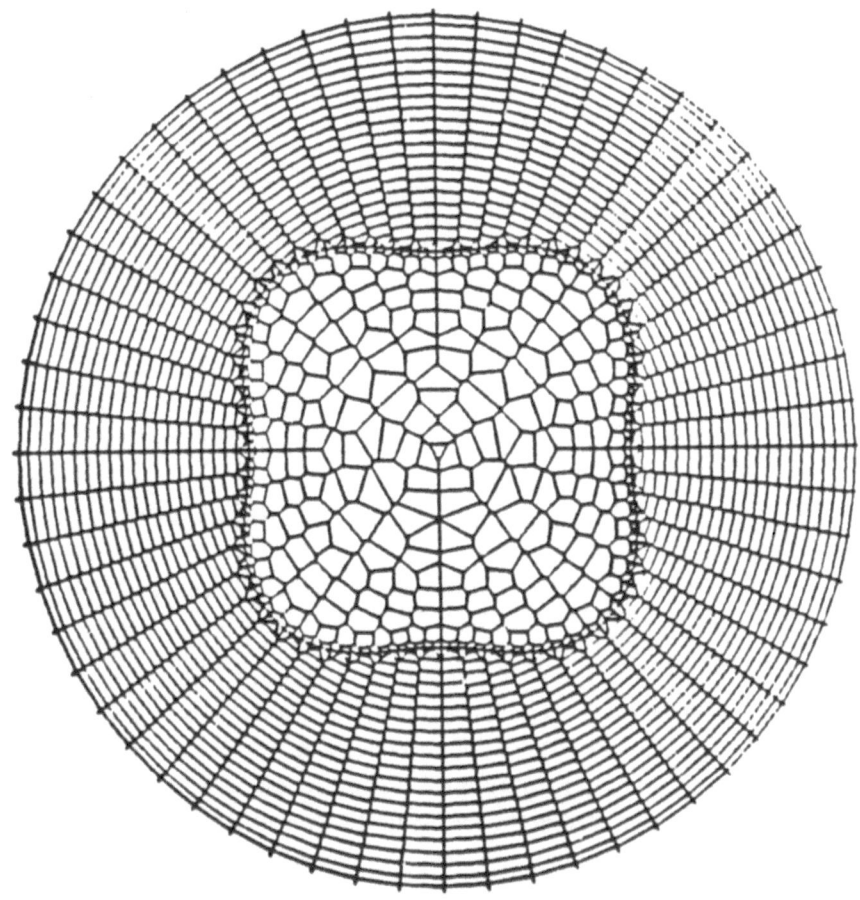

Figure 3.   2-dimensional projection of a 3-dimensional Voronoi mesh.

represent an interface as a (N-1) construct in a N-dimensional space.  Taking these
two   concepts into   consideration   we came up with an interface tracking   construct
called   a   ZOC.   An   example   of   a   ZOC is   shown   in   Figure 4,   where   several
observations   can be made.   First, we can see that the interface separating the two
fluids is distinct.   Also, we notice that zoning away from the interface in the two
regions is discontinuous with respect to the other region and the interface.

Most of the technical details of maintaining a ZOC will not be discussed, but
some   of   the   more   general   aspects   of   this   interface   tracking   concept   may   be
interesting.   These are listed below:

A) A   ZOC   is   essentially   a   special   free   Lagrange   region   that   uses   its
   connectivity matrix to connect to the surrounding regions.

B) Points can be added or   subtracted from a ZOC   to   maintain the interface.

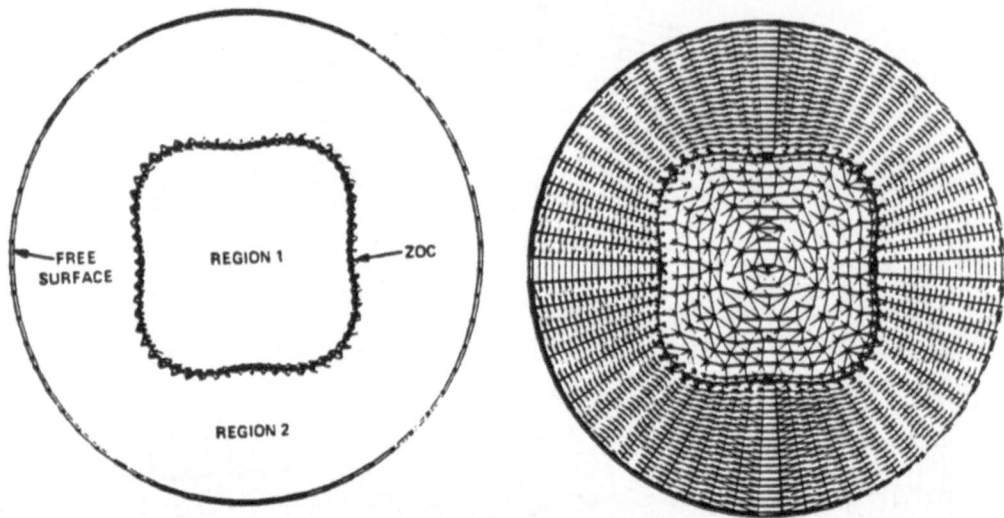

Figure 4. Example of ZOC. The left figure shows the ZOC in relation to the free surface. The right figure shows the corresponding grid.

This process is made especially easy since a connectivity matrix is used to connect points and thus the mesh reconnections account for the fact that a point has been added or subtracted. The process of adding new points is trivial because the Voronoi mesh indicates when a new point should be added and where the new point should be located. The rest of the details of adding a point involves the redistribution of mass, momentum, and energy in a local region of space.

C) A ZOC will work in two dimensions, to maintain a line interface, just as well as in three dimensions.

D) The treatment of slip-lines should be automatic with a ZOC since there is no restriction on the tangential velocity of the fluid on either side of the interface.

E) This interface will be used to connect regions that use different hydro algorithms. This means a free Lagrange algorithm can be used on one side of the interface and a fixed mesh algorithm on the other side.

F) A ZOC makes a natural communication buffer for connecting two separate algorithms that are running as distributed processes. This is where the dimensionality of the ZOC becomes important because the data transfer between processes must be kept to a minimum. The separate region processes are N-dimensional data structures and a ZOC is a (N-1) data structure, which means the amount of information being communicated between the regions is small compared to the regions themselves.

## APPENDIX A
## THE HYDRODYNAMIC EQUATIONS AND THEIR
## FINITE DIFFERENCE REPRESENTATION

The hydrodynamic equations that are solved on this free Lagrange system of mass points are given below. These equations represent the conservation of mass, momentum, and specific internal energy, respectively;

Continuity Equation;

$$\frac{1}{\rho}\frac{D\rho}{Dt} = -\vec{\nabla}\cdot\vec{u} \quad , \tag{Eq. 1}$$

Conservation of Momentum;

$$\rho\frac{D\vec{u}}{Dt} = -\vec{\nabla}p - [\vec{\nabla}\cdot\vec{\vec{T}}] \quad , \tag{Eq. 2}$$

Conservation of Internal Energy;

$$\rho\frac{D\varepsilon}{Dt} = -p(\vec{\nabla}\cdot\vec{u}) - (\vec{\vec{T}}:\vec{\nabla}\vec{u}) \quad , \tag{Eq. 3}$$

where,

$\rho$ = fluid density ,

p = fluid pressure ,

$\varepsilon$ = internal energy/unit mass ,

$\vec{u}$ = fluid velocity vector ,

$\vec{\vec{T}}$ = total stress tensor , and

$= \vec{\vec{\tau}} + \vec{\vec{Q}}$ ,

where,

$\vec{\vec{\tau}}$ = stress tensor and

$\vec{\vec{Q}}$ = artificial viscosity tensor .

$$\vec{\vec{Q}} = \begin{cases} \ell^2\rho \; div(\vec{u})\{(\vec{\nabla}\vec{u}) - \frac{1}{3}div(\vec{u})\vec{\vec{e}}\} & , \text{ if } div \; (\vec{u}) < 0 \\ \\ 0 & , \text{ if } div \; (\vec{u}) \geq 0 \end{cases} \tag{Eq. 4}$$

where,

$div(\vec{u})$ = divergence of velocity vector $\vec{u}$ ,

$\vec{\nabla}\vec{u}$ = dyadic product of the differential operator $\vec{\nabla}$ and the velocity vector $\vec{u}$ ,

$\overset{\rightrightarrows}{e}$ = unit tensor , and

$\ell$ = constant * local grid spacing .

The algorithm used to solve these equations can be outlined as follows. First, Equations 1 through 3 are integrated over an arbitrary volume (in reference to the code we integrate over a computational cell). The volume integrals are transformed to surface integrals by using the divergence theorem. The mean-value-theorem from calculus is used to obtain average quantities. These resulting equations are then cast into finite difference form as shown below. The following notation will be used in writing the finite difference representation of the fluid flow equations;

$i$ = mass point i, spatial position at $(XI, YI, ZI)$ ,

$j$ = $j^{th}$ nearest neighbor of mass point i, spatial position at $(XJ, YJ, ZJ)$ ,

$J$ = total number of nearest neighbors associated with mass point i ,

$n$ = present time step (time = t) ,

$n + 1$ = next time step (time = t + $\Delta$t) ,

$R_{i,j}$ = distance from mass point i to nearest neighbor j ,

$A_{i,j}$ = area of polygon face separating mass point i from nearest neighbor j ,

$V_{i,j}$ = volume of the polyhedron associated with mass point i and nearest neighbor j ,

$M_i$ = mass of fluid associated with mass point i ,

$P_{i,j}$ = fluid pressure at the face associated with mass point i and nearest neighbor j ,

$\rho_i$ = fluid density in cell i ,

$\vec{U}_{i,j}$ = fluid velocity at the face associated with mass point i and nearest neighbor j , and

$\hat{n}_{i,j}$ = normal vector to the face associated with mass point i and nearest neighbor j .

$$V_i^{n+1} \left(\frac{D\rho}{Dt}\right)_i = -\rho_i^n \sum_{j=1}^{J} \hat{n}_{i,j} \cdot \vec{U}_{i,j}^{n+1} A_{i,j} \qquad \text{(Eq. 5)}$$

$$M_i \left(\frac{D\vec{U}}{Dt}\right)_i = -\sum_{j=1}^{J} P_{i,j} \hat{n}_{i,j} A_{i,j} - \sum_{j=1}^{J} \hat{n}_{i,j} \cdot \overset{\rightrightarrows}{T}_{i,j} A_{i,j} \qquad \text{(Eq. 6)}$$

$$M_i \left(\frac{D\varepsilon}{Dt}\right)_i = -P_i \sum_{j=1}^{J} \hat{n}_{i,j} \cdot \vec{U}_{i,j}^{n+1} A_{i,j} - \sum_{j=1}^{J} \hat{n}_{i,j} \cdot [\overset{\rightrightarrows}{T} \cdot \vec{U}]_{i,j} A_{i,j} +$$

$$+ \vec{U}_{i,j} \cdot \sum_{j=1}^{J} \hat{n}_{i,j} \cdot \overset{\rightrightarrows}{T}_{i,j} A_{i,j} \qquad \text{(Eq. 7)}$$

APPENDIX B

MESH CONNECTIVITY

(NEAREST NEIGHBOR CALCULATIONS)

The purpose of this Appendix is to describe the manner in which the connectivity matrix for the free Lagrange algorithm is calculated. The connectivity matrix contains the "nearest" neighbors for all the mass points. These connections are used for calculating surface areas and volumes of the Voronoi cells that make up the computation mesh. In addition to describing the geometry of the cells the connectivity matrix indicates which cells will interact hydrodynamically.

Each Voronoi cell is made of an arbitrary number of intersecting planes. These planes construct a convex polyhedron with an arbitrary number of "faces" about each mass point. Each face forms a polygon with an arbitrary number of "edges". The trick is to come up with an algorithm that can calculate the connectivity matrix from an arbitrary distribution of points. As we proceed through this discussion the two following definitions should be kept in mind:

A) "face" neighbor: Any two points that are "nearest" neighbors are separated, in 3 dimensions, by a polygon shaped "face". A "face" neighbor therefore refers to the "nearest" neighbor that is across a given "face" from a given central point "I". There is a one-to-one correspondence between the number of "face" neighbors that are associated with a point and the number of "faces" that make up the polyhedral cell surrounding that point.

B) "edge" neighbor: Each of the "faces" of a Voronoi cell is a polygon. Each vertex of a given polygon "face" is found to be the center of a sphere that passes through four points. The four points are: the central point "I", its "face" neighbor "j", and two other points that are called "edge" neighbors (they are referred to as "k-1" and "k"). Also, since each polyhedron is a closed figure each "edge" neighbor must also be a "face" neighbor. It should be noted that there is a one-to-one correspondence between the number of "edges" on a given polygonal "face" and the number of "edge" neighbor associated with that "face".

The main idea in discovering the "nearest" neighbors of a point is to identify the list of "edge" neighbors for each "face" of the polyhedron surrounding that point. As each set of "edge" neighbors is discovered they are put on a stack of "face" neighbors, then the next "face" neighbor to be looked at is pulled from this stack. When the stack of "face" neighbors is empty the Voronoi polyhedron is complete and the connectivity for this given point has been found.

The process of constructing the connectivity matrix for a given point is described more completely in the following steps:

A) Assemble a list of "possible" neighbors. This process depends on whether or not the point has been calculated before.

a)  If the point hasn't been calculated before.

    1)  Use any "logical" neighbor information to select as many potential neighbors as possible.

    2)  Use a proximity rule to select a set of closest points based on filters such as distance between and material type.

b)  If the point has been calculated before.

    1)  Recall the old connectivity matrix for a given point.

    2)  Form the list of possible "nearest" neighbors from the old "nearest" neighbors plus the "nearest" neighbors of the old "nearest" neighbors.

B)  From this list of "possible" neighbors we identify the $1^{st}$ "face" neighbor by the procedure outlined below.

a)  First all points are translated so the coordinate axes are centered at the central mass point, point "I", by using the following equation,

$$\sum_{\ell=1}^{nkin} \{\vec{X}_\ell = \vec{X}_\ell - \vec{X}_I\}$$

(Eq. 8)

where,

$\vec{X}_\ell$ = coordinates of a possible neighbor "$\ell$" ,

$\vec{X}_I$ = coordinates of point "I" , and

nkin = number of points on the list of "possible" neighbors for point "I" .

b)  Select the first "face" neighbor by using the following equation,

$$N_1 = \text{Index of } [\min_{\ell=1}^{nkin} \ \{|\vec{X}_\ell|\}]$$

(Eq. 9)

where,

$|\vec{X}_\ell|$ = the distance to point $\ell$ and

$N_1$ = index of the $1^{st}$ "nearest" neighbor (this is the index into the global mesh arrays) .

c)  Select the first "edge" neighbor for the first "face" by using the following equation.

$$E_{1,1} = \text{Index of } [\min_{\ell=1}^{nkin} \ \{W_1|(\vec{X}_V)_1 \times (\vec{X}_V)_\ell| +$$

$$+ W_2 \cos ((\vec{X}_N)_1, \vec{X}_\ell) +$$

$$+ W_3 |\vec{X}_\ell|\}]$$

(Eq. 10)

where,

$$E_{1,1} \neq N_1 \quad,$$

$E_{k,j}$ = the $k^{th}$ "edge" neighbor of the $j^{th}$ "face" neighbor ,

$(\vec{X}_N)_j$ = the coordinates of the "face" neighbor "$N_j$" ,

$W_i$ = weighting parameters
$$W_1 > W_2 > W_3 \quad.$$

$(\vec{X}_v)_\ell$ = the coordinates of the $\ell^{th}$ Voronoi point. This point is found as the center of a circle which passes through three points. These points are: the central point "I", the "face" neighbor "$N_j$" and the next possible neighbor "$\ell$" ($N_\ell \neq N_j$). The equations, in matrix form, that determine the (X,Y,Z) - coordinates of the $\ell^{th}$ Voronoi point are:

$$\begin{bmatrix} (\vec{X}_N)_j \\ \\ \vec{X}_\ell \\ \\ (\vec{X}_N)_j \times \vec{X}_\ell \end{bmatrix} \bullet \begin{pmatrix} \\ (\vec{X}_v)_\ell \\ \\ \end{pmatrix} = -\tfrac{1}{2} \begin{pmatrix} |(\vec{X}_N)_j|^2 \\ \\ |\vec{X}_\ell|^2 \\ \\ 0 \end{pmatrix}$$

NOTE: This step describes the "nearest" neighbor algorithm in 2 dimensions (Z=0). Here we just use it to "start" the 3-dimensional algorithm described below.

d) The process of selecting the rest of the "nearest" neighbors for point "I" serves a dual purpose. First, we finish calculating the rest of the "face" neighbors and we also discover the list of "edge" neighbors that make up the polygonal "face" that separates Points "I" and "j". An important point to notice is that the list of "nearest" neighbors for point "I" are contained within the sets of "edge" neighbors, i.e., the "nearest" neighbors are a subset of the "edge" neighbors. Therefore, by sifting the "edge" neighbors we obtain a list of unique points that represent the "face" neighbors for point "I". The "edge" and "face" neighbor lists will bootstrap each other to completely describe the polyhedral cell surrounding point "I". The 3-D "nearest" neighbor selection algorithm is described below.

1) We already know the first "face" neighbor, $N_1$, and the first "edge" neighbor, $E_{1,1}$, for face one for point "I". These were found in Steps (B.b) and (B.c).

2) Now we calculate the "edge" neighbors for "face" j.

$$E_{k,j} = \text{Index of } [\min_{\ell=1}^{nkin} \{W_1 \, (\vec{X}_N)_j \, \bullet$$

$$\bullet \, [[(\vec{X}_V)_{k-1} - \tfrac{1}{2}(\vec{X}_N)_j] \times$$

$$\times \, [(\vec{X}_V)_\ell - \tfrac{1}{2}(\vec{X}_N)_j]] +$$

$$+ \, W_2 \, \cos \, ((\vec{X}_N)_j, \, \vec{X}_\ell) +$$

$$+ \, W_3 \, \cos \, (\vec{X}_{k-1}, \, \vec{X}_\ell) +$$

$$+ \, W_4 \, |\vec{X}_\ell|\}]$$

(Eq. 11)

where,

$$E_{k,j} \neq N_j \quad,$$

$$E_{k,j} \neq N_{1,j} \quad,$$

   j = index of the current face ,

   k = 2 to K (until $E_{k+1,j} = E_{1,j}$, where K=number of "edge" neighbors) ,

   $W_i$ = weighting parameters

$$W_1 > W_2 > W_3 > W_4 \quad,$$

$(\vec{X}_N)_j$ = coordinates of nearest j ,

$(\vec{X}_V)_\ell$ = coordinates of the $\ell^{th}$ Voronoi point and ,

$$\begin{bmatrix} (\vec{X}_N)_j \\ \\ \vec{X}_{k-1} \\ \\ \vec{X}_\ell \end{bmatrix} \bullet \left( (\vec{X}_V)_\ell \right) = -\tfrac{1}{2} \left( \begin{array}{c} |(\vec{X}_N)_j|^2 \\ \\ |\vec{X}_{k-1}|^2 \\ \\ |\vec{X}_\ell|^2 \end{array} \right).$$

3) From the list of "edge" neighbors, $(E_k, \, k=1, \, K)$, we add the unique indices to the list of "face" neighbors, $N_j, \, j=1, \, J)$.

4) Increment j and continue with Step (d.2). This continues until all the "face" neighbors have been calculated, i.e., J > j. J is the number of "face" neighbors that are associated with point "I".

REFERENCES:

1.  Trease, H. E. (1981), A Two-Dimensional Free Lagrangian Hydrodynamics
    Model, Ph.D. Thesis, University of Illinois, Urbana-Champaign.

2.  Crowley, W. P. (1971), "FLAG: A Free Lagrange Method for Numerically Sim-
    ulating Hydrodynamic Flow in Two Dimensions," Proceedings of the Second In-
    ternational Conference on Numerical Methods in Fluid Dynamics, Lecture Notes
    in Physics, Vol. 8, pp. 37-43, Springer-Verlag, New York.

3.  Fritts, M. J. and Boris, J. P. (1979), "The Lagrangian Solution of Transient
    Problems in Hydrodynamics using a Triangular Mesh," Journal of Computational
    Physics, Vol. 3, pp. 319-343, Academic Press, New York.

4.  Kirkpatrick, R. C. (1976), "FREE FLOW HYDRO," Internal Report (TD-2), Los
    Alamos National Laboratory, Los Alamos, New Mexico.

A Vectorized "Near Neighbors" Algorithm of Order N
Using A Monotonic Logical Grid

J.P. Boris and S.G. Lambrakos
Laboratory for Computational Physics
U.S. Naval Research Laboratory, Code 4040
Washington, DC  20375

ABSTRACT

In free Lagrangian representations of fluid dynamics, the fluid is assigned
to discretized parcels which are defined throughout the flow by a large number of
nodes moving with the local fluid velocity. These Lagrangian nodes define a
finite difference or finite element grid for calculating fluid dynamic averages
and driving gradients in the vicinity of the fluid parcels. Because the nodes
move with the fluid, the convective terms in continuity equations governing the
flow are transformed away. Thus unwanted numerical diffusion is reduced greatly
or eliminated. The price for this improved numerical accuracy is having to
compute derivatives in a complicated shifting geometry and having to keep track of
which of the many Lagrangian nodes are nearby.

When N nodes move essentially randomly in space, $N*(N-1)/2$ interactions might
be important in determining how a given node moves. The exact positions and
velocities of the neighboring nodes must be known. Knowing statistical averages
and the general properties of the Lagrangian fluid parcels currently nearby does
not provide enough data to compute local interactions accurately. At any instant
a given Lagrangian node may interact strongly with only a few of the others.
Unfortunately, keeping track of the other nodes with which it interacts or
recomputing them each timestep is computationally very expensive. The goal is
efficient, simple algorithms which select only the important near neighbor without
having to check all N-squared interactions. Effort on the near neighbors problem
has persisted in computational physics for several decades. To date the best
algorithms scale nominally as N, or N log N rather than $N^2$, but they are scalar
algorithms and address memory randomly using linked lists.

This report introduces a simple three-dimensional nearest-neighbors algorithm
whose cost scales as $N(1+ \varepsilon \log N)$ with $\varepsilon$ a small coefficient, not as the square
of N, and which vectorizes using data from contiguous memory locations. A compact
data structure to store the object data, called a Monotonic Logical Grid (MLG), is
defined dynamically so that nodes which are adjacent in real space automatically
have close address indices in the MLG data arrays as well. As two nodes move past
each other in space, their data are exchanged or "swapped" in the MLG data arrays

to maintain a strictly monotone mapping between the geometric locations and the corresponding storage location indices. The MLG is logically regular so many algorithms which normally apply only to Eulerian calculations, such as multigrid elliptic solvers, should be applicable.

To construct an MLG the data values for each node are stored at location $(i,j,k)$ in the grid such that the X positions of all the nodes increase monotonically with index $i$, the Y positions increase monotonically with index $j$, and the Z positions increase monotonically with index $k$. Such a well-structured mapping from the real positions to regular, compact data arrays can always be found. An O (N-log N) constructive algorithm for one such MLG has been found.

Further, when motion of the nodes results in a local violation of some of the monotonicity conditions on which the original MLG was based, another MLG can be found nearby. This means that local changes in the Lagrangian node positions and hence spatial ordering do not trigger global changes in where the fluid parcel data have to be stored in the MLG data arrays. The data relocations to restore the MLG as nodes pass each other in space can be vectorized without inefficient gather-scatter oprations or variable-length (scalar) linked lists. The MLG data structure and algorithms allow contiguous-data vector operations which are long enough to be efficient for fluid dynamic calculations and for the nodal data "swapping" which restores the MLG.

An efficient vector solution of the nearest-neighbors problem also advances many important applications in addition to Lagrangian fluid dynamics. For an important class of molecular dynamics problems, the candidates to enter into chemical reactions. Many important physics problems in gases, liquids, solids, and transitions among these phases require detailed manybody calculations where the close encounters are most important.

For graphics based on vertex-and-edge representations of complex 3D shapes, local relationships and orientations of nearby vertices determine which surfaces are visible. It is clearly advantageous to be able to construct a 2D image of a complex 3D scene, for example, using the parallelism made possible by Very Large-Scale Integration (VLSI). Terrain management simulation models and also multidimensional radiation transport models are currently limited in their ability to compute geometric obscuration. For controlling airline traffic over crowded airports, collisions with nearby planes are the most immediate danger - and demand shorter timescales for detection and corrective response.

These applications all require rapidly updating many distinct local configurations as the nodes move. A commercial distributed processing system, a minicomputer host with modest array processors, would be fast enough using an MLG to integrate 10,000 interacting nodes and restructure the data base a thousand times in about 15 minutes, useful for realtime applications as well as fluid dynamic applications. The MLG also permits partitioning to take advantage of asynchronous multi-processor parallelism in VLSI-based distributed processing architectures of the future.

## I. INTRODUCTION

This report introduces a simple three-dimensional nearest-neighbors algorithm whose cost scales as N, not as the square of N, and which vectorizes easily using data from contiguous memory locations. A compact data structure to store the node data, called a Monotonic Logical Grid (MLG), is defined dynamically so that nodes which are adjacent in real space automatically have close address indices in the MLG data arrays as well. As two nodes move past each other in space, their data are exchanged or "swapped" in the MLG data arrays to keep a strictly monotone mapping between the geometric locations and the corresponding storage location indices.

To construct an MLG the data values for each node are stored at location (i,j,k) in the grid such that the X positions of all the nodes increase monotonically with index i, the Y positions increase monotonically with index j, and the Z positions increase monotonically with index k. Section III describes the algorithm in some detail.

It is not obvious but it is true that such an organized logical ordering of even random locations can always be found. In Section II an order (N log N) constructive algorithm for one such MLG is provided-proving existence. Generally more than one MLG meeting all the monotonicity conditions seems to be possible so the technical problem of selecting the optimum MLG for a particular application has to be addressed. In one case, minimizing average distances to neighbors in the MLG may give the best grid. In other problems it may be best to maximize the shortest distance to any point which is not a near neighbor in the logical grid.

Further, when node motions result in a local violation of the monotonicity conditions on which the original MLG was based, another MLG can be found nearby. This means that local changes in the node positions and hence spatial ordering do not trigger global changes in where these node data have to be stored in the MLG. The data relocations to maintain the MLG as nodes pass each other in space can be vectorized without inefficient gather/scatter operations or variable-length (scalar) linked lists. The MLG data structure and algorithms allow contiguous-data vector operations which are long enough to be efficient for physical force sums, for $F = Ma$ orbit integrations, and for the object data "swapping" used to restructure the MLG whenever the monotonicity conditions are violated.

The cost to execute a simple test version of the model is one hour on a DEC VAX 11/780 for one thousand particles for 1000 timesteps. We used a power series force law for the 124 nearest neighbors, assuming that the average particle separation distance is smaller than the cutoff radius $R_c$ of the force law. A commercial distributed processing system, a minicomputer host with modest array processors, would be fast enough using an MLG to integrate 5000 interacting nodes and restructure the data base thousands of times in about 15 minutes, useful for realtime applications where current supercomputers using other algorithms will be inadequate. The MLG also permits partitioning to take advantage of asynchronous multi-processor parallelism in VLSI/VHSIC-based distributed processing systems (e.g. Fox and Otto, 1984).

Section II contains a description of the MLG itself, an O(N log N) sort algorithm to find a starting MLG from arbitrary initial data, and simple algorithms which restructure the grid dynamically as the objects move. Section II also presents a few simple tests of the method. Section III considers several extensions. Section IV contains a summary and conclusions.

## II. THE MONOTONIC LOGICAL GRID ALGORITHM

A Monotonic Logical Grid (MLG) is a simple, compact way of indexing and storing the data describing a number of nodes moving in space. For N points in three dimensions, the arrays of locations, $X(i,j,k)$, $Y(i,j,k)$, and $Z(i,j,k)$, constitute an MLG if and only if

$$X(i,j,k) \leq X(i+1,j,k) \quad \text{for} \quad 1 \leq i \leq NX-1,$$
$$Y(i,j,k) \leq Y(i,j+1,k) \quad \text{for} \quad 1 \leq j \leq NY-1, \text{ and} \quad (2.1)$$
$$Z(i,j,k) \leq Z(i,j,k+1) \quad \text{for} \quad 1 \leq k \leq NZ-1.$$

Given N = NX*NY*NZ random locations, the spatial lattice defined by an MLG is irregular. However, the cells defined by logically neighboring locations are distorted cubes and thus form a useful consistent partitioning of the spatial volume. When the N locations satisfy Eqs. (2.1) and any additional constraints or relations specifying other than infinite-space boundary conditions, they are in "MLG order". This ordering is useful because the direction for going from one node to another in space and in the MLG are the same. Further, other nodes which are between two locations in space will also be between them in the MLG. Thus neighbors in real space have neighboring address indices in the MLG as well.

Figure 2.1 shows three different spatial configurations of 16 nodes in the lower three panels. The nodes are ordered into four rows and four columns in each of these configurations corresponding to regular storage of the node data in the two-dimensional MLG shown above. The cells of the MLG move with the nodes and thus always have exactly one node in them. When all the nodes move to the upper left of the region, as in the lower right panel, the MLG is just as regular as when the nodes are uniformly spaces. This mapping of irregular locations onto a very regular data structure is what permits optimal use of vector and multiprocessor hardware.

Figure 2.2 illustrates several different MLG mappings of the same 16 node locations. The upper left panel shows the 16 locations in a regular spatial lattice. The obvious numbering of the locations into four rows of four nodes each is an MLG because the X and Y components of all the node locations (dots) increase monotonically with the X and Y indices, i and j. In the remaining three lower panels the locations have been displaced from the regular spacing. Each of these panels contains the same data, but the MLGs for storing these data, as indicated by the logical mesh lines, all differ. The lower left panel is a recognizeable distortion of the regular grid above. The indexing of the nodes would be identical in both cases although the actual data stored would differ somewhat since the nodes have moved away from their regular locations. The lower center and lower right panels show different logical indexing for the same physical data giving two other Monotonic Logical Grids (MLGs). In the center the connections to points in the second row from above and from below have all been displaced to the left. On the right, the connections to the second column from points located logically in columns one and three have been displaced downward.

These figures show that there can be a number of MLGs with the same Lagrangian data, all satisfying the required monotonicity conditions from Eqs. (2.1). These spatial monotonicity conditions constitute $3N - NX{\times}NY - NY{\times}NZ - NZ{\times}NX$ numerical comparisons which can be performed to determine if a particular organization of the locations is in MLG order. For each dimension of the desired data structure such a monotonicity condition can be defined.

In space the coordinates can be rotated or redefined and this corresponds to a different family of MLGs. The monotonicity conditions may not change in the new coordinate system but the node coordinates will. Even if the coordinate system is held fixed, the node motions will quickly invalidate at least some of the relations (2.1) requiring a reorganization of the node data in the arrays to store a completely monotone mapping. Using the monotonicity conditions, a given data structure can be checked efficiently to see if the locations are in MLG order. However, additional algorithms are needed when MLG order is violated.

If the nodes are not in MLG order, the following algorithm using a vector sort routine $O(N \log N)$ can be used to rearrange them. First sort all N locations

into the order of increasing Z. The first NX x NY of them, should be indexed k =
1, and sorted into the order of increasing Y. The first NX of these, should be
indexed j = 1 and then sorted into the order of increasing X. These locations are
indexed from i = 1 to i = NX. The next NX locations, indexed j = 2 but still k =
1, are again ordered and indexed from i = 1 to i = NX. This procedure is
continued until the first NX x NY plane of locations has been arranged. Since the
locations were initially ordered in Z, the subsequent reorderings within the k = 1
plane cannot disturb the monotonicity conditions relating the first plane to any
subsequent reordering of the second and subsequent planes. Similarly, all the
locations will satisfy the monotonicity conditions in Y and X as well.

Once the first plane is ordered, the next NX x NY locations are indexed k =
2, and the MLG ordering within this plane is constructed just as for the first
plane. All NZ planes are organized this way. The process requires of order

$$NZ \times NY \times NX \times (\log NZ + \log NY + \log NX)$$
$$+ NZ \times (NY \times NX \times (\log NY + \log NX) + NY \times NX \times \log NX) \quad (2.2)$$
$$= NZ \times NY \times NX \times (\log NZ + 2 \log NY + 3 \log NX)$$

operations to construct the MLG. This sort algorithm could be repeated every
timestep as necessary to restructure the MLG when node motions in one of the three
coordinate directions cause some of the conditions (2.1) to be violated.

The existence of this constructive algorithm proves that at least one MLG for
even random locations always exists and that it is not hard to find. As a
consequence, data manipulation and summation algorithms in the MLG can always
assume the rigorous spatial monotonicity of the MLG. When several node locations
are identical, any ordering the sort procedure comes up with is correct as the
conditions (2.1) are satisfied. Locally degenerate grids are possible when
several locations overlap.

Although this algorithm is fast, it has two limitations, it always shuffles
all the data, and it may move data a long distance in index space to correct even
small changes in position. To counter these objections, an order (N log N)
algorithm is described which executes local but vectorizeable exchange or
"swapping" operations on the MLG data to restore monotonicity everywhere. The
extra factor log N is present because small monotonicity upsets from the previous
set of locations sometimes require information from the other side of the grid for
their correction. Empirically the numerical coefficient of this N log N term is
small.

If two nodes move less than a typical separation distance per timestep, a
condition generally required for accurate integration of the equations of motion,
a few iterations are usually enough to restore MLG order. A "swap" is executed by
testing the conditions in Eqs. (2.1), and then, when the corresponding
monotonicity condition is violated, exchanging the locations in the logical grid
of all data pertaining to the two nodes involved. Each direction is checked

separately. A red-black algorithm (Adams and Jordan, 1984) would allow at
least half the tests in a given direction to be performed simultaneously and thus
vectorized while converging as fast as a scalar iteration. This general procedure
would also be very useful in studying physical systems with Cellular Automata
because swapping is a natural bit-by-bit operation.

No more than five arithmetic operations are required to test for monotonicity
and to prepare to swap any amount of data in a real/floating point representation.
A floating point 0.5 is logically "or"ed with the sign bit of the coordinate
difference to obtain a number S which is 0.5 if the two coordinates are in MLG
order and -0.5 if they are not. This takes three operations. Two more operations
give the weights w and $(1 - w)$ where

$$w = S + 0.5, \qquad (1 - w) = S - 0.5. \qquad (2.3)$$

The weight w is unity when the coordinates are in order and zero when they are
not. If the test $X(i,j,k) \leq X(i+1,j,k)$ is being performed, the X components of
the node locations can be exchanged using the following formulae (six
operations):

$$T(i,j,k) = w \times X(i,j,k),$$
$$U(i,j,k) = (1 - w) \times X(i,j,k), \qquad (2.4)$$
$$X(i,j,k) = T(i,j,k) + (1 - w) \times X(i+1,j,k), \text{ and}$$
$$X(i+1,j,k) = w \times X(i+1,j,k) + U(i,j,k).$$

When the locations are in MLG order, the swapping formulae change nothing. When
two locations are out of order, these formulae interchange the node data on the
MLG so they will be in order for the next iteration. All node data at every grid
point can be treated identically with this procedure. The algorithm vectorizes
easily even though some of the pairs of nodes have to trade places in the MLG and
others do not.

These six operations must be repeated to swap each data variable stored in
the MLG. As a minimum these include the three components locations and an
identification number, ID# $(i,j,k)$, to mark which of the N particles currently is
at $i,j,k$ in the MLG. To vectorize the complete algorithm, the velocity components
$VX(i,j,k)$, $VY(i,j,k)$, $VZ(i,j,k)$, the mass $M(i,j,k)$, and another force law constant
$FC(i,j,k)$ must also be moved about dynamically. These nine variables require 54
operations to be moved between adjacent cells for each swapping iteration. Thus
$N_{os} \sim 60$ operations are required for each iteration in each direction for each
node. This is about as much work as calculating three components of the force
acting between two nodes which are near neighbors in the MLG. With $N_{si} = \sim 4$
swapping iterations being performed in each direction, the total cost of restoring
the MLG every timestep is about the same as calculating forces from 12 neighbors.
When timesteps are short, this can be reduced even further.

The operation count for the overall MLG algorithm is problem dependent.
Using typical simulation variables, the cost of the MLG in vector floating-point

operations to execute a timestep, exclusive of the relatively inexpensive orbit calculations, is

#Flops for the Monotonic Logical Grid algorithm = $F_{mlg}$

$= N \times (N_{nn} \times N_{ot}$ for neighboring object interactions

$+ 3 \times N_{si} \times N_{os})$ for swapping interations in X, Y, Z

$= 2.25 \times 10^7$ flop$\sigma \rightarrow -0.5$ seconds/step at 50 megaflops.

Here $N_{si} = \sim 4$ is the number of iterations of vector swapping performed over the entire grid to restructure the MLG after the object positions change each timestep. $N_{os} = \sim 60$ is the number of floating point operations to execute a single swap of two objects in the MLG. Here also $N_{nn} = \sim 60$ is the number of near neighbors included in the force calculations and $N_{ot} = \sim 60$ is the number of operations per interaction.

The speedup expected using this algorithm is large, approximately a factor greater than thirty for 5000 nodes. Not only is the $N^2$ dependence removed but the actual near neighbor interactions can be computed with very high efficiency, comparable to the best order-N scalar algorithms. Only about a fifth of the computation is expended on maintaining the MLG data structure. The rest is used in computing pairs of interactions at full vector efficiency.

The random motion of point particles in a cubical domain is taken as a test problem to illustrate the concepts. A topologically regular 8 × 8 × 8 3D grid is defined for storing the position and velocity components of 512 randomly located points. The domain is doubly periodic in X and Y and is bounded in Z by two reflecting end walls at Z = 0 cm and $Z = Z_{max} = 8$ dZ. A number of short calculations have been performed using this system to test and develop various aspects of the model. Figure 2.3 shows the first of eight planes of this 3D MLG, plotting the X and Y locations of the 64 nodes currently on that plane. The initial conditions for the calculation are shown in the upper right, regularly-spaced locations with random velocities uniformly distributed in each coordinate from $-10^7$ cm/sec to $+10^7$ cm/sec. The three remaining panels show plots of the 64 locations in the same MLG data plane at three times. As the nodes move in the plane and between planes, a complicated but clearly structured MLG is always maintained.

Under a number of different physical circumstances and numerous different initial conditions the model has been able to find an MLG after only a few swapping iterations. The average near-neighbor separations increase somewhat at first over their almost minimal initial values. Rather quickly, however, random swapping halts the increase of this average distance to the near neighbors. Figure 2.4 displays the frequency distribution for the number of swapping iterations required to restore the MLG after relative motion of the objects has disrupted it. Three cases were run from the same physical initial conditions and zero-sized non-colliding particles, with timesteps $\delta t = 2.5 \times 10^{-16}$ sec, 1.0 ×

$10^{-15}$ sec, and $4.0 \times 10^{-15}$ sec. The lightly shaded bars in Figure 2.4a correspond to the intermediate case with $10^{-15}$ sec as the timestep. For this case $dl_{max} = .1\Delta$ meaning that the fastest particles traverse 1/10 of the regular initial spacing of $\Delta = 10^{-7}$ cm per timestep. The data with unshaded bars, $dl_{max} = .025\,\Delta$, show the results when $\delta t$ is smaller by a factor of four and the data depicted with dark bars show results when $\delta t$ is a factor of four larger, i.e. $dl_{max} = .4\Delta$.

To interpret the figure consider $dl_{max} = 0.1\Delta$. About 40% of the timesteps (frequency 0.38) required 4 iterations of swapping to restore the MLG. Less than 10% of the timesteps required 6 or more iterations. The average number of iterations required is 4.0 for $dl_{max} = 0.1\Delta$. When $dl_{max} = 0.025\Delta$, the average number of swapping iterations is 2.85, about $2\sqrt{2}$. When $dl_{max} = 0.4\,\Delta$, the average is 5.0 swapping iterations per timestep. Thus the actual computational work decreases per unit integration time with longer timesteps because the number of swapping iterations increases much more slowly than the timestep increases.

A great deal of swapping goes on in the first few iterations out to the average number for the particular timestep chosen. For timesteps with relatively large numbers of iterations I, the likelihood of this extra work being required decreases by a factor of two or three for each extra iteration. These timesteps requiring a relatively large amount of work contribute little to the average computation load needed to restore the MLG because they occur infrequently but constitute the log N factor in swapping.

In test calculations, with nonzero particle size, forces were calculated between a given particle and the $5 \times 5 \times 5$ cubical nearest neighbors interaction template of 125 neighboring particle centered on it in the MLG. Since the interaction has to be computed only once for a pair and can be ignored for self interactions, the tests had the following number of near neighbors

$$N_{nn} = (5 \times 5 \times 5 - 1)/2 = 62 = {}^{\sim}60. \qquad (2.6)$$

When many particle are within the cutoff distance $R_c$, the interaction template should be extended, perhaps to $7 \times 7 \times 7$. An appreciable fraction of the forces calculated will be beyond the cutoff distance but this extra work is compensated by the fact that all the work can now be performed by vector operations working from contiguous locations in the computer storage. This gain is typically an order of magnitude or more in speed and is still worthwhile even if a factor of two or three is wasted calculating unnecessary interactions.

When particles are far apart compared to the cutoff radius $R_c$, only the 13 neighbor interactions from the $3 \times 3 \times 3$ interaction template need be considered. This number 13 is the same as the number of chaining cells which have to be considered in Hockney's PPPM data structure to find all nodes within the cutoff radius $R_c$. Figure 2.5 shows a schematic rendition of these different interaction templates. Only the half of the template with index offset larger than zero has to be considered since all interactions with nodes having a lower storage address index will have been calculated previously. As shown, shells of interaction can be defined which will correspond approximately to neighbors at different physical distances. The 16 neighboring nodes indicated with grey squares form the closest shell. The 30 triangle nodes are a bit further away, on average, and the 16 circle nodes form the furthest shell of the near neighbors template.


## SECTION III. ADDITIONAL ASPECTS OF MONOTONIC LOGICAL GRIDS

### III.A  Possibilities for Further Optimization

By a fully vectorizeable process of exchanging or "swapping" between adjacent logical cells every few timesteps, the near neighbor MLG ordering is kept intact even though the nodes move from cell to cell. Thus the method can be applied to gas, solid, and liquid systems using the same logical structure for problems of interesting size, i.e. 1,000 - 10,000 particles. The MLG algorithms forego a regular grid in space with a variable number of nodes in each cell for an irregular spatial grid which has exactly one node per cell by construction. This logical simplification, brought about by the MLG mapping, permits extensive optimization under current and planned supercomputer architectures (e.g. Fox and Otto, 1984) without sacrificing the generality needed to make it useful. The technique can also be discretized completely for use as Lagrangian Cellular Automata.

Optimization of near neighbor algorithms for particle dynamics is both machine and problem dependent. Vectorization techniques to achieve very high rates of computation require that all logical and arithmetic operations be performed on organized arrays of independent data. Distributed processing approaches to massive parallelism rely on a number of self-controlled processing centers operating asynchronously, but according to fixed rules of cooperation, on an evolving data base. To take advantage of both approaches simultaneously requires being able to define a number of vectorizeable segments of the problem which can be calculated independently. Furthermore, the vectors must be long enough to be computationally efficient but short enough that the memory needed in each asynchronous processing center is not prohibitively expensive. The MLG algorithms presented in the previous section can be partitioned for multi-tasking across a number of independent processors.

To maximize the length of vectors within each partition when the typical MLG dimension, NX ≈ NY ≈ NZ ≈ $N^{1/3}$, is only about 20 (8,000 objects) requires treating a substantial fraction of a plane as a single vector. In the 8 × 8 × 8 test problem, vectors of length 64 can be used throughout except for the X-direction monotonicity tests where vectors half as long would result. This is accomplished by collapsing several indices into one index and by paying careful attention to the boundary conditions.

Optimum computational efficiency results when the last few swapping operations are performed only for the grid points which might have become non-monotone due to adjacent swaps taking place during the previous iteration. Reductions of up to a factor of two in computer work to maintain monotonicity might be obtained by reducing the number of inactive vector swap attempts greatly at the cost of considerably increased program complexity. The scalar program to perform the few remaining swaps and keep track of which few nodes might have had their monotonicity conditions affected by the previous swaps is complicated. To date, convergence of the number of vector swapping iterations required has been so fast that this extra work has not been indicated. In the future it may be worth the effort for production calculations.

The same kind of gain can be obtained by trimming the near neighbors template defining which logical neighbors are likely enough to be close spatial neighbors that they should be included in the vector interaction calculations automatically. When a scalar "clean-up" portion is added to the vector force summing algorithm, the number of logically neighboring nodes which are always considered can be reduced significantly below that required to ensure no close uncounters. Figure 2.5 shows three shells of logical interactions in the near neighbors template, each succeeding shell taking neighbors which are logically, and usually physically, farther away.

By keeping track of maximum X, Y, and Z displacements along each row, column, and plane in the near neighbors template, a smaller template can be used with assurance that there will be no close "uncounters". After performing the interaction analysis on the neighbors within the vector shell, the boundaries of the shell can be checked to ensure that nodes on the logical boundary are far enough away spatially. These checks over all the nodes in the grid can also be vectorized. If the template has been pared sufficiently to ensure a worthwhile reduction in the number of nodes that have to be considered most of the time, there will generally be a fraction for which one or more of the neighbors in the vector shell were not far enough away to ensure that the next neighbor, which is outside the vector shell, can safely be neglected.

For the few nodes which may have spatially close neighbors which are removed more than two or three locations logically, a scalar calculation can be performed.

It would probably save a factor of two or so in overall computational cost to reduce the vector shells of the near neighbors template until the scalar cost competes with the significantly reduced vector cost. The expense of keeping track of $X_{max}$, $Y_{max}$, and $Z_{max}$ are minimal, three vector operations per node per near neighbor interaction. The test to determine which few nodes require extra (scalar) work is even cheaper, a few vector operations per node. Once a node has been found to require extra work, the scalar search can be extended to whatever logical distance is necessary to ensure that physically nearby nodes do not go uncounted.

A 5 × 5 × 5 cubical near neighbors template has 62 interactions which will be considered for each node. From empirical evidence to date this is adequate provided the critical radius of consideration is somewhat less than the average separation, here taken to be the original object spacing. Relatively few close uncounters can occur because neglected objects are logically at least 3 and generally 4, 5, or 6 nodes away.

Holes can be added to the MLG, locations which move or stay fixed in space but which don't contain a node. Any node neighboring one of these holes has one fewer real nodes in its interaction template because of the hole but this obvious disadvantage is balanced by the fact that hole locations can be updated any way necessary to improve the local structure of the MLG. By adding or shifting holes about judiciously it may be possible to avoid highly distorted MLGs. The holes would be subject to swapping with regular nodes just as if they were nodes but their equations of motion can be different and their interactions with regular nodes zero.

Figure 3.1 was computed using the 512 particle model with point non-interacting particles and the complete 5 × 5 × 5 interaction template shown in Figure 2.5. The volume around each particle was divided into shells of thickness 1 Angstrom and the number of particles in each radial shell was counted for particles logically outside the 5 × 5 × 5 template to determine how often "close uncounters" occur. A close uncounter occurs when a node gets close physically to another node without coming within the near neighbors template and thus ensuring that the interaction is "counted" in the vector sum. Concentrating first on the common features of the two physically identical calculations shown in the figure, we see that probability of an uncounted particle penetrating the interaction volume drops off very rapidly as the distance becomes small, that is, when the physical interaction would be important. The algorithm is one hundred times less likely to find an undetected particle coming within 10 Angstroms than to find one coming within 20 Angstroms. It is another one hundred times less likely to find one coming within 5 Angstroms and no close uncounters were ever found less than about 3 Angstroms.

The problem was repeated with a ninth plane of 64 locations added to the calculation for holes. The holes were given the average location of their six nearest logical neighbors as a propagation law and rapidly mixed throughout the volume with the particles. The probability of close uncounters was essentially unchanged. In this test, at least, holes don't seem to help much. This is probably because the propagation law used took no detailed account of local grid irregularities signalling a possible problem. Clearly research is needed to refine this generalization to the point where it appreciably optimizes the MLG representation.

## III.B  Fluid Dynamic Applications of the MLG

A few words about the application of the MLG to Lagrangian fluid dynamics is appropriate here. Each node of the grid can be identified with a fluid or vortex element. The advantage is in having a regular grid available to solve the physical evolution equations. Elliptic equations, for example, become amenable to highly efficient, vectorized multigrid methods (e.g., DeVore, 1984) on regular $N_x \times N_y \times N_z$ grids even though the fluid elements themselves move randomly. Two-dimensional and four-dimensional problems can be handled just as easily by the same methods.

Work is needed telling how to evaluate spatial derivatives accurately on the distorted MLG. When points are far apart spatially, the fluid cannot be as accurately represented as when they are close. To keep the resolution more nearly uniform than the specific fluid flow may be capable of, it can become necessary to remove nodes where they are crowded and to inject them elsewhere to better resolve some regions. To do this in the MLG involves finding a fluid element which can be merged with a larger one nearby in a manner which conserves mass, momentum, and energy. This frees up a location which can be "shifted" to the correct row, column, and plane to improve a deteriorating local resolution. The process in 2D is quite analogous to the operations needed to order sliding tiles numbered 1 to 15 in the 4 × 4 spaces of a popular child's game. The hole is shifted to the place where it is most needed.

After a shift operation, the same as an ordered series of swaps, it is likely that local swapping may be necessary to reestablish monotonicity. Shifting a line in one direction may well trigger swaps in the other two directions, so the real cost of adding and subtracting Lagrangian nodes locally to control resolution has to be measured (or estimated) for each configuration being considered. Certainly the specific additions, deletions and shifts will have a significant scalar component of computation. It does not make sense to execute a vector swapping

iteration over the whole grid unless a significant fraction of the objects are
being swapped. However, as long as at least 5% of the nodes are being moved, the
overall vector swapping iteration probably pays.

For comparison consider another free Lagrangian approach, the Lagrangian
Triangular (Tetrahedronal) Grid (Crowley, 1972; Boris and Fritts, 1975; Fritts and
Boris, 1979). In this approach the logical grid structure varies in time as the
nodes move. The number of near neighbors can vary from node to node and the
number and identity of these neighbors can vary at a given node as the Lagrangian
configurations change. This extra freedom, not allowed in the MLG, is used to
maintain a local grid structure optimized to guarantee diagonal dominance of the
simplest conservative finite—difference elliptic operator. The price is the loss
of local order in the grid and hence no vectorization.

Generalizing this Lagrangian Triangular Grid (LTG) to 3D is straightforward
but operationally very complicated. The grid is composed of adaptively
restructuring arrangements of tetrahedra in this case. The local grid structure
can still be changed as needed to maintain diagonal dominance of the elliptic
operator and resolution can be increased or decreased locally as needed. Scalar
linked lists become a necessary evil to keep track of near neighbors though the
resulting algorithms are still of order N.

Clearly the local spatial structure of the MLG is not as "good" as in the
generally structured LTG but the global structure compensates for this. The
monotonicity conditions specify a meaningful and useful relationship between
spatial derivatives and grid differences. As a result, fluid flows with long
range correlations, unlike the random particle motions used in earlier tests, may
lead to an additional computational expense at specific
times. In the smooth flow of large rotating and translating vortices, an
initially rectilinear grid might survive many timesteps before any of the local
monotonicity constraints are violated. Nearby points would move in almost the
same way. Once the fluid rotates far enough, however, monotonicity violations
would have to occur. Because of the long range correlation of the motions, a
number of swapping iterations may be necessary to reset the MLG.

In a turbulent flow with coherent flow structures at several spatial scales,
we can expect intermittant bursts of swapping activity from different scales at
different temporal frequencies. Small coherent structures in the flow require
fewer swapping iterations to restore monotonicity but will require them more often
than large structures when the rotation rates are higher. Though the integrated
number of these swapping iterations is large, it is unlikely to be larger than the
number of swaps required for random object motion with the same typical distances
traversed.

The strong possibility exists that statistics on the number and frequency
spectrum of grid swaps may prove to be a very good diagnostic of the progress of

turbulent mixing and flow interpenetration. Each swap can be viewed as a single quantum of geometrical upset. The object swapping process is necessarily intermittant and thus presents the possibility of easy integral measures of local intermittancy and turbulence spectrum. Studying the object swapping time series might provide a very direct way of diagnosing such fluid simulations for the onset and character of chaos.


III.C  Other Applications of a Monotonic Logical Grid

   The MLG suggests itself for use in multiphase fluid problems. Each grid node could be used to represent a droplet in a spray or a grain of sand in a sandstorm. Droplets could have varying sizes which increase or decrease in time due to local surface effects like condensation, evaporation, or abrasion and all the droplets would not have to be simulated. The accumulation of temporal averages over times and distances short compared to changes in the background flow means that only a small fraction of all the droplets or particles would have to be followed in the MLG to get a good estimate of the interactions of the whole distribution of particles with the background gas.

   Collisions of two drops could occasion merging or fragmentation. The MLG can accomodate either by shift operations which transport nodes from where they are no longer needed to new sites where fresh droplets are formed or enter the system. The gaseous background could be represented on a Eulerian mesh to facilitate swapping of mass, momentum and energy back and forth between particles and gas. The volume overlap of MLG cells with cells of the Eulerian grid can be used to circumvent a major complication of Monte Carlo methods, choosing the values of continuum functions at places where there are no particles or Lagrangian nodes. The MLG cells provide a natural way to interpolate back and forth between the two representations.

   When insolation of dust or droplet clouds is important, the MLG provides a simple way to assess the radiation opacity along any particular direction. The grid axes can be chosen in a given direction and the swapping algorithm used to resort the points along that direction. There are a number of line-of-sight obscuration problems where this flexibility will be useful.

## IV.  SUMMARY AND CONCLUSIONS

This report introduced a simple, fully vectorized algorithm to determine near neighbors whose cost scales as $N(1+\varepsilon \log N)$ where N is the number of independent nodes and $\varepsilon$ is a small number of order .02.  This was accomplished by defining a Monotonic Logical Grid (MLG) for storing the object data dynamically so that objects which are adjacent in real space are automatically close neighbors in the logical grid as well.  As a simple geometric test problem, a regular $8 \times 8 \times 8$  3D grid was used to store the position and velocity components of 512 randomly located particles in a cubical domain.  For this idealized system the points were given random velocities and the MLG was evolved for many transits of the system by the faster particles.  Statistics on near encounters of logically far away points and on the number of restructuring operations required were presented.

It was found that the reconnections of the dynamically changing MLG can generally be computed locally in a very few vectorized iterations without using inefficient gather or scatter operations.  Almost all of the grid restructuring occasioned by particles passing each other occurs in the first two or three vectorized iterations.  Further optimization is possible by changing to local scalar swapping after a few iterations.  It is also found that the spatially closest nodes are nearby in the MLG as well.  Two or three logical grid locations effectively defines the spatial near neighborhood except for a vanishing small number of cases which can be detected and corrected inexpensively.

The MLG differs from previous near neighbor algorithms.  It effectively removes the constraint of having to associate a cell of the logical grid with a fixed region of real space, but introduces the constraint of only one node per computational cell.  When many of the nodes cluster somewhere, a corresponding fraction of the storage locations in the MLG are automatically associated with that region.  This means that substantial variations in node density are adaptively gridded by the MLG and large regions of space, as well as computer memory, are not occupied by empty cells.

This algorithm gives regular global orderings of the node data and so allows efficient contiguous vector operations which are longer than the relatively small number of neighbors considered for each node but can be much shorter than the total number N.  The algorithm will execute efficiently in small array processors and permits direct partitioning to take advantage of massive asynchronous parallel- ism in VLSI/VHSIC-based distributed processing systems.  The cost to execute the simplest version of the model is one hour on a DEC VAX 11/780 for one thousand particles for 1000 timesteps when a simple force law for the 124 nearest neighbors is used.  With a commercially available Distributed Processing System, 10,000 in- teracting nodes could be monitored and the data base updated and then restructured thousands of times in about 15 minutes, fast enough for realtime applications.

A number of potential applications were discussed briefly. Obviously other uses will suggest themselves as the good properties and restrictions of the Monotonic Logical Grid (MLG) mapping between real space and relative (logical or computer storage) space becomes better understood. These problem-independent properties will be necessary to the successful
application of the representation to practical problems with other mathematical, logical or physical constraints. Swapping and shifting operations and holes were introduced to allow efficient local and global grid readjustments. Practical experience with the MLG is still minimal, however, so major pathologies may yet be uncovered in some applications.

Many MLG configurations may be possible for the same physical node arrangements and simple examples suggest that the best configurations are much better than the worst. Thus efficient methods of optimizing local and global structure within the monotonicity constraints will eventually be imperative. Additional work is needed on the following questions:

1. What is the mathematical nature of the simple representations for spatial derivative operators and integral conservation operators and how can they be optimized computationally?

2. Is there an algorithm to optimize the grid structure using holes and/or adaptively varied local modifications of the monotonicity functions?

3. What is the cost of not reaching monotonicity every cycle?

4. What is the geometric or information theoretic meaning behind the ambiguity of possible representations, i.e. what kind of an uncertainty principle does this represent?

5. What is the ultimate scaling of cost with system size?

## Acknowledgements

We would like to acknowledge years of informative and rewarding discussions with Dr. Martin Fritts on topics ranging over all aspects of this subject. His diligent and creative efforts on the LTG approach have provided an information base for the development of the MLG algorithm. This work was supported by the Office of Naval Research projects in Large-Scale Scientific Computing (44-1909, RR014-03-05), Computational Hydrodynamics (44-0573, RR01403-02) and Molecular Dynamics (44-1950-0-5, 61153N), and by the Naval Research Laboratory.

REFERENCES

1. R.W. Hockney and J.W. Eastwood, "Computer Simulation Using Particles", Chapter 8, pp267-304 (McGraw-Hill Inc, New York, 1981).

2. W.F. Gunsteren, H.J.C. Berendsen, F. Colonna, D. Perahia, J.P. Hollenberg, and D. Lellouch, "On Searching Neighbors in Computer Simulations of Macromolecular Systems", Journal of Computational Chemistry, Vol. 5, No. 3, pp272-279 (1984).

3. G.C. Fox and S.W. Otto, "Algorithms for Concurrent Processors", Physics Today, pp50-59, May 1984.

4. L.M. Adams and H.F. Jordan, "Is SOR Color-Blind?", ICASE Report No. 84-14, NASA Langley Research Center, May 1984.

5. J.P. Boris, M. Fritts, and K.L. Hain, "Free Surface Hydrodynamics Using a Lagrangian Triangular Mesh", Proceedings of the First International Conference on Numerical Ship Hydrodynamics, Gaithersburg, MD, October 20-22, 1975.

6. M.J. Fritts and J.P. Boris, "The Lagrangian Solution of Transient Problems in Hydrodynamics Using a Triangular Mesh", J. Comp. Phys., Vol 31, No 2, p173, May 1979.

7. W.P. Crowley, "FLAG: A Free Lagrange Method for Numerically Simulating Hydrodynamic Flows in Two Dimensions", Proceedings of the Second International Conference on Numerical Methods in Fluid Dynamics, (Springer-Verlag, New York, 1971).

8. C.R. Devore, "Vectorization and Implementation of an Efficient Multigrid Algorithm for the Solution of Elliptic Partial Differential Equations", NRL Memorandum Report, October, 1984.

**Space**

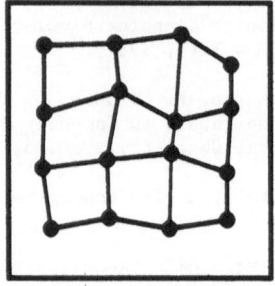

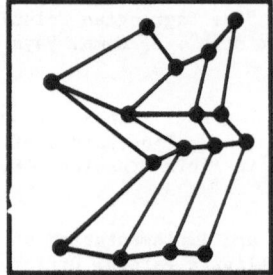

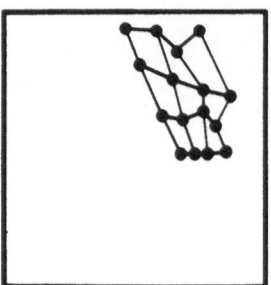

Figure 2.1   Three Different Spatial Configurations and the
Corresponding Monotonic Logical Grid

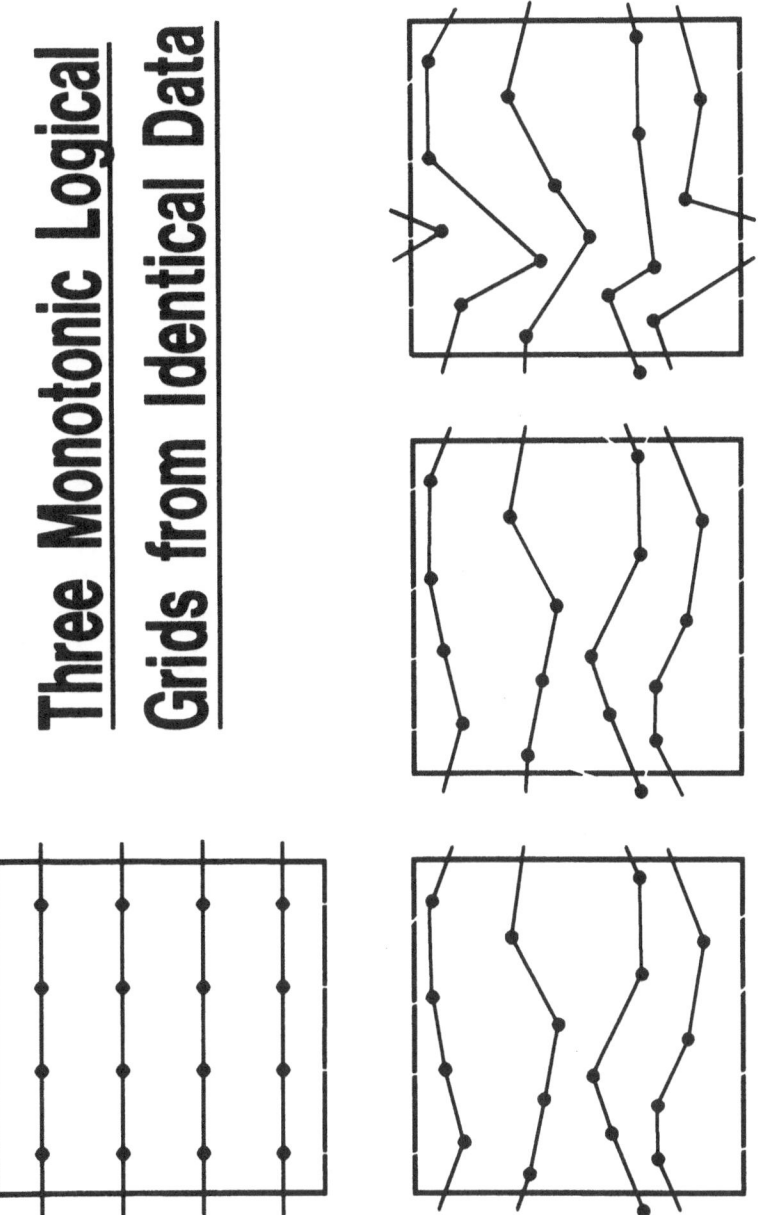

Figure 2.2   Three Monotonic Logical Grids from Identical Data

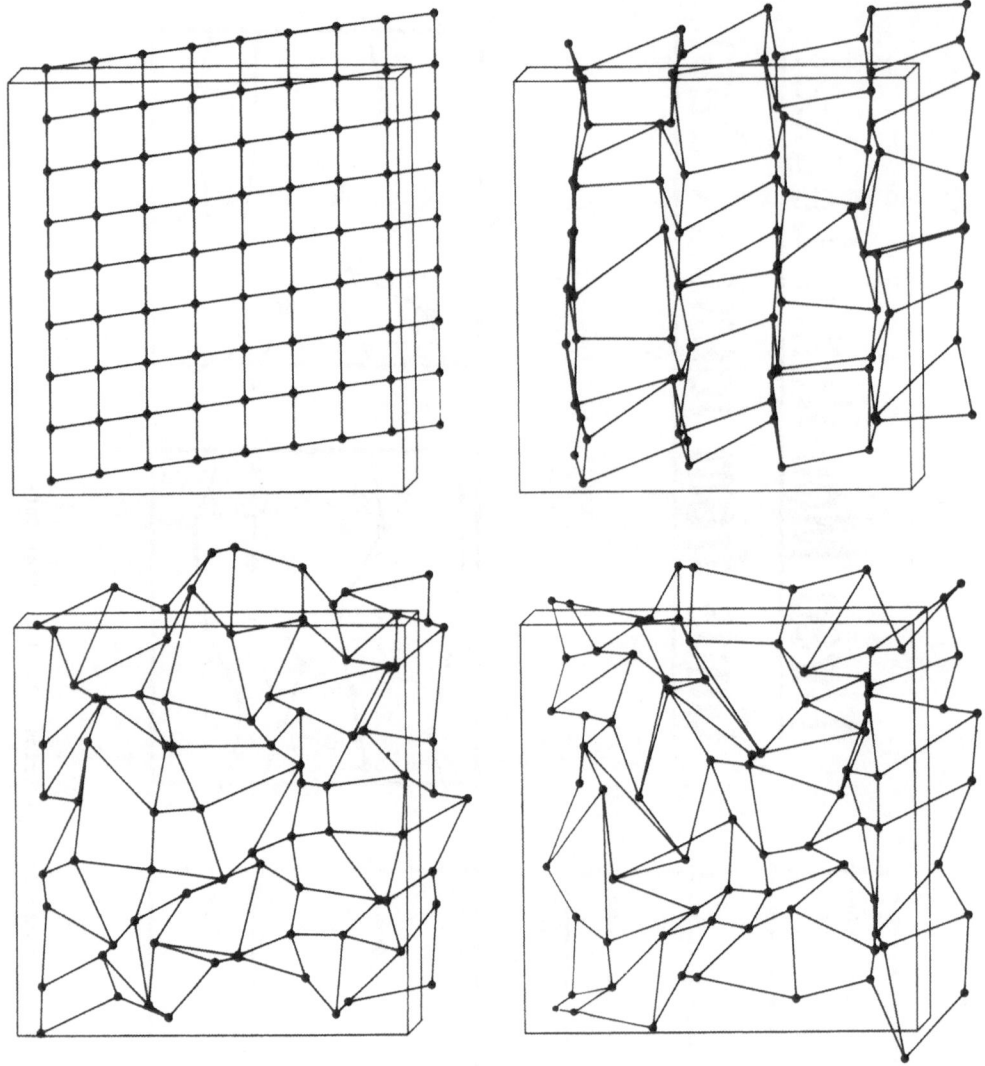

Figure 2.3  The MLG for One Plane of the 512 Particle Test Problem

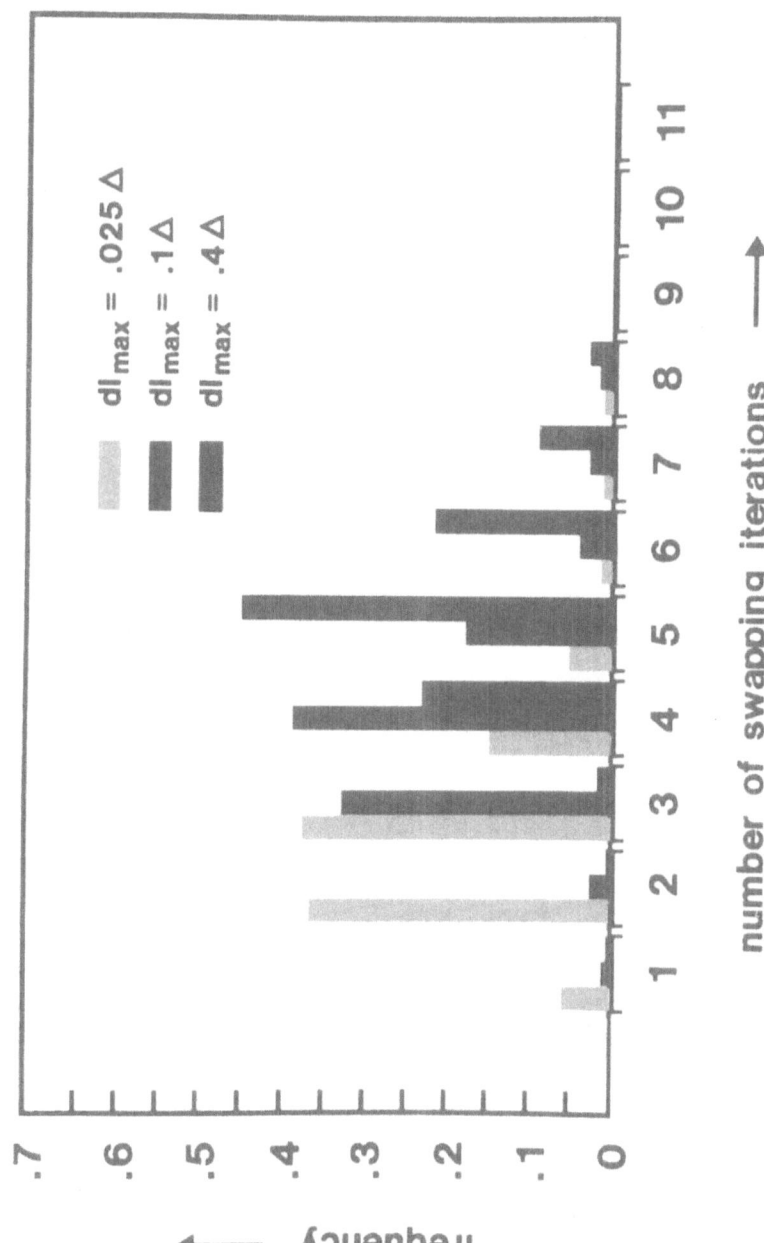

Figure 2.4  Frequency of Requiring N Swapping Iterations to Construct a
Monotonic Logical Grid

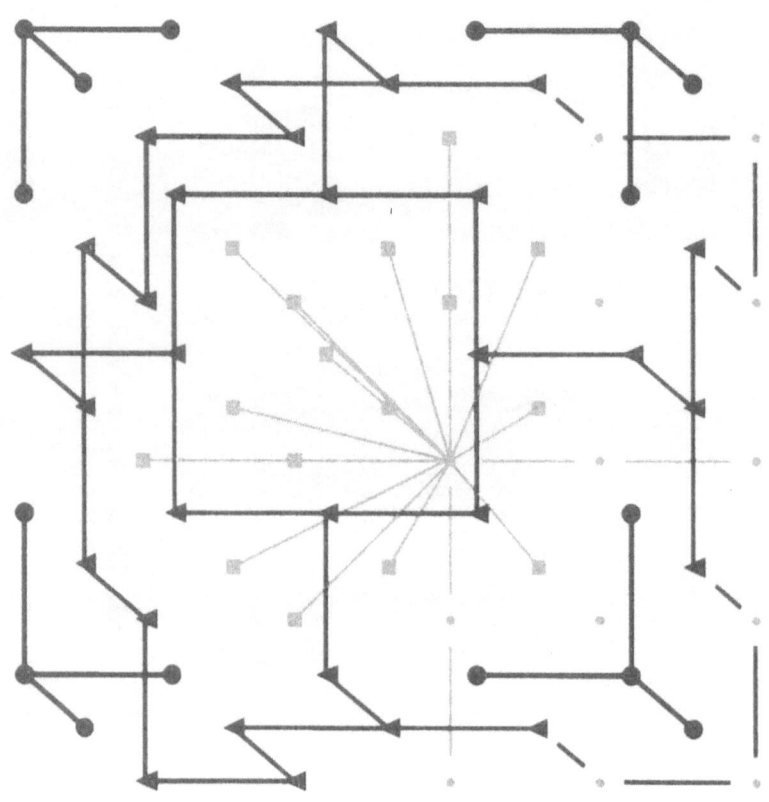

■  16 pts
▲  30 pts
●  16 pts

# Rc = 4Å 5Å 6Å     (8 × 8 × 8)

| | 4Å | 5Å | 6Å | | | (8 × 8 × 8) |
|---|---|---|---|---|---|---|
| 4Å | | .0039 | | | .0051 | |
| 5Å | .0004 | .0058 | | .0008 | .0160 | |
| 6Å | .0027 | .0121 | .0016 | .0019 | .0238 | |
| .0012 | .0055 | .0585 | .0678 | .0499 | .0226 | |
| .0016 | .0273 | .1540 | .1587 | .1286 | .0756 | .0008 |
| | .0448 | .2928 | .3396 | .2090 | .1458 | |
| .0008 | .0511 | .7610 | 2.1150 | .5980 | .0316 | |
| .0035 | .1380 | 1.8610 | 4.4490 | 1.3080 | .1415 | |
| .0094 | .2339 | 3.0650 | 7.1080 | 2.1180 | .2858 | |
| | | | Target | 2.2770 | | |
| | | | Cell | 4.7060 | | |
| | | | | 7.1910 | | |

Table 3.1   Probability of Unrecorded Close Encounters

Data Structures and Vectorization in a
Two-Dimensional Free-Lagrangian Code

Robert E. Cooper
Lawrence Livermore National Laboratory
Livermore, Ca. 94550

## ABSTRACT

The data structures used to define the mesh and drive the physics
calculations of a 2-D Free-Lagrangian hydro code are described. The
mesh contains three and four sided zones but the data structures can
be generalized to support n-sided zones. The mesh is redundantly
described by three different data structures: zone oriented, node
oriented, and edge oriented. Physics algorithms, graphics routines,
and reconnection algorithms use the data structure that is easiest to
use or that results in the best execution rate. On the Cray
computer, these data structures are sufficient to vectorize most of
the physics algorithms. The basic vectorization methods will be
discussed. From experience, given a physics algorith, a data
structure can be constructed that will vectorize the algorithm.

## BACKGROUND

The physics code that is the topic of this paper is a large
two-dimensional finite difference free-lagrangian code. Because
these calculations require a considerable amount of computer time,
substantial effort has been spent by Computer Scientists to make the
physics run efficiently. Assembly language numerical libraries were
used on the CDC 7600; compiler vectorization and specialized assembly
language routines are being used on the Cray-1 and Cray-XMP.

The code is really a code system with three major parts: generator,
physics, and interactive graphics. Smaller utilities and
post-processors complete the system. The physics section is the only
part that has intentionally been optimized though the graphics has
also benefitted from the optimizations.

Currently, several Computer Scientists and Physicists are working on
the code. A system of alter decks allow everyone to change different
sections of the code concurrently. We use two different compilers:
CFT (Cray Research Inc. Fortran compiler), and LRLTRAN (LLNL's
Fortran compiler). Both compilers produce vector coding from Fortran
do-loops. Approximately, 100 line of Fortran (six subroutines) have
been converted to assembly language subroutines. Physics do-loops
that consume an appreciable amount of time are either vectorized by
the compilers or have been converted to one of the six assembly
language subroutines.

The Mesh

The mesh consists of either three
or four sided zones; dangling zones
(Fig. 1) and interior holes are not
allowed. The code could be changed
to allow zones with more than four
sides and interior holes. Neither
of these features has seemed
important enough to warrant the
work necessary to implement them.
Mesh arrays come in several
different types: zone, node, edge,
zone-name, and node-name.

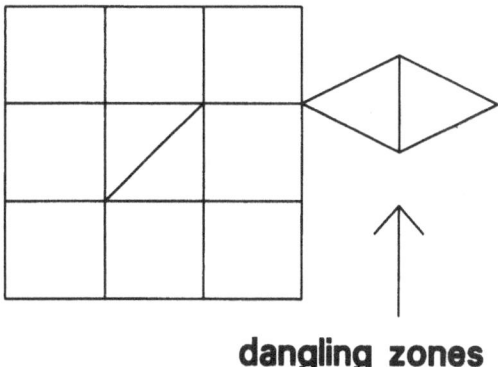

**dangling zones**

Figure 1

Zone Arrays

Zone arrays contains zone-centered physics variables (density, etc.)
or zone-centered mesh data (neighboring zone numbers, node numbers on
the zone boundary, etc.). To facilitate vectorization, data for
triangular zones are first and data for quadrilateral zones last.
Triangular zones are in no particular order and quadrilateral zones
are also in no particular order. If a zone has been eliminated from
the mesh as the result of a mesh reconnection, the storage for that
zone will be squeezed out of the zone arrays.

Node Arrays

Node arrays contain node-centered physics variables (coordinates,
velocities, etc.) or node-centered mesh data (number of neighboring
mesh nodes, mesh boundary flag, etc.). Nodes are in no particular
order. If a node has been eliminated from the mesh as the result of
a mesh optimization, the storage for that node will be squeezed out
of the node arrays.

Node-Name Array

Every node in a mesh has both a "name" and a "number". The name is
the integer that people associate with a particular node; the number
is the integer Fortran coding uses to fetch node data for that node.
The name of a node does not change — once people remember the name of
their favorite node they do not want that name to change. The number
of a node may change when the code repacks node arrays after a node
is removed from the mesh. Node numbers are stored in the Node-name
array. The name of a node is the position in the Node-name array
where its number is stored i.e., the second word of the Node-name
array contains the node number for node name two.

During a calculation, node data for a node may be moved to a different position in the node arrays. When this happens the Node-Number for the moved node is changed to the new position. Also during a calculation, a node may be removed from the mesh by a mesh reconnection. When this happens, the Node-Number for the removed node is modified so it indexes a node still in the mesh that is spatially close to the removed node.

The code's mesh is originally generated from a logically rectangular mesh. People are accustomed to referring to nodes by their two indexes, "K" and "L". The code allows people to refer to a node by its "K,L" indexes by storing the "K" and "L" into the Node-Name array. People can refer to a node by supplying either the "K,L" or the Node-Name. To summarize:

    Node-Name-Array(Node-Name) = K / L / Node-Number

(The above notation indicates that three quantities, K, L and Node-Number, are packed into one computer word. Packing in the code is a matter of practical efficiency. The meshes are large enough compared to the computer memory that packing is necessary.)


Zone-Name Array


The Zone-Name array is the analog to the Node-Name array but for Zone arrays. To summarize:

    Zone-Name-Array(Zone-Name) = K / L / Zone-Number


Edge Arrays


Edge arrays contain edge-centered physics variables (e.g., differences between zone centered variables) or edge-centered mesh data (e.g., zone numbers on each side of the edge). To facilitate vectorization and to simplify the graphics routines edges are ordered:

    1) Edges that have the same material on each side.
    2) Edges that have different materials on each side.
    3) Edges that have material on only one side. These edges are
       ordered to trace the boundary of the mesh.

Within edge groups 1 and 2, the order is random. If an edge has been eliminated from a mesh because of a reconnection, the storage for the edge data is squeezed out of the edge arrays.

An Example

So far this paper has given a general descriptions of the arrays used
in this code; the following section will give the data structures for
a simple mesh (Fig. 2). The code carries three separate independent
descriptions of the mesh. Each description is from a different
perspective: zone, edge, and node. Physics, graphics, and
reconnection algorithms use the mesh description that is easiest to
use or that results in the best execution rate.

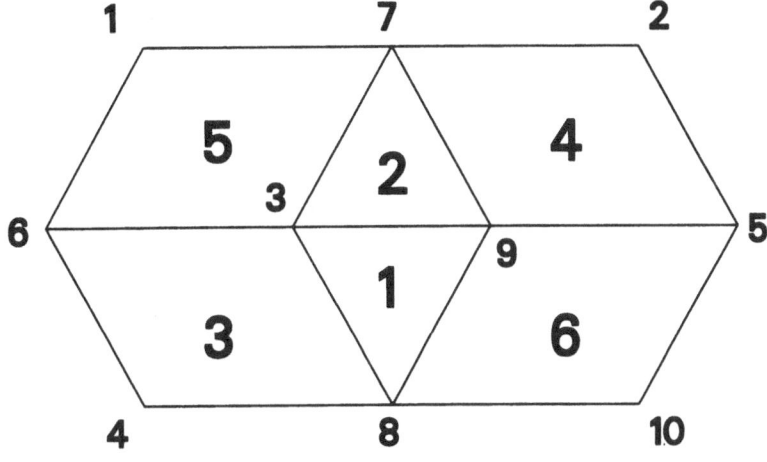

Figure 2.

Zone Perspective

The mesh can be completely described by describing each individual
zone in terms of its zone number and the nodes that are on the
boundary of the zone. Zone array, ZONE, contains such a destription:

    ZONE(zone number) = L / K / J / I

where I, J, K, and L are the node numbers of the nodes on the
boundary of the zone. If the zone number is a triangle, L is zero.
To make this description unique, I is the minimum node number and I,
J, K and L are in counter-clockwise order. Without uniqueness ZONE
could describe the same mesh in many different ways and numerical
round-off would be different for each description. For the example
mesh:

    ZONE(1) = 0 / 9 / 8 / 3        ZONE(4) = 5 / 9 / 7 / 2
    ZONE(2) = 0 / 7 / 9 / 3        ZONE(5) = 7 / 3 / 6 / 1
    ZONE(3) = 8 / 4 / 6 / 3        ZONE(6) = 10 / 8 / 9 / 5

Edge Perspective

The mesh can be completely described by describing each individual
edge in terms of the edge number, the zones numbers of the zones on

each side of the edge, and the node numbers of the nodes at each end of the edge. Edge array, EDGE, contains such a destription:

EDGE(edge number) = Z2 / Z1 / N2 / N1

N1 and N2 are the node numbers of the nodes that define the edge. If you are looking from N1 to N2, Z1 is the zone number of the zone to the left of the edge and Z2 is to the right. For uniqueness, N1 is less than N2 for edges interior to the mesh. For the example mesh:

```
EDGE( 1) = 2 / 5 / 7 / 3          EDGE( 9) = 0 / 4 / 2 / 5
EDGE( 2) = 6 / 1 / 9 / 8          EDGE(10) = 0 / 4 / 7 / 2
EDGE( 3) = 2 / 4 / 9 / 7          EDGE(11) = 0 / 5 / 1 / 7
EDGE( 4) = 3 / 1 / 8 / 3          EDGE(12) = 0 / 5 / 6 / 1
EDGE( 5) = 1 / 2 / 9 / 3          EDGE(13) = 0 / 3 / 4 / 6
EDGE( 6) = 5 / 3 / 6 / 3          EDGE(14) = 0 / 3 / 8 / 4
EDGE( 7) = 4 / 6 / 9 / 5          EDGE(15) = 0 / 6 / 10 / 8
EDGE( 8) = 0 / 6 / 5 / 10
```

## Node Perspective

The mesh can be completely described by listing the neighboring nodes and zones around each node in the mesh. For each node in the mesh there is an entry in array NBR. The entry lists the neighboring nodes and zones in counter-clockwise order. The first word of each entry also contains the node number of the center node and the length of the entry. This information is used to repack NBR. Node array, NODE, contains an index into NBR, the length of the entry, and a flag which indicates whether the center node is on the mesh boundary or not. To summarize:

NODE(node number) = flag / length-of-NBR-entry / index-to-NBR

NBR(index) = length / center-node-number / zone number / node number

Three NBR entries are given below for the example mesh.

```
NODE( 8) = 1 / 4 / 69
NODE( 9) = 0 / 4 / 73
NODE(10) = 1 / 2 / 101

NBR( 69) = 4 / 8 / 6 / 10
NBR( 70) = 0 / 0 / 1 / 9
NBR( 71) = 0 / 0 / 3 / 3
NBR( 72) = 0 / 0 / 0 / 4

NBR( 73) = 4 / 9 / 4 / 5
NBR( 74) = 0 / 0 / 2 / 7
NBR( 75) = 0 / 0 / 1 / 3
NBR( 76) = 0 / 0 / 6 / 8

NBR(101) = 2 / 10 / 6 / 5
NBR(102) = 0 / 0 / 0 / 8
```

After generating and before running, the zone(node) names and numbers are usually the same. The Zone-name and Node-name arrays for the example mesh are given below. (The K's and L's are not shown.)

```
Zone-name(1) = / / 1      Node-name( 1) = / /  1
Zone-name(2) = / / 2      Node-name( 2) = / /  2
Zone-name(3) = / / 3      Node-name( 3) = / /  3
Zone-name(4) = / / 4      Node-name( 4) = / /  4
Zone-name(5) = / / 5      Node-name( 5) = / /  5
Zone-name(6) = / / 6      Node-name( 6) = / /  6
                          Node-name( 7) = / /  7
                          Node-name( 8) = / /  8
                          Node-name( 9) = / /  9
                          Node-name(10) = / / 10
```

To demonstrate how the data structures change when a reconnection
occurrs, lets merge nodes 3 and 9, removing node 3 from the mesh
The reconnect will also remove zones 1 and 2 from the mesh. The
resultant mesh is in Figure 3.

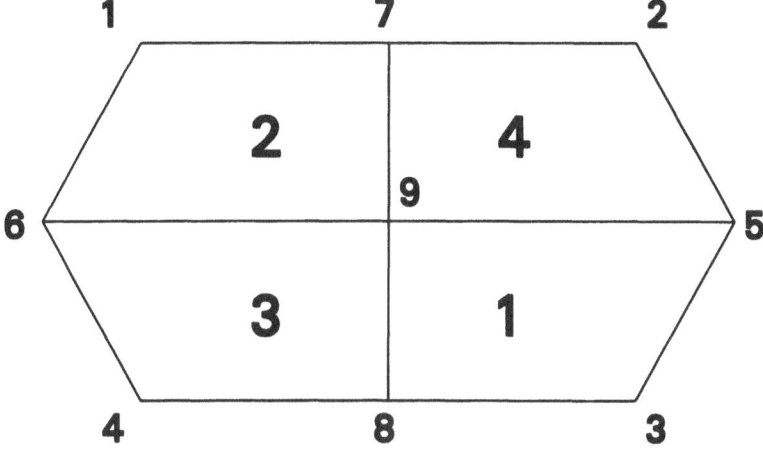

Figure 3.

Data in Node arrays will be rearranged in the following manner:

Data in Zone arrays will be rearranged in the following manner:

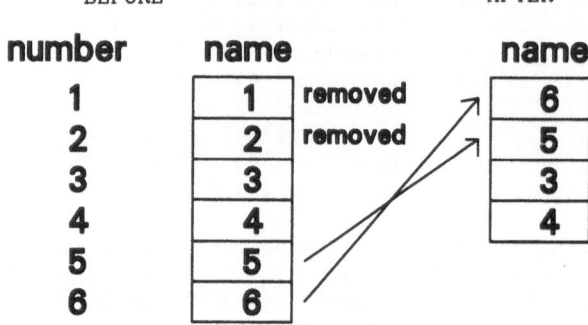

|  | BEFORE |  | AFTER |
| :-: | :-: | :-: | :-: |
|  | **number** | **name** | **name** |

ZONE, EDGE, NBR, NODE, Zone-name and Node-name:

```
ZONE(1) = 8 / 9 / 5 / 3          EDGE( 1) = 2 / 4 / 9 / 7
ZONE(2) = 7 / 9 / 6 / 1          EDGE( 2) = 1 / 3 / 9 / 8
ZONE(3) = 6 / 9 / 8 / 4          EDGE( 3) = 3 / 2 / 9 / 6
ZONE(4) = 5 / 9 / 7 / 2          EDGE( 4) = 4 / 1 / 9 / 5
                                 EDGE( 5) = 0 / 1 / 5 / 3
                                 EDGE( 6) = 0 / 4 / 2 / 5
NODE(3) = 1 / 2 / 55             EDGE( 7) = 0 / 4 / 7 / 2
NODE(8) = 1 / 3 / 65             EDGE( 8) = 0 / 2 / 1 / 7
NODE(9) = 0 / 4 / 23             EDGE( 9) = 0 / 2 / 6 / 1
                                 EDGE(10) = 0 / 3 / 4 / 6
NBR(23) = 4 / 9 / 4 / 5          EDGE(11) = 0 / 3 / 8 / 4
NBR(24) = 0 / 0 / 2 / 7          EDGE(12) = 0 / 1 / 3 / 8
NBR(25) = 0 / 0 / 3 / 6
NBR(26) = 0 / 0 / 1 / 8

NBR(55) = 2 / 3 / 1 / 5          Zone-name(1) = / / 3
NBR(56) = 0 / 0 / 0 / 8          Zone-name(2) = / / 5
                                 Zone-name(3) = / / 3
NBR(65) = 3 / 8 / 1 / 3          Zone-name(4) = / / 4
NBR(66) = 0 / 0 / 3 / 9          Zone-name(5) = / / 2
NBR(67) = 0 / 0 / 0 / 4          Zone-name(6) = / / 1

Node-name( 1) = / / 1
Node-name( 2) = / / 2
Node-name( 3) = / / 9
Node-name( 4) = / / 4
Node-name( 5) = / / 5
Node-name( 6) = / / 6
Node-name( 7) = / / 7
Node-name( 8) = / / 8
Node-name( 9) = / / 9
Node-name(10) = / / 3
```

Vectorization


Over the years a large amount of time has been spent making the code
run efficiently. On the 7600 optimal assembly language routines were
used that performed arithmatic operations on vectors. Also, several
time consuming do-loops were coded in optimal assembly language.
Originally, NODE was the only data structure used by the physics.
ZONE and EDGE were added which allowed the same physics equations to
be calculated much more quickly. When the Cray-1 computer arrived at
LLNL, we learned to use vectorizing Fortran compilers. When nodes
and zones that had been removed from the mesh were squeezed out of
node and zone arrays, many "if" tests were eliminated from otherwise
vectorizable do-loops. Again, coding time consuming do-loops into
optimal assembly language routine is still used as a last resort.
The number of assembly language routines we have written is small and
the size of each routine is also small. When we run diagnostic
routines to tell us what parts of the code the CPU is spending most
of its time, we find that our equations account for less than half of
the running time. The rest of the time is spent in library routines
we call but otherwise have no control over.

On first impression, vectorization in a general mesh might seem
difficult. In reality, most physics equations are very easy to
vectorize. The large majority of do-loops deal strictly with zone
centered quantities and invariably the compiler vectorizes them.
These are Strictly Zonal loops i.e., their inputs and outputs are
zone centered quantities. In addition, the equations for Strictly
Zonal loops are usually identical for triangular and quadrilateral
zones, so all the zones can be processed in the same do-loop.


```
            DO M=1,NUMZONES
               ZE(M) = ZE(M) + ZDELTAE(M)
            ENDDO
```

Figure 4. Example of a Strictly Zonal do-loop


There are several do-loops that deal strictly with node centered
quantities and they vectorize similarly. These are Strictly Nodal
loops i.e., their inputs and outputs are node centered quantities.
There are also a few examples of do-loops that deal strictly edge
centered quantities.


```
            DO M=1,NUMNODES
               R(M) = R(M) + DELTAT * RVEL(M)
            ENDDO
```

Figure 5. Example of a Strictly Nodal do-loop


The situation is much more interesting when physics equations have
have more than one input type (e.g., nodal and zonal) and one output
type (e.g., nodal). The coding in Figure 6 calculates a node
centered mass from a zone centered mass. The inputs are PMASS
(initialized to 0.) and ZMASS, the zone centered mass. The output
is PMASS. This type of loop, Zone to Node, only occurs four time in

our code and does not comsume appreciable computer time. It can be vectorized.

```
DO M=1,NUMTRI
    I=AND(          ZONE(M),      77777B)
    J=AND(SHIFTR(ZONE(M),15),77777B)
    K=AND(SHIFTR(ZONE(M),30),77777B)
    PMASS(I) = PMASS(I) + ZMASS(M)
    PMASS(J) = PMASS(J) + ZMASS(M)
    PMASS(K) = PMASS(K) + ZMASS(M)
ENDDO
```

Figure 6. Example of Zone to Node loop (triangles only)


The next type of do-loop, Node to Zone, will be dealt with in detail because it demonstrates the most common method of vectorizing a do-loop that the compiler cannot vectorize. The inputs are nodal and zonal; its outputs are zonal. The nodal inputs are ordered into temporary zonal arrays. Once this is done, the do-loops become Strictly Zonal and vectorization is easy.

The Fortran in Figure 7 calculates the centroids of the zones in the mesh. Figure 8 gives a sample mesh and the values of temporary and output arrays in Figure 7. Because the algorithm is different for triangles than for quadrilaterals, one subroutine will calculate centroids for triangles and another for quadrilaterals. Only the first will be included. The coding uses eight temporary zone arrays, RI, ZI, RJ, ZJ, RK, ZK, RL, and ZL. In practice to conserve memory, the temporary arrays are only 64 words long. A mesh calculation is actually processed 64 zones at a time. Variable NUMTRI is the number of triangles in the mesh and NUMQUAD is the number of quadrilaterals. Subroutine FETCH3 orders the input data. In the code an assembly language routine is used but the equivalence Fortran is listed here. On the CRAY-XMP, FETCH3 vectorizes completely.

There are other varieties of this kind of physics loop, but each can be characterized by the types of inputs and its output. More importantly, the do-loop can be vectorized after the input variables have been reordered into the same type as the output type. If the output variable type is "edge" and the inputs are nodal and zonal, the nodal and zonal inputs can be reordered into a temporary array of type "edge". The compiler will do the rest. Variable ZONE is used to reorder a nodal variable into a zonal temporary; EDGE is used to reorder zonal or nodal variables into an "edge" temporary. There are subroutines similar to FETCH3 that perform the reordering.

If a physics algorithm requires more or different data than ZONE and EDGE can produce, then a new data structure can be constructed which will produce the necessary data. Several algorithms have been implemented in our code that have required new data structures to drive the reordering of input data. Once the data structure was created, the algorithm was easily vectorized.

```
                              .
                              .
                              .
C           CALL ROUTINES TO CALCULATE CENTROIDS

      CALL CALCENT3(CENTR,CENTZ,RI,ZI,RJ,ZJ,RK,ZK,R,Z,ZONE,
     1              NUMTRI)
      CALL CALCENT4(CENTRI(NUMTRI+1),CENTZ(NUMTRI+1),RI,ZI,RJ,
     1              ZJ,RK,ZK,RL,ZL,R,Z,ZONE(NUMTRI+1),NUMQUAD)
                              .
                              .
                              .
      SUBROUTINE CALCENT3(CENTR,CENTZ,RI,ZI,RJ,ZJ,RK,ZK,R,Z,
     1                    ZONE,NUM)

C           CALCULATE CENTROIDS OF TRIANGLES IN MESH

      DIMENSION CENTR(1),CENTZ(1),RI(1),ZI(1),RJ(1),ZJ(1),
     1          RK(1),ZK(1),R(1),Z(1),ZONE(1)

C           FETCH NODE COORDINATES INTO ZONE ORDER

      CALL FETCH3(RI,RJ,RK,R,ZONE,NUM)
      CALL FETCH3(ZI,ZJ,ZK,Z,ZONE,NUM)

C           CALCULATE CENTROIDS

      DO FOR I=1,NUM
         CENTR(I)=(RI(I)+RJ(I)+RK(I))/3.
         CENTZ(I)=(ZI(I)+ZJ(I)+ZK(I))/3.
      ENDDO

      RETURN
      END
      SUBROUTINE FETCH3(AI,AJ,AK,A,ZONE,NUM)

C           GATHER ELEMENTS OF ARRAY A INTO A1, A2, AND A3
C           BASED ON INDEXES IN ARRAY ZONE

      DO FOR M=1,NUM
         I=AND(        ZONE(M),      77777B)
         J=AND(SHIFTR(ZONE(M),15),77777B)
         K=AND(SHIFTR(ZONE(M),30),77777B)
         AI(M)=A(I)
         AJ(M)=A(J)
         AK(M)=A(K)
      ENDDO

      RETURN
      END
```

Figure 7. Vectorizing a Node to Zone loop (triangles only)

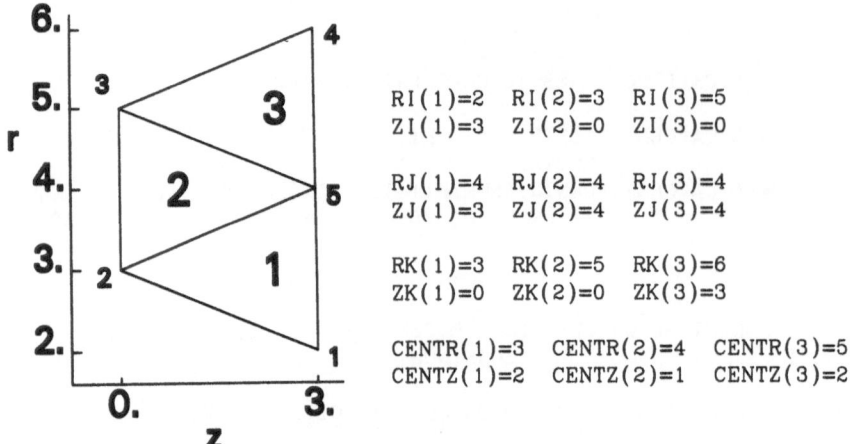

RI(1)=2    RI(2)=3    RI(3)=5
ZI(1)=3    ZI(2)=0    ZI(3)=0

RJ(1)=4    RJ(2)=4    RJ(3)=4
ZJ(1)=3    ZJ(2)=4    ZJ(3)=4

RK(1)=3    RK(2)=5    RK(3)=6
ZK(1)=0    ZK(2)=0    ZK(3)=3

CENTR(1)=3    CENTR(2)=4    CENTR(3)=5
CENTZ(1)=2    CENTZ(2)=1    CENTZ(3)=2

Figure 8.

There are other types of physics do-loops in the code, but all of
them can be vectorized in the same fashion as the Node to Zone loop.
They depend on a "fetch" routine to reorder the input data types into
the same order as tye output data type.  The computer scientist needs
only to generate a data structure that has the correct indexes to
perform the reordering.  This scheme of vectorization requires a fast
"fetch" routine.  High-end scientific CPU's are implementing such
"fetches" in hardware.  Optimal assembly language routines should be
used on other CPU's.

This method of vectorization has not lead to a proliferation of data
structures but has generated data structures that tend to be
generally useful.  Data structures, like ZONE and EDGE, are used not
only to vectorize do-loops, but are also used frequently in parts of
the code that are logic-intensive and scalar.  The data structures
that redundantly describe the mesh are worth their memory space
because of their use in vectorization and use in producing cleaner
code.

*This work was performed under the auspices of the U.S. Department of Energy
by Lawrence Livermore National Laboratory under contract No. W-7405-Eng-48.

MR. ARMSTRONG: A lot of things have been said about the Voronoi mesh. It seems to me, that they've been made to appear quite unnecessarily complicated. The way I look at them, for any triangular mesh, if the triangulation is proper, there is no line in it such that any two angles subtended by each side of the line sum to more than 180 degrees. From this we find that the Voronoi mesh is obtained immediately by simply locating the circumcenters of each triangle. This seems by far the easiest way to construct the Voronoi mesh from the triangulation. But, the Voronoi mesh has some distinct advantages over the triangulation. One is that you can direct lines from the three sides of a triangle and show that they all coincide, at the circumcenter. A more important advantage, in my view, is that when two circumcenters coalesce, which happens precisely when these two angles add up to 180 degrees, then your angle criteria tells you to swap the two diagonals. Swapping the diagonals at that point guarantees that there is no change in the allocation of the areas of the two Voronoi cells. This seems to be the crucial advantage of the Voronoi mesh over the triangular mesh. In passing, I would like to say that any method which will give you radically different answers depending upon which way you draw the diagonal cannot be assumed to be very accurate.

I have said that there is one obvious drawback in the Voronoi mesh. It is that we insist on drawing the perpendicular bisector of the line joining two mesh points. For example, we take an exaggerated case where one mesh point has an area of 100 and the second mesh point has an area of 1. Then if we draw the perpendicular bisector the result is obviously ridiculous from my point of view. However, it is very easy to generalize the Voronoi mesh. We draw another locus of points which are not equally distant from those two nodes, but instead, are such that the distance to those nodes is equal to the difference of the sectors associated with two nodes. If the area of this node is $\pi r_1^2$, and the other is $\pi r_2^2$, then we draw the locus of a point such that the ratios of the squares of the distances from end one and end two is equal to our $\pi r_1^2/\pi r_2^2$, we get a more rational position of the line. It goes down here somewhere (see Figure 1). Moreover, it's just as easy to construct a mesh in this way.

The sort of thing you end up with consists of nodes that have a different area associated with them. By dividing the corresponding sides shown here, obviously not by bisectors, but by what I call perpendicular non-bisectors, they are concurrent at a certain point off-center from the present circumcenter. This has the same advantage as the Voronoi mesh in that the perpendicular which divides a given line in this ratio performs the same function for masses assigned to it. As opposing angles get larger, these two points will move towards one another, just as circumcenters did. When they cross over or when they coincide, that is when the preference for diagonals is completely invariant with respect to the proper diagonal. There is one snag I have to point out, and that is if you get extreme differences in

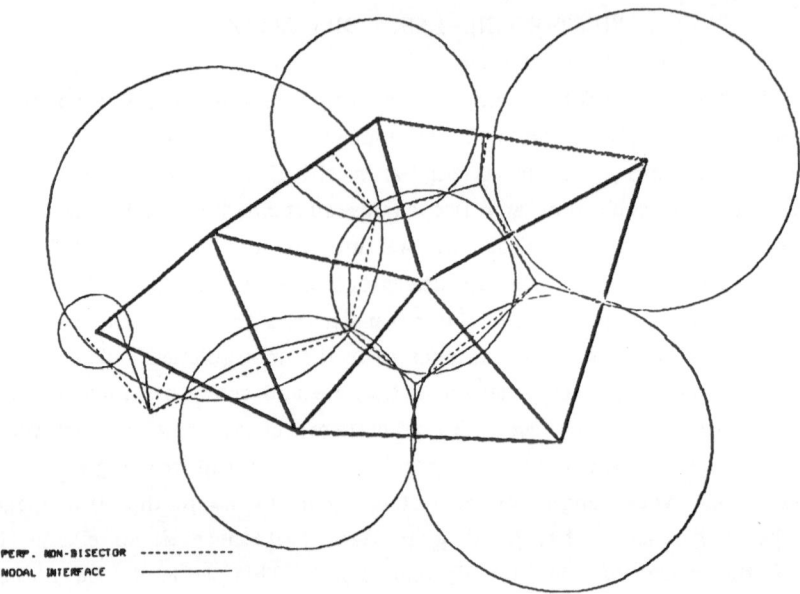

PERP. NON-BISECTOR ------------
NODAL INTERFACE ‾‾‾‾‾‾‾‾‾‾

Figure 1. Viewgraph Showing an Example of Mr. Armstrong's Generalized Voronoi Mesh.

the $r_1^2$ and $r_2^2$, this could cause the perpendicular to fall outside the line, which is obviously very undesirable. So in fact, having constructed the "circum-off-center" in the way I have described, we must give up something. What I would suggest doing is joining that "circum-off-center", not by perpendiculars to the sides, but to the points on the sides which divide the lines in the ratio of the radii in the two areas. I think that this would retain all of the advantages of the Voronoi mesh.

MR. GLENN: Have you tested that concept on an irregular set of points where you can see the differences between a given state on one side and that same state on the other side?

MR. ARMSTRONG: I have not tested it in a Free-Lagrange program. I set up all my meshes such that all the nodes have the very same mirror masses.

MR. GLENN: You can't do that in a lot of practical cases.

MR. ARMSTRONG: I suggest that in those cases the generalized Voronoi mesh would be better than the Voronoi mesh.

MR. WINSLOW: We use the Voronoi mesh as the Voronoi mesh. The mesh you constructed has point centered quantities which you wish to use to divide space into the regions which would carry some physical significance.

MR. ARMSTRONG: If you draw the circumcenters for all the triangles and join them by perpendicular bisectors across each side, then you have the Voronoi mesh. Getting back to the facts between the Voronoi mesh and the median based mesh, I would imagine that you may get different results with different strategies for flipping diagonals. You seem to be losing the flexibility of the Voronoi cells. As one very specific example; if you were to solve a shock problem which had a pressure surface with reasonable continuity, you would like the edges of the cells to follow the shock. The Voronoi mesh could not possibly guarantee that. You may very well end up with an edge that might be perpendicular to the shock. This means that it would cut through the pressure surface. Therefore, you would question the accuracy of the solution. Is there anyone who would like to expand on that?

MR. GLENN: By aligning the edges of the cells with a shock you would be introducing dissipation. This, of course would cancel any of the advantages of using the Voronoi mesh.

MR. TREASE: It's been found that the cell faces do align themselves with a shock. Take a regular mesh and propagate a shock at a 45° angle to the mesh. We find that the mesh aligns itself such that the cell edges are perpendicular to the shock propagation direction. Or take a spherically diverging shock. The orientation of the mesh to the shock turns out to be roughly spherical.

MR. BORIS: I would like to draw a picture of a test problem. For the one that I came up with, a diagram will be helpful (see Figure 2). Take a single point and surround it by a ring of other points, and then a second ring further out. Some of these points are moving, but the outer points in the system are not moving, such that we have a vertex and a closed ring vortex around it. Now, if you triangulate this system, point "I" will try and move through the long side. The test problem that Marty Fritts came up with really has to be done with these codes, in order to establish whether or not they handle this problem accurately.

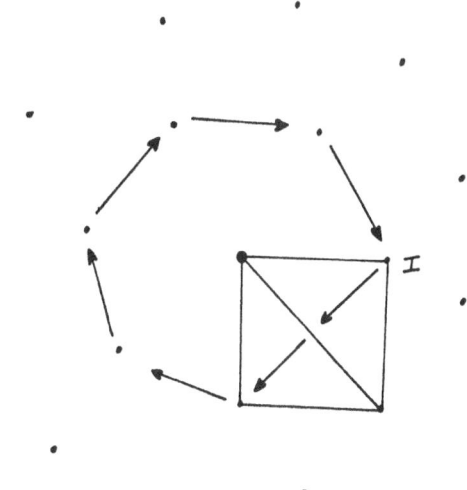

Figure 2.  Viewgraph of Mr. Boris' Vortex Flow Test Problem.

Another problem that these codes should do equally well, is: you have a point vortex. Regardless of the structure, if there is only one point with vorticity in the system, that vertex shouldn't move itself. If there are problems with walls, it's a different issue. In free space, a single point vortex, regardless of the structure around it should not start migrating. Both of these things are combinations of the kind of test problems that all of these schemes should be tried on. If you eliminate those kinds of problem areas, you would be able to have much better behavior of your solutions.

MR. TREASE: I don't understand the problem of moving an interface or testing an interface scheme. The cell compressions we've always been able to handle accurately; it is the interface movement that gives us problems.

MR. BORIS: It is a rotation about a point that is the problem, I believe. Perhaps the Voronoi grid would do this problem correctly. I don't know what the Voronoi mesh looks like. It might rotate without hanging up. What we have seen in some of the calculations is that as one cell gets squeezed up, a neighboring cell gets stretched out as the neighboring point is moving away. The mesh gets screwed up because two points are bunched across the interface with no real ability to relieve each other.

MR. WINSLOW: Let's squeeze them a little bit in the other direction, where these points are closer, so it is like this. If I followed Armstrong's strategy here, that would happen very naturally, the way I exchange (diagonals) as soon as the sum of those angles is greater than 180 degrees.

MR. BORIS: I think that for this kind of thing, if you have a Voronoi grid, it may not happen, but I haven't worked on it. In some formulations of a median based mesh, this is a problem Marty and I worked very hard to get rid of. I would like to quit there, if I can.

MR. WINSLOW: I would like to ask one more question about that. Does it matter in your experience whether the triangles have quantities associated with the nodes or the centers?

MR. BORIS: That is what it is all about. It does matter. We found one scheme which gets around this problem; that isn't to say there aren't others.

MR. WINSLOW: You have a velocity in the zone and other quantities at the nodes, and then you get good behavior?

MR. BORIS: Yes, but you have constraints on how you can update adjacent points in order to maintain vorticity conservation.

MR. FRITTS: I would like to make one note on that example. I think that this test problem is a particularly good example (see Figure 3). It shows two problems with any mesh. The first is that given a problem with an enclosed vortex, you must have a self-consistent velocity field that is rotational. The point is that it doesn't matter where the triangle is connected, as long as you have a velocity which is forcing a point to go through the opposite side. Whether it is a good triangle or bad triangle, it makes no difference. If you have a pressure

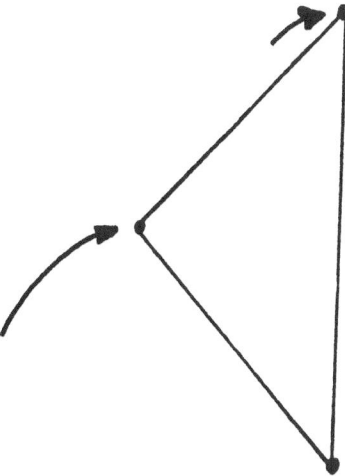

Figure 3.  Viewgraph of a Triangle in a Vortex Flow.

generated anywhere in the mesh that will oppose that particular flow, then you have got a problem with the formulation. It is not a problem of geometry, or of the triangle being badly oriented or badly shaped. It is a problem of the difference schemes and where the variables are centered. You oppose the flow and you have locked yourself into solid body rotation. That is one point.

The other point is that even if you had a scheme good enough to get exactly the right velocities, the vorticity would have to be advanced to the next time step, so that it is consistent with the velocities at that next time step. Even if you can rotate that triangle exactly right, or any triangle within this vortex exactly correctly, you still have the problem that when you update the vertex positions, the velocities carried by the updated vertices will now be in the wrong positions to sum to the correct vorticity. Although you have the positions updated correctly, you cannot correctly conserve all the other problem variables. This is really a good test problem. It covers both aspects at once.

MR. CROWLEY: I think what I've heard from this is we have a test problem that has been suggested by Marty and Jay that may help us sort out the differences of the advantages and disadvantages between different methods.

MR. DUKOWICZ: I think that accuracy is really the issue in this field. After all, we are working with highly irregular grids, and there is a tremendous source of inaccuracy associated with that. We can illustrate this, if we have a problem where we expect the flow to be smooth, a low mach number flow, for example. Invariably, the mesh points or grid points become scrambled, irregular, and the result is that calculations are rather hashy. I'm sure most everyone in this field has had results of that type. We should see if we can do anything about this kind of inaccuracy. I would like to hear any comments from mathematicians in the group on the accuracy of Voronoi versus other meshes. If there are any comments to be had without spilling the beans on talks that might be coming up. Maybe we have worn this topic out!

MR. EISEMAN: I have one question about the generalized Voronoi mesh regarding controlling the cell size. If I'm not mistaken, you are just shifting the interface by moving cell sides. Is there any way to control cell size by doing this? Perhaps if you could control cell sizes, you could have one cell that is very large, and the body of another cell very small, and sort of smooth this?

MR. ARMSTRONG: There are other factors. I'll tell you what cell areas to associate with those nodes. I should admit there is a danger in using the generalized Voronoi mesh, where you can get into a positive feedback loop. This is where you decide you have certain cell areas for your nodes, then you move the interface toward the smaller node and recalculate the area of the smaller node. The cell area can get smaller and then the interface moves further, continuing to recalculate the cell area sets up a positive feedback loop.

MR. GLENN: I want to direct a question to you again, Pat, relative to the mesh that Bob Cooper had up there. We are talking about aspect ratios being very, very large. What do you know about the accuracy of a problem like that; any rules of thumb when the aspect ratio is so and so? Typically, when we see that type of mesh in a Quasi-Eulerian Code, the calculation is garbage.

MR. CROWLEY: I don't quite know how to answer that. Equilateral triangles provide us with the most accurate results. We know how to do the types of problems we do. They would be impossible to run with aspect ratios near unity. We accept what we have, and what I have learned in this meeting that we should be starting to consider some error estimates based on the actual mesh during execution. We should try to optimize the mesh to reduce the error; to try to adapt to a minimum error.

MR. WINSLOW: I can imagine a special test problem with triangles where you know what the answer is. With such a test problem we could evaluate accuracy for different methods and compare with theoretical estimates of accuracy.

MR. ERLEBACHER: It is a very interesting point here, concerning equilateral triangles. You can show most of the time the maximum accuracy is found, but the actual error could be less than maximum estimated errors. Now, an interesting thought in going back to this shock problem. For any surface problem, possibly, the best triangles would be equilateral on the physical surface, not on the plane. After all, if you look at structural analysis, look at people constructing domes with triangle structures, the triangles are all equilateral. If you projected them onto a plane, you could very well end up with screwed up triangles and that comes back to the part of what people are using. You see that the grids are very, very screwed up near the boundary line. If you look at the physical surface and draw the mesh on that surface, you would probably find that the mesh is a little bit more square everywhere.

MR. CROWLEY: I would like to respond to that. I had a further thought in answer to Lou Glenn's question; it depends on what is happening in the problem. If you have large aspect ratio zones, and you have some motion that is going in at an angle to them, you have a questionable solution. But if you have large aspect ratio zones, and the motion is perpendicular to the long axis, you might want to believe that solution. After all, when we think of a nice problem, it has square zones.

MR. WINSLOW: What I'm saying, and it is kind of unambiguous if you do so, there are a lot of simple problems in which you know what the answer is, and it is easy to construct test cases to see what happens.

MR. CROWLEY: I think what I hear you saying, is that if you see nice zoning, you will believe the answer; if you see ratty zoning as Bob Cooper showed, you will question the answer.

MR. WINSLOW: I don't know what the answer is.

MR. CROWLEY: We need help, don't we?

MR. WINSLOW: What I'm saying is that it is easy to check.

MR. EISEMAN: The surface is something we will talk about in more detail tomorrow. But basically, if you have a surface, and you have points on the surface as observers, think of the tessellation of the surface as planar sections that are put together. With that surface you should be doing better than you would otherwise do, considering the surface as projected straight down to the plane below. If you extract a salient property of the solution, and you have a piecewise rendition of that, you would think in terms of what would be the most accurate representation. For example, make a least-squares-fit to that surface so the question is very well posed, and then just about any difference formula you could come up with would be very accurate. I think it is very important to ask the question, where am I looking for my error. I should be looking at the error of the object I wish to approximate, in the physical space.

MR. LÖHNER: In finite elements, we make shape functions, and the question is then how good can the shape functions approximate the function I'm trying to obtain. If nothing happens in the perpendicular direction, why not use long aspect ratio triangles in that direction. We are trying to somehow approximate our function, and if my shape function can be accurate, it doesn't matter whether I have an aspect ratio of one to a thousand. The typical example is the shock; if the shock is completely straight, I don't need to have equilateral elements there. I could have tremendously elongated elements, and it would still be adequate.

MR. WINSLOW: If you send shock waves in the long direction of the element, what happens in your finite elements?

MR. LÖHNER: It starts generating elongated elements, but I could have very thick elements there. The important thing is to keep a small distance across the shock.

MR. COUET: You are saying it's a matter of resolution?

MR. LÖHNER: I have to approximate the actual solution of the shock with my shape functions. That is all I need to do.

MR. ARMSTRONG: Let us ask what effect this subdivision has on accuracy. I have seen problems set up by other people where they start off with a rectangular mesh and divide it into triangles. There is an equally valid choice of which way to triangulate. How much are the results varied according to which diagonal is chosen?

MR. CROWLEY: We eliminated that problem by going to quadrilaterals.

MR. COUET: I've talked about this case with Marty Fritts. The problem that's most difficult to test is that of a point vortex. If you want to go for a 3-D method to follow vortex rings in a periodic domain you can try spectral codes. There is a lot of analysis for 2-D or 3-D, I think, that tells you what kind of accuracy you have on any kind of mesh you are dealing with.

MR. WINSLOW: There are people here who use the Voronoi mesh and people who use the median centered mesh. Is anybody here who has had experience with both types of meshes?

MR. ARMSTRONG: Yes, I have. I performed the calculations with generalized Voronoi meshes, and I must admit getting nicer results with the median mesh.

MR. WINSLOW: Are you talking about solving dynamic problems?

MR. ARMSTRONG: I'm talking about highly dynamic problems.

MR. AUGENBAUM: I have only used Voronoi and from what I've seen of the analysis of the difference operators, the gradient and Laplace operators seem to be all first order. I know from experience the difference operators with the Voronoi mesh are not first order; they are first order in a weak sense. A lot of the problems one has to address have to do with finding more accurate ways of centering the difference operators and possibility finding an element that will give us accuracy. Another comment is that in nonlinear problems, generally, you have oscillating solutions. In Lagrangian methods, you don't have the advection term. The extra degrees of freedom that a point can have, can give high frequency oscillations, which turns out to produce very noisy contours. I don't know if anybody has, and I certainly haven't, developed filters to filter out high frequency noise; to smooth out some of these artifacts, and to get better results.

MR. CROWLEY: Has someone developed a high frequency filter on an irregular mesh?

MR. ERLEBACHER: I generalized gradients, to first order accuracy, as a part of my thesis. I was about to mention a test problem. This test problem seems to be a Voronoi test problem. It is part of my thesis to test my irregular triangle grid. It's one I found while using second-order on a triangular grid. I needed these high frequency filters I developed. I just don't understand why people haven't used that very same problem as a test. It's an ordinary problem, conservative, and it seems obvious people would test that out.

MR. AUGENBAUM: I just wanted to say we are really not interested in Eulerian methods. We are working with Lagrangian methods on problems that I think would be typically solved by Eulerian methods. The other point I would like to make is about the filters. I think that is a very useful idea for some of the calculations that we do; they are hashy, and we know it has to come from nonlinear effects. One way to do that is with a convolution type of a filter. An easy way to implement that type of filter is to realize that convolution is the same as an application of a diffusion equation. You can filter by diffusing for a very short time on the time scale of the diffusion. This is equivalent to a length scale filter.

MR. CROWLEY: Let me add to that. The artificial viscosity is the actual filter.

MR. AUGENBAUM: It is used in every time step. If you want streamlines, you don't want artificial viscosity as a high filter. You want one that can sort of be tuned.

MR. CROWLEY: You can think of the artificial viscosity when you transform as a k-squared effect where k is the frequency. Maybe we should try to hit some of these other topics. Anybody have any strong feelings?

MR. N. JOHNSON: I would like to bring up one point about the Voronoi mesh that is very bothersome to me. It basically deals with numerical techniques that construct a mesh while ignoring the orientation of material interfaces. After you have constructed the mesh, you still have interfaces and you must use the orientation of those interfaces to calculate certain quantities; this is especially true in calculating viscous stresses or elastic stresses. The Voronoi surface that is constructed between two mesh points is not a material interface; the surface always remains perpendicular to the line between the two mesh points. In calculations that I have done, for non-Newtonian fluids or elastic solids, the Voronoi cell produce stresses that are not realistic.

MR. WINSLOW: How about a median based mesh? Would it be more accurate?

MR. N. JOHNSON: I don't know. I have not used that. Do people have experience with three points versus four points versus five points?

MR. SINZ: I would like to respond to the three versus four versus five points. We found four points are faster and more accurate, but we do get mesh tangling with them. I think that when you go to five points, then there is some problem in the acceleration calculation.

MR. N. JOHNSON: It's only when motion becomes severely deformed and starts stretching, then you start having problems.

MR. EISEMAN: I would like to make one comment on that. I think Jay's swapping algorithm just might bring quite a bit of life back in your codes.

I know it will for adaptive curvilinear grids. In your codes, you already have the uniform contributions of the features. You don't have adaptive curvilinear grids when you get to the point near disaster. There you can use swapping algorithms. I would second the motion.

MR. LÖHNER: I would like to bring up another point on this issue. Have you ever experienced locking or other related phenomena when you have compressible flows? When you go to very low mach numbers it is known that the triangular element has certain problems with the locking and so you need to do something about that, whereas the four-node element hasn't got that problem. That would be an answer, from another point of view, to your question.

MR. FRITTS: We solved that problem.

MR. CROWLEY: He solved it in the incompressible case.

MR. FRITTS: Another comment on nodes. To me, it's a little bit unimportant how many nodes are in the polygon. What's more important is the template that the nodes are representing, and how well and in which direction you need information gathered in using that template for a particular problem. And that I think is true for either finite element or finite difference codes. You really have to have the information about neighboring nodes available to you when you advance the nodes. For the nodes in higher order polygons, where you must increase accuracy, the templates are very highly distorted. I think really the attention should focus on what each node "sees".

MR. CROWLEY: Are there other comments?

MR. ROCKENBACH: We have used three- and four-noded elements and n-sided polygons on the diffusion operator. John Dukowicz's four-noded finite element method seems to work very nicely. Harold and I have looked at this to see if there was a definite difference and neither one of us could find any difference. Since there is a little bit of competition going on as to which method we would like to use, you can bet we looked at it very, very closely. I'm just now getting ready to look at doing the finite element pressure gradient terms using Lagrangian coordinates, but I am not far enough along in that.

MR. BORIS: What was your conclusion?

MR. ROCKENBACH: There was no difference. They both work very well.

MR. WINSLOW: Didn't Harold say he found that using the triangular mesh and the Voronoi mesh to be effective, maybe?

MR. TREASE: We were considering using the Voronoi mesh to designate the nearest neighbors, and the median mesh for the hydro.

MR. ROCKENBACH: The idea of using the Delaunay mesh to define your finite elements would be kind of useful.

MR. TREASE: Use of the Voronoi mesh changes when used in conjunction with the finite element formulation. We don't use it during the hydro calculation. We only use it to update the nearest neighbors for the mesh. I might make another comment about that; because we use the Voronoi mesh, we are required to swap neighbors when the mesh says change neighbors. Otherwise, the corners of my polyhedrons start to overlap. With a finite element formulation on the median mesh, you can stretch these triangles a little bit farther than I normally would with the finite difference formulation on the Voronoi mesh.

MR. N. JOHNSON: I would like to switch to the 3-D discussion if nobody objects.

MR. KIRKPATRICK: I could come up and give a brief presentation.

We are talking about the business of the sum of the opposite angles of a quadrilateral being 180 degrees as a reconnection test. It turns out that this notion is for 2-D, but can be expanded to 3-D. Another test one can perform is, first, if you have three points, as Armstrong said, a circle that passes through these three points defines the center of that circle. Then you draw the perpendicular bisector, between some fourth other point, that you are interested in, and each of the other three points. This perpendicular bisector will pass on one side or the other of the center of the circle. That tells you whether or not the other point is an interloper. For the three points that define the circle in two dimensions, you have a simple test by taking a dot product between two vectors. These vectors are drawn from the center of the circle to the midpoint on the line between the fourth point and each of the other three points of the triangle. This test determines whether or not you are going to have another point that is going to cause a reconnection. Another way of looking at this is simply to ask whether or not the point is inside or outside of the circle.

This generalizes for three dimensions, too. Here we have four points on the surface of a sphere, and then having again a bisecting plane in relation to the center of the sphere. In 3-D the test for changing neighbors depends on whether or not other points are inside or outside of the sphere defined by the other four points. That again gives you a very simple test. If you already have a connection between the four points, it gives you the apexes of a possible polyhedron. All you have to do is find the distance from the center of the sphere to each of the apexes of the polyhedron and the distance from the center point to some other point. If it is greater than the distance to any of the points defining the sphere, then it cannot be an interloper. This simple test should be very efficient. I was surprised to see the complicated test involving angles and normals to surfaces, and so forth. I was not aware that a 3-D test could be done involving angles the same way as you have for the two dimensional case. In any case, this seems to be considerably simpler than going through the other type of case to find out if a point was inside or outside the sphere. You only have to calculate where the center of the sphere is. Frequently, many other points can be determined to be known interlopers because they are now inside of the sphere. In the initial generation phase, it is still rather simple. I wholeheartedly endorse Dr. Armstrong's idea of initial mesh generation by what he calls growing crystals, that is, finding a first tetrahedron and continuing to find other tetrahedrons. You can do this in just about any order and then go back and sort the neighbors in a fairly efficient fashion.

EDITOR'S NOTE: The rest of Mr. Kirkpatrick's comments on nearest neighbor calculations can be found in his paper that has been included in the proceedings as an appendix.

# SOLUTION ADAPTIVITY USING A TRIANGULAR MESH

Peter R. Eiseman
Department of Applied Physics and Nuclear Engineering
Columbia University
New York, NY  10027

## INTRODUCTION

Solution adaptivity will first be discussed from a general perspective and then from the specific vantage point of triangular meshes. The generality first establishes a common format from which a variety of adaptive methods can be considered. The specific use of triangular meshes provides a very flexible means for simulating phenomena with dynamically changing configurations. The basic work with triangles stems from joint work with Gordon Erlebacher who has presented different but related material in his paper at this conference. Rather than a presentation of only past work, I have taken the liberty here to suggest new strategies for improvement and to describe the consequent operations.

The numerical simulation of rapidly varying phenomena is considered with meshes which adapt to the disturbances in the phenomena. With clear limitations on the total number of mesh points, the adaptation is some strategy which places the points in locations from which the phenomena is accurately observed and hence well-resolved. The locations for enhanced resolution conform to the disturbances. These typically occur at unknown places, have varying intensities, and may also evolve temporally.

To establish a general perspective from which a suitable adaptive mesh strategy can be developed, the basic adaptive data is first consolidated into the form of an abstract surface which we call a monitor surface. It is formed by extracting solution quantities which contain the disturbances and then by evaluating the assembled result at each existing mesh point to trace out a surface that suitably monitors the solution behavior.

Once established, the monitor surface must be accurately represented to obtain a concurrent accurate representation of the solution. While the monitor surface can often be taken as the solution itself, we note that the monitor surface is usually more simply defined, can be more readily adjusted to sensitive parameters, and can even be formed from error estimates arising out of a solution procedure. With the objective of accurately representing the monitor surface, all adaptive mesh strategies are reduced to mesh generation strategies for surfaces and their possible temporal evolution. Whether the surfaces are viewed by looking up at them from physi-

Research supported by NASA Grant NAG1-479

cal space or by directly examining them as geometric entities, the adaptive strategies are now posed as entirely geometric strategies. The resolution then depends
upon geometric quantities.

To resolve a given geometry, the mesh is altered by pointwise movement, by local changes in the number of points, or by a combination of both movement and such
changes. Each method of alteration then corresponds to a particular adaptive strategy — albeit, the many possible variations in coupling the altered mesh to a solution procedure. Pointwise movement can be applied to both regular and irregularly
structured meshes. On application to curvilinear grids, the inherent regularity
must be preserved for otherwise it folds and thus produces a discrete singular
transformation. As a consequence, the movement of such grid points must be executed
carefully. The degree of care is further amplified by demands for a good structure
which often translates into a demand for only small deviations from orthogonality.
By contrast, when movement is applied to a general connectivity triangular mesh that
can be dynamically restructured, the level of care is substantially reduced. This
is, of course, balanced by the need for a suitable restructuring procedure. The
distinctive advantage here is that the connectivity pattern can dynamically adjust
to a dynamically changing topology while maintaining a good structure in each local-
ity — while not increasing the complexity of the data base. The price for the data
base is assumed initially and is not altered by movement. While changes in the total number of points may appear to more significantly cause extra expense, the use
of linked lists help to minimize such expenses. As a consequence, the general connectivity triangular mesh provides us with a good tool for dynamically changing topologies in two-dimensional simulations. A similar resolution scheme can be done
with curvilinear grids by providing local refinement in the form of locally embedded
fine subgrids or more generally overlapping local grids. While the embedded case
has dangling coordinate curves, the overlapping case generally has no alignment
whatsoever with the major grid or grids should there be multiple layers. In either
case, the boundary conditions must be treated to appropriately transfer solution
data between grids. When the data transfer problem is done well, the curvilinear
grids also provide a good tool for dynamically changing topologies. For a
discussion of this strategy, see Steger et al. [1].

To maintain a single common data base for such problems, we shall emphasize the
use of a general connectivity triangular mesh. Our discussion will concentrate on
adaptivity as outlined above and will start with the development of monitor surfaces
and their geometric properties. These will be expressed relative to the triangular
mesh. Then we will proceed to movement, to restructuring, to changes in the number
of points, to locally regular structures, and finally to show some computed results.

THE MONITOR SURFACE

Rapid spatial variations in physical processes often occur in only a small number of physical quantities which can be readily identified. The quantities are not necessarily selected only from the solution vector of the system of partial differential equations used to describe the process. More generally, they are functions of the solution vector and its derivatives. With each selected quantity, there is usually a clear and direct physical interpretation which gives the observer a firm intuitive base from which to work. If there is more than one such quantity, then the observer must simultaneously view each to see the separate evolution of disturbances. In order to simplify matters, it is desirable to view only one object which contains all of the disturbances. This can be established by forming some function of the various quantities which provides a reasonable representation for each and which is appropriately scaled relative to the physical region. When the function is evaluated at each point in the physical region, a surface is formed over the physical region. This surface monitors all of the possible severe solution behavior and is appropriately called a monitor surface [2].

In a direct manner, the monitor surface can always be generated by a function which is just a vector of the appropriately scaled or normalized physical quantities under consideration. The surface has the same dimensionality as the physical space and is embedded in a Euclidian space with an increase in dimensionality equal to the number entries in the vector. The increased dimensionality is also the dimensionality of the field of normal vectors to the surface. Since the normal fields are the key elements employed in the examination of the surface geometry, a reasonable simplification can be expected when the normal field dimension is as small as possible.

To reduce the dimensionality, we combine distinct physical attributes by forming suitable scalar functions of the corresponding physical quantities that were considered as the components of the single vector above. The functions are suitable when all of the spatial disturbances in the given quantities are preserved, when the relative scaling provides a proper balance between the various strengths of the disturbances, and when the aggregate scaling is commensurate with the overall lengths of the physical region. The simplest such functions are linear and have coefficients that satisfy the desired scaling conditions. When the monitor surface is generated from a vector with the scalar functions as entries, a collapse in dimensionality is clearly obtained for the embedding space. When all of the physical quantities can be suitably combined by means of a single scalar function, the collapse is maximal and the embedding space is only one dimension higher than the monitor surface. This maximal collapse is possible in most physical situations and produces a surface geometry that can be measured by a single unit normal vector field. As the surface is traversed in uniform increments of arc length, the single normal field gives the motion of the corresponding tangent planes attached to the succes-

sion of surface points in our path. For two-dimensional surfaces, the motion determines normal curvatures which in turn yield mean and Gaussian curvature.

With the definition of the monitor surface, the problem of solution adaptivity
is reduced to the geometric problem of most accurately representing the surface and
its evolution by using a finite collection of mesh points. When a triangular mesh
is employed in physical space and the normal field is fully collapsed, the monitor
surface is defined by a scalar function and is represented by evaluations at the
physical mesh points together with the connectivity pattern vertically lifted from
the physical space. An illustration of this relationship is displayed in Figure 1.
Here the vertical projection of the surface mesh (1a) is just the physical space
mesh (1b).

Since the evaluations use the data generated from a solution algorithm, a
smoothing step or steps (in the form of a numerical filter) may be required to provide a smooth surface. This would prevent small solution-generated fluctuations
(i.e., wiggles) from appearing and would yield a monitor surface where only the fundamental physical disturbances are present. In cases where the solution is actually
discontinuous, the smoothing would slightly smear the discontinuities and represent
them as continuous disturbances in the monitor surface. Here, it is important to
note that smoothing is _not_ applied to the solution itself for that may lead to an
inaccurate simulation. Smoothing is simply applied, when needed, to ensure that the
monitor surface is continuous enough to permit calculations of derivatives and curvatures which are accurate and smoothly distributed. As a consequence, we can always assume a smooth monitor surface, and thus, can test the various strategies for
pointwise placement by using virtually any smoothly defined function in place of the
assembled solution data. This was done in the case of Figure 1.

With curvilinear grids, similar smoothly defined functions were also used to
examine the effect of curvature attraction. In an example, an artificial disturbance was placed in front of a biconvex airfoil and was defined to temporally deform
from a vertical line into a parabolic arc over the airfoil. This was simply given
by $z = \frac{1}{2} \tanh 3r$ with $r = x - ty^2 + 1.5$. From a physical perspective, hyperbolic
tangents arise in analytical solutions to Burger's equations which provide elementary views on the effect of viscosity (Reynolds number). Noting that the center of
such disturbances lie at the inflection point $r = 0$, similar disturbances in more
general spatial configurations are readily prescribed with r as a function of x and
y as done here. It should also be noted that multiple disturbances can also be specified by linear combinations of such hyperbolic tangent forms. In parallel trials,
the grid was adapted to the evolving parabolic disturbance. The first trial sequence contained no curvature attraction and is displayed in Figure 2. The second
sequence used normal curvatures for attraction and is displayed at corresponding
frames with the first in Figure 3. In each sequence, only the physical space grid
is viewed: the surface grid is not. The tracking and clustering were accomplished

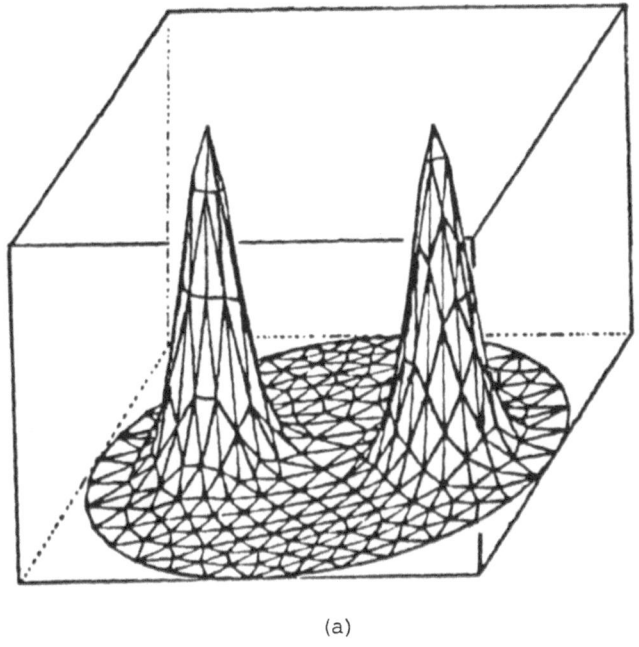

(a)

(b)

<u>Figure 1</u>:  A simulated monitor surface (a) and its projection (b)

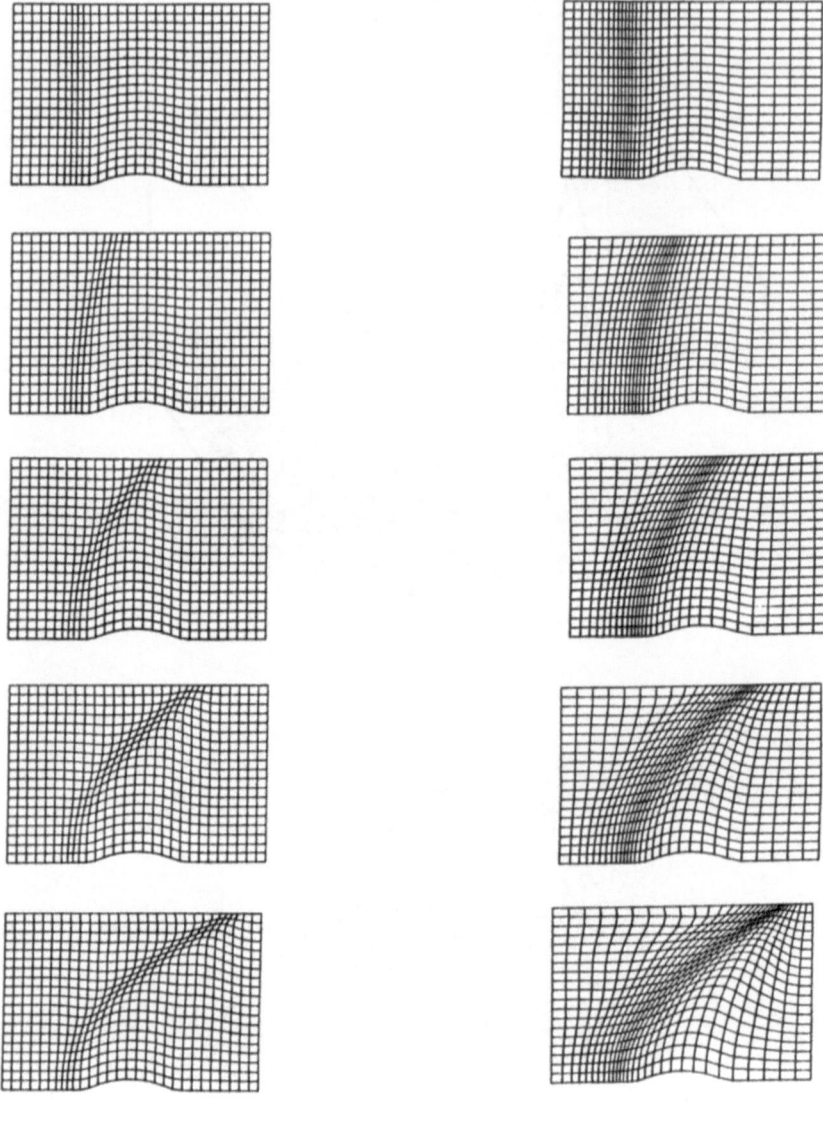

Figure 2: The resolution of
a moving disturbance without
curvature attraction

Figure 3: The resolution of
a moving disturbance with
curvature attraction

with the alternating direction adaptive grid method [3]. On examining the two cases, the effect of curvature attraction is explicitly seen to be a broader band of resolution together with a greater degree of grid alignment. The broader band comes from the high curvature regions adjoining the high gradient region. While the first sequence resolves only the gradients, the second also picks up the adjoining regions where the surface folds into and out of the high gradient. The alignment comes because curvature attraction tends to pull coordinate curves along such folds. The alignment, however, is limited by the choice of grid topology. In a similar case with a circular disturbance and a Cartesian topology, the alignment occurred with coordinate curve segments rather than entire coordinate curves. In the context of a general connectivity triangular mesh, a full alignment should generally be available. This will be accomplished by using a curvature sensitive restructuring algorithm together with movement.

MESH POINT MOVEMENT

The discrete representation of the monitor surface can be improved with the currently available points by moving them into positions from which the observation of surface geometry is also improved. The concurrent improvement in the numerical simulation of the physical problem then follows from the definition of the monitor surface. To accurately represent the surface, curvature clustering must be applied relative to a uniform distribution of surface points. In fluid dynamics, the consequent motion is neither Lagrangean or Eulerian but instead is an adaptive movement that is generally somewhere in between these two cases.

In the context of a general connectivity triangular mesh, a movement strategy is developed by creating a local weight balance about each point and then by applying it iteratively. The balance is executed about physical space mesh points and is determined only by the directly connected neighbors. For a point P, the direct neighbors form the vertices $v_1, v_2, \ldots, v_n$ of a surrounding polygon which is illustrated in Figure 4. Also illustrated is the connectivity that defines the triangles attached to P. Each triangle i is considered to have a weight $w_i$ which for the purpose of balancing against the other triangles is considered to be concentrated at its barycenter.

$$b_i = \frac{1}{3} \left( P + v_i + v_{i+1} \right) \tag{1}$$

The various barycenters are displayed in Figure 4. The relative force balance between the barycentric weights gives the change in position.

$$\Delta P = \left\{ \sum_{i=1}^{n} (b_i - P) w_i \right\} / \left\{ \sum_{\ell=1}^{n} w_\ell \right\} \tag{2}$$

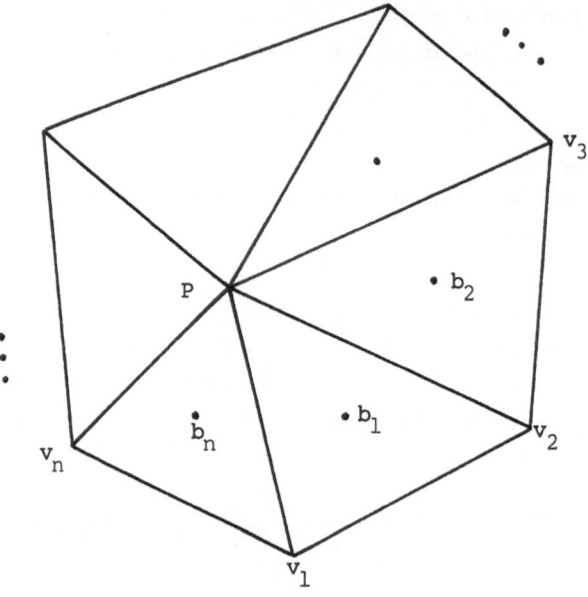

Figure 4: The movement stencil

Upon addition to P, this gives the new position

$$P_{new} = \frac{\sum\limits_{i=1}^{n} b_i w_i}{\sum\limits_{\ell=1}^{n} w_\ell} \qquad [3]$$

which is recognized as a center of gravity expression for the polygonal disk.

The control over the motion comes with the choice of weights. Since clustering is generally desired relative to a uniform distribution of points on the monitor surface, the weight is assumed to contain a factor of incremental surface area and to reduce to such areas when no clustering is specified. For simplicity, the weight is assumed to be a linear combination of attracting quantitites $M_i$. For an arbitrary triangle, it assumes the form

$$w = \left(1 + c_1 M_1 + \cdots + c_m M_m\right)A \qquad [4]$$

where A is the area of the triangle on the <u>surface</u> and the coefficients $c_j$ give the relative level of importance for each quantity $M_j$. By using the surface triangles, the gradients that are viewed from physical space are implicitly contained and are thereby resolved. Globally, this resolution comes from w = A and the iterative relaxation into a uniform surface mesh which projects down onto the mesh in the physical region. Because of the linearity, the coefficients in Eq. 4 can also be readily converted into fractions of the total number of points [3] – [4]. This permits a rational way to decide upon their magnitude [2].

To cluster points in regions where the surface bends or folds, a measure of curvature is needed. The basic measure comes from normal curvatures which were directly used in the alternating direction adaptive method for curvilinear grids [3]. For triangles, it is simpler to consider either the mean or Gaussian curvature which are respectively the mean and product of the principle normal curvatures. The selected measure must increase for the steep fronts which commonly occur and which are typically characterized by a large principle curvature in the direction of propagation and a small or vanishing one along its length. In these occurrences, mean curvature increases while Gaussian curvature is very small or vanishing. As a consequence, the mean curvature is preferred. Using the scalar function f and the Monge patch $(x,y,f(x,y))$ to represent the monitor surface, the mean curvature is given by

$$H = \frac{(1+f_y^2)f_{xx} - 2f_x f_y f_{xy} + (1+f_x^2)f_{yy}}{2\left[1 + f_x^2 + f_y^2\right]^{3/2}} \qquad [5]$$

where subscripts have been employed to denote derivatives. Since gradients and Laplacians are most easily constructed in the physical space triangular mesh, an approximation using these two operators is sought. This leads to the curvature measure

$$K = \frac{\nabla^2 f}{2\left[1 + (\nabla f)^2\right]^{3/2}} \qquad [6]$$

which agrees with H when f is 1, x, y, $x^2$, and $y^2$ but not xy because of the missing cross-term in the numerator. With m = 1 and $M_1 = |K|$ in Eq. 4, the weight for the resolution of surface curvature is given by

$$w = (1 + c|K|)A \qquad [7]$$

The barycentric value of K is used here and is taken by averaging the vertex values. When the boundary of the physical region needs resolution, a boundary curvature $M_2$ must also be included by taking m as 2 in Eq. 4.

In addition to attracting quantities such as curvature, weight terms can be inserted to improve the structure of the triangles. Erlebacher [5] – [6] has shown that such improvements tend to enhance the accuracy of the discrete operators. In particular, the error bounds are smaller when no vertex angle is permitted to become excessively small. As a consequence, a structural restoring force is desired when an angle becomes too small. One method to apply the force is to include the weighting quantity

$$
M_3 = \left\{ \begin{array}{ll} (\theta_c - \theta)/\theta_c & \text{for } 0 < \theta < \theta_c \\ \\ 0 & \text{for } \theta_c < \theta < \pi \end{array} \right\} \qquad [8]
$$

by increasing m to 3 in Eq. 4. This causes a linearly increasing force which goes into effect when an angle from P in the movement stencil of Figure 4 falls below a critical smallest permissible angle $\theta_c$. A reasonable choice for $\theta_c$ would be $\pi/3$ or some other nearby angle. To enforce the angle condition without the effect of area weighting, the corresponding coefficient $c_3$ must include a factor of 1/A to remove the area multiplication of Eq. 4.

While the error bounds were derived only with respect to physical space, the major enhancement of accuracy may actually come from an improved structure on the monitor surface rather than in the underlying physical region. One reason for this suspicion comes from the observation of boundary layers attached to a wall. Here, the monitor surface would locally be determined by velocity magnitudes that fall from free stream conditions to zero over a small distance. In physical space, clustering requirements towards the wall would be much more intense than along the wall. The result would be many long thin triangles slammed against the wall, and from our previous analysis, a sequence of meaningless error bounds. By contrast, the corresponding triangles on the monitor surface would not have that severe thinness. Instead, the rapid vertical rise in the monitor surface would be covered by reasonably shaped triangles. As a consequence, the quality of the surface mesh should be emphasized, and in correspondence, the angles for the structural quuantity of Eq. 8 should be taken from the surface triangles about the lifted P.

Wherever it is applied, the structural part of the movement acts only for an improvement when the mesh connectivity pattern is fixed. To obtain further structural advantages, the pattern must be dynamically changed. With such changes, the reliability of the center of gravity movement scheme is enhanced because the pattern continually adjusts to maintain convexity in the gross features of the polygonal cells about each mesh point. This causes each new position to fall within the polygonal boundary. In the case of Voronoi cells, such convexity is automatically assured.

RESTRUCTURING

One of the major advantages to using triangular meshes in free Lagrangian simulations has been the capability to dynamically restructure the connectivity pattern among the moving nodes. With the restructuring, the simulations could continue without the diffusive transferral of data between poorly structured Lagrangian meshes and well-structured Eulerian meshes. Crowley [7] introduced the idea of restructuring by considering the quadrilaterals formed by adjacent triangles and then by flipping the diagonal if that would improve the structure. Fritts and Boris [8] continued the development and arrived at a flipping condition which provided them with a diagonally dominant discrete Laplacian operator. With additional discrete Laplace operators, Erlebacher [6] and Erlebacher and Eiseman [9] considered yet another flipping condition. This was applied to maintain the same structural integrity for the more general adaptive motion.

In all of the previous studies, the restructuring operation has been confined to diagonal flips that are determined by the quadrilateral geometry in the physical region. As such, no information about the solution was used. To obtain a potentially substantial advantage over previous developments, a restructuring strategy that adapts to the solution is suggested. The objective is then to improve the structure of the monitor surface mesh rather than its projection down onto the physical region.

On the surface, a good structure is determined not only by the relative lengths but also by the accuracy with which the surface is geometrically represented. With the choice of diagonal, there are two possible surface patches that are separated by the tetrahedonal volume determined by the four vertices. A geometric selection would choose the one which conforms most closely to the surface. The situation of primary concern would be the accurate representation of sharp fronts. These appear as sharp folds in the monitor surface where, for example, a sudden drop occurs across some rather mildly varying curve. A quadrilateral on the edge of the drop is depicted in Figure 5 along with the two possible diagonals. In Figure 5a, the diagonal clearly cuts through the front while that in Figure 5b follows it. The best

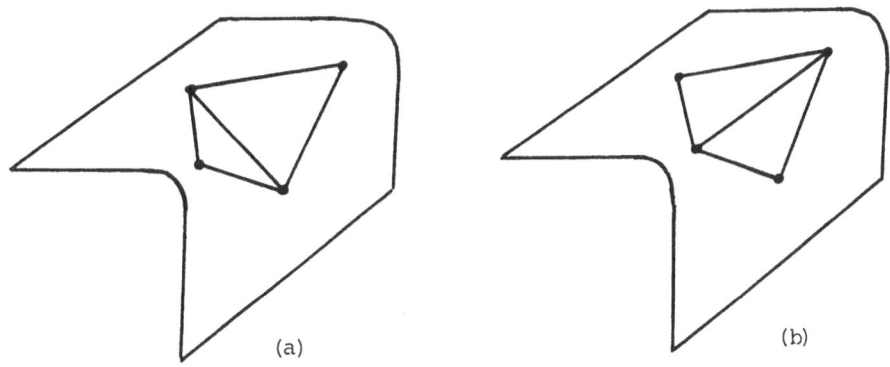

(a)                                                      (b)

Figure 5: A geometric choice of surface diagonals

surface representation is clearly given by Figure 5b which is also more properly aligned. To automatically make this choice, the change in surface direction is computed along each diagonal and then the diagonal with minimal change is selected provided that there is a sufficiently large disparity between the changes. If no such disparity arises, then there is no concern about cutting through or aligning with some front of a disturbance. As a consequence, the choice of diagonals is then determined solely by comparisons of relative lengths.

For the geometric part of the diagonal selection process, the surface direction must be computed at each vertex of the quadrilateral frame on the surface. The surface direction of any point is defined by the inclination of the tangent plane attached to it. With surfaces embedded in three dimensions, each tangent plane is determined by a single normal vector. To examine all vertices on an equal basis, the normal vectors must also be of unit length and be oriented to point on the same side of the surface. By computations in the mesh on the physical region, the normals are computed by evaluating the gradient with a numerical operator and then by scaling it to unit length. The total change in surface direction between the endpoints of a diagonal is just the difference between the respective unit normals. A major part of this change, however, could be merely a rotation about the given diagonal, and thus, would not contribute information to determine whether or not the diagonal cuts through a front. Instead of the total change, the appropriate measure is then its projection onto the associated diagonal direction to give the change <u>along</u> the diagonal. This is computed by forming the dot product between the difference of surface normals and a unit vector along the diagonal. The geometric comparison between diagonals is then based upon the absolute values of these dot products. When the difference between values exceeds a specified critical amount, the diagonal with the smallest value is selected provided that the corresponding diagonal in physical space lies in the quadrilateral. This method of selection is essentially based upon taking the diagonal with the smallest normal curvature. A more full discussion of normal curvature and its application in grid generation is given by Eiseman [2], [3].

When the disparity in normal curvatures falls below the above critical amount, the judgment is made on comparisons between lengths. The first step here is to compute the parametric locations where the two surface diagonals are most closely spaced. This occurs at the points connected by the line segment which is perpendicular to both diagonals. To maintain a uniform comparison, each diagonal is parameterized by a variable which is linear and assumes values from 0 to 1. Next, each diagonal is partitioned into a generous mid-range and its outer extremities. For a parameter $\lambda$, a reasonable definition for the mid-range is given by $0.2 < \lambda < 0.8$. In a symmetric fashion, this leaves 20% of the diagonal length on each end for the extremities. The selection procedure is now given by the two conditions:

1. If the parametric locations of closest spacing fall respectively in the mid-

range of one and the outer extremity of the other, then select the diagonal with
the outer extremity location.

2. Otherwise, if the diagonals have different lengths, then choose the shortest. If the lengths are equal, then keep the current diagonal.

Implicit in the application of these conditions is the same feasibility requirement
noted earlier that the chosen diagonal in the physical plane must lie in its quadrilateral. The first condition here is just a check to prevent an excessively thin
triangle caused by a diagonal which crosses through the extremity of the other in a
well-centered manner. When either both cross at extremities or in the mid-range,
this strategy does not work. The result is the improvement obtained with shortest
lengths in the second condition. These comparisons of relative lengths are lifted
versions of previous comparisons executed in the physical region by Erlebacher [6]
and Erlebacher and Eiseman [9]. By lifting this part of the restructuring process
to the monitor surface, the mesh quality is improved on the surface where it would
seem to be of more benefit rather than on the physical region where certain error
bounds may be improved. It must be noted that the error bounds are done for arbitrary functions rather than specific ones. The benefit is then anticipated because
the mesh is improved on the specific object in need of resolution.

Altogether, a solution adaptive restructuring process is formed by the successive comparisons of normal curvatures and then relative lengths. When further
coupled with the previous adaptive movement and with methods to adaptively alter the
number of points, an extremely high degree of conformity is obtained between the
mesh and the essential solution behavior.

CHANGES IN THE NUMBER OF NODES

The number of nodes in a triangular mesh has been changed by various investiga-
tors for the expressed purpose of providing a suitable refinement for a physical
simulation. The resulting sizes of the triangles are typically adjusted to give
resolution only in certain desired places while other regions are still relatively
coarse. In the pure Lagrangian context, Fritts and Boris [8] consider node addition
by putting the new node on an existing edge and then by connecting it to the opposing vertices of the adjacent triangles which share the edge. As a consequence, new
edges and triangles are established and may need some restructuring to yield good
shapes. An illustration is given in Figure 6a. In the general adaptive setting,
Erlebacher [6] and Erelebacher and Eiseman [9] considered node addition by putting
the node at a triangle barycenter when refinement was required. An illustration is
given in Figure 6b. By restructuring with an adjacent triangle, the result can become close to that of Figure 6a. The closeness, however, is not exact. The basic
distinction is that the angles from the barycentric addition are smaller than those

from the midpoint addition. A single diagonal flip does not remove this distinction since two of the smaller angles would remain. While two additional flips would be necessary, the effect on other adjoining triangles may not be desirable. A further method of refinement is to connect triangle midpoints to form a uniform dissection into four similar triangles of the same size. This type of refinement has occurred with finite element applications, and is illustrated in Figure 6c. To fit this into the general connectivity pattern, an additional vertex connection is required for

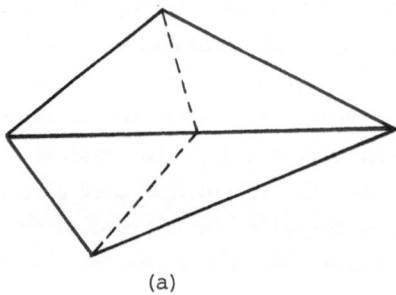

(a)

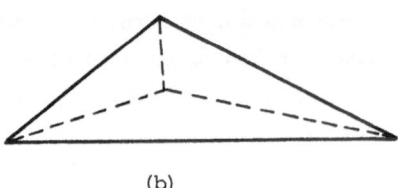

(b)

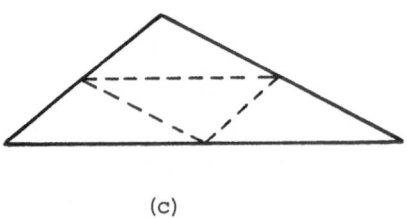

(c)

Figure 6: Mesh refinement by
        (a) vertex connections to a new edge point
        (b) vertex connections to a new barycenter point
        (c) connections between three new edge points

each edge, thereby adding an extra link from each opposing vertex.

In the previous developments, the number of points were changed to increase the resolution rather than to improve the mesh structure. In fact, the mesh structure often suffered since small angles would result and sometimes could not be successfully restructured. This can happen because the associated diagonal flips may cause smaller angles elsewhere. What is then desired is to be able to provide resolution while maintaining a decent structure. With this in mind, attention is first shifted from triangles to cells. Since cells are centered about existing mesh points, the operation under consideration must be some sort of action concerning these points and certain connections from them. When the number of connections from a point deviates from six, the possible angles for the attached triangles must also have a deviation from 60° which means that there must be a deviation from a decomposition into equilateral triangles. The equilateral triangles are preferred because the accuracy of difference operators is usually enhanced [5]. While the computation of angles is directly executed in the physical region, the computation on the monitor surface involves an extra step: the entire cell must first be projected onto the tangent plane attached to the cell center. In either the physical space or the tangent plane, there is only 360° to subdivide which naturally leads to the concern about deviations from six connections. As a consequence, a method to alter the number of connections is sought. Once established, the objective is then to minimize the deviations from six.

The method proposed here is based upon the successive application of two elementary operations. These shall be respectively called the fission and fusion of cells. They are simply the inverse operations of breaking up a single cell into two cells and of combining two adjacent cells into one cell. In correspondence, the fission operation of breaking up cells is a method of point addition while the fusion operation of combining cells is a method of point subtraction. As an illustration, an octagonal polygon is depicted in Figure 7 in two configurations that are related by the fission and fusion operations. In Figure 7a, there is a single central cell about the point c together with partial cells about vertex points 1 - 8. In Figure 7b, there are two central cells about points a and b. The respective cellular boundaries are depicted in Figures 7c and 7d.

To break a cell into two pieces, a divide line must first be established for the break. This is represented by the dashed line 1c5 in Figure 7a which separates the vertices 2, 3, and 4 from 6, 7, and 8. By viewing the divide line as two coincident curves, the point c is the common location of the respective segment centers. Under the assumption that the centers are always connected, the line segment between them continuously increases from zero length as they are pulled apart. This gives a continuous evolution into the form displayed in Figure 7b where the two centers at c have continuously migrated into the new locations at a and b. Under the given action, a cut along the divide line has been made and then opened up to produce two

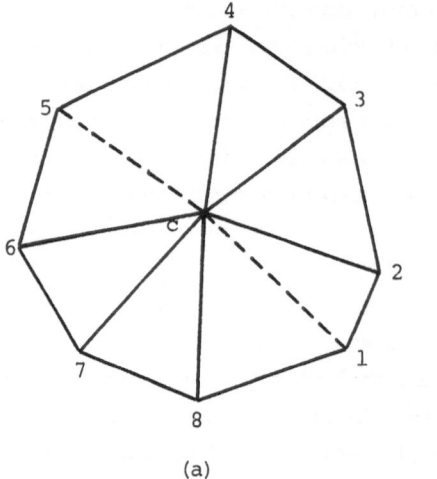

(a)

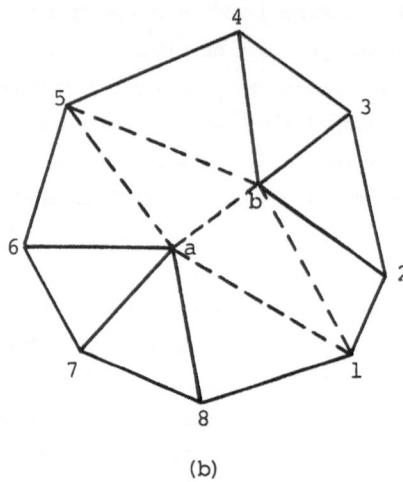

(b)

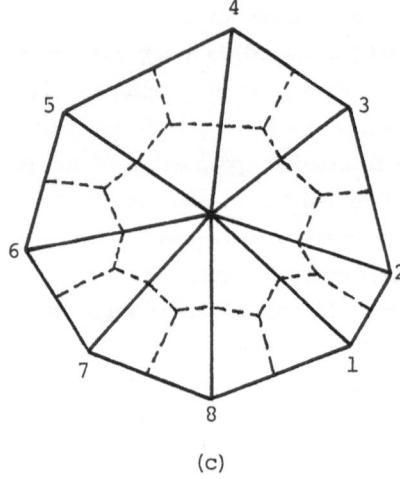

(c)

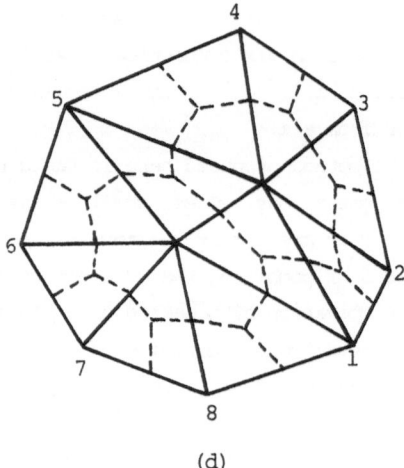

(d)

Figure 7: A polygon decomposed with
(a) one central cell
(b) two central cells
(c) cellular boundaries for (a)
(d) cellular boundaries for (b)

additional triangles, three additional edges, and one additional point. The extra point yields an extra cell which is clearly seen as the splitting of one cell into two. The corresponding transition in going from Figure 7c to 7d is clearly a continuous process with respect to areas. Accordingly, the transition process yields an increase in cell size for each side of the divide line and this diminishes the disparity in size when compared with the original unsplit cell. To minimize the size disparity, the divide line must be chosen to cut the cell into two parts of almost equal size. The approximate size matching for each side produces a more gradual change in cell sizes which can be expected to enhance solution accuracy since the solution procedure is then more smoothly coupled with neighboring cells.

In addition, to control the number of connections to the new cell centers, the orientation of the divide line must also be chosen to adequately separate the number of vertices in the original stencil. With a separation into an equal number of vertices for each side, the divide line 1c5 in Figure 7a was opened up to produce two hexagonal stencils from a single octagonal one. In so doing, it should also be noted that the vertices 1 and 5 at the ends of the divide line each acquire an extra connection. As a consequence, the selection of a dividing line should also biased towards end point vertices that have a small number of other connections. In the more general case with 2n vertices surrounding a central vertex, a divide line that equally separates the vertices will open up to produce two interlocked stencils that each have n+2 vertices surrounding a central one. With 2n+1 surrounding vertices, the most even splitting will similarly produce stencils that have n+2 and n+3 surrounding vertices for the two respective centers. In the most general splitting, if a divide line separates the number of surrounding vertices into i and j for the respective sides, then the two resulting stencils have i+3 and j+3 surrounding vertices respectively. The addition of 3 represents the uncounted endpoints of the divide line together with the opposing center. By iteration, an arbitrarily large number of connections to a vertex can clearly be reduced to a small number. In a global sense, an iterative cycle can be expected to produce node connections that vary in number from five to seven, and thereby, leave open the opportunity to produce a nearly equilateral tesselation.

Upon viewing the operation of opening up slits along dividing lines, note is given to the arbitrary nature with which the operation is accomplished. This nature arises because the old center and the two new centers are given arbitrary locations. As a consequence, when a slit is opened up and then collapsed, the original configuration is generally altered by a simple translation of its center. Such translations imply that the opening and collapsing operations are not inverseses of each other unless the centers are suitably determined in each instance. The natural and adaptively consistent determination comes from the placement of each center at the center of mass of its corresponding stencil. In the opening operation, an iteration of the adaptive movement algorithm (Eq. 3) is required to obtain the locations a and b in Figure 7b or more generally to obtain the two new centers for arbitrary cases. In the collapsing operation, a single application of movement from Eq. 3 is all that is necessary. Under the plausable assumption that centers of gravity are the present locations and are to be maintained, the opening and collapsing become specific elementary operations which are inverses of each other and which can be reasonably called cellular fission and fusion respectively. The inverse is clear since now the successive application of one and then the other does not change the mesh.

In the context of a general connectivity polygonal mesh, the opening and collapsing operation was also suggested by Crowley [10] at this conference. He called them CREP and MERGE respectively and applied them in a different manner.


LOCALLY REGULAR STRUCTURES


While a general connectivity triangular mesh is advantageous for simulations with dynamically changing configurations of disturbances, it is disadvantageous with regard to the required data structure. This greatly limits the number of mesh points that can reasonably be employed; and thereby, depreciates the basic advantage as the configurations demanding resolution extend beyond a certain amount of complexity.

The use of locally regular structures is one method to offset the data structure limitation and to thereby obtain an enhanced range of application. This is important since a solution algorithm could then be used in cases which may start with a relatively modest amount of complexity and end with an exceedingly detailed pattern of disturbances. The basic strategy is to maintain a modestly sized moving mesh which should be enough to detect a wide range of patterns and then to insert within each moving triangle a regularly structured fine mesh to resolve the detected disturbances. Here, the unrefined moving mesh is considered to be the coarse mesh. With the fine mesh being defined relative to the coarse mesh in a regular manner, a rather modest addition to the data structure is required. The main expense is minor and only integer arithmetic is required to identify the nearest neighbors within each coarse mesh triangle. Unlike the general connectivity coarse mesh, there are no arrays to identify the constitutive parts of the connectivity pattern.

In the most regular instance, each local refinement is accomplished by successive connections between midpoints of triangle edges. The process starts with the coarse triangles, connects midpoints to reach the next refinement level, and continues until the desired level is reached. For any triangle, the midpoint connections are linear as illustrated in Figure 6c and the consequent subdivision results in four identical triangles, each with one-quarter of the original area. As a consequence, a seemingly modest three-level refinement decreases the area by a factor of 1/64 which is clearly not that modest. Moreover, with a modest three hundred triangles for the coarse mesh, the three-level refinement produces a 19,200 triangle

fine mesh without the price of detailed connectivity relationships that would otherwise be stored and used.

Aside from the dramatic reduction in area, the intrinsic structure of the refinement is also noteworthy. This shall be observed by examination of the two-level
refinement depicted in Figure 8 which is displayed within two coarse triangles. The
coarse triangles are abc and abd and share the common edge ab. The first observation is that the (nearest neighbor) stencils about all non-coarse vertices are hexagonal. This is clear for both edge vertices such as 1, 2, 3 and internal vertices
4 - 9. The other edges such as ac also yield the same hexagonal structure unless
they represent physical boundaries in which case they each truncate to a half-hexa-
gon. The hexagonal structure, however, cannot be assumed for the coarse vertices
because of the more general connectivity pattern of the coarse mesh. The next observation is that the hexagonal structures are precisely identical within each
coarse triangle and are partially so along the edges. This is witnessed by noting
the hexagonal congruence of the stencils centered at vertices 4, 5 6 and similarly
for 7, 8, 9, while half of each congruence appears on its corresponding side of the
edge ab for points 1, 2, 3. As a final structural observation, the predominant
stencil is hexagonal. If the mesh section displayed in Figure 8 is entirely internal, then there are twenty-one hexagons compared to at most four non-hexagons
which can possibly occur at a, b, c, or d. With an additional level of refinement,
the number of points increases to eighty-one, of which seventy-seven are known to be
hexagonal. In general, with k levels, the two coarse triangles contain $(2^k+1)^2$
points which is a further indicator of the rapid shift towards hexagons as the number of levels increases.

Regardless of coarse mesh motion, we note that the predominance of hexagonal
stencils, the congruence properties of such stencils, and the even subdivisions of
area are all preserved. The fine mesh motion is clearly dictated by the coarse mesh
in the rigid manner resulting from the successive linear connections to various edge
midpoints. It is this rigidity which yields the preservation of mesh properties and
correspondingly determines the pointwise locations within coarse mesh triangles. As
a further consequence, there is no need to store the determined locations since they
are trivially reconstructed. Altogether, this most regular local structure may be
viewed as a case where a basic relatively coarse mesh carries along local pads of
resolution within its triangles.

The regularity of the pads leads to an ideal setting for multigrid methods.
The essential characteristic of multigrid methods is a cycling between fine and
coarse grids which is done to more rapidly converge upon the desired solution. It
is generally the case that robustness and efficiency are substantially enhanced when
the mesh is highly regular as it is with our most regular local structure. The
cycling process, however, is restricted to a coarse mesh which may appear to be
somewhat fine relative to the entire physical region. Using the strategy suggested

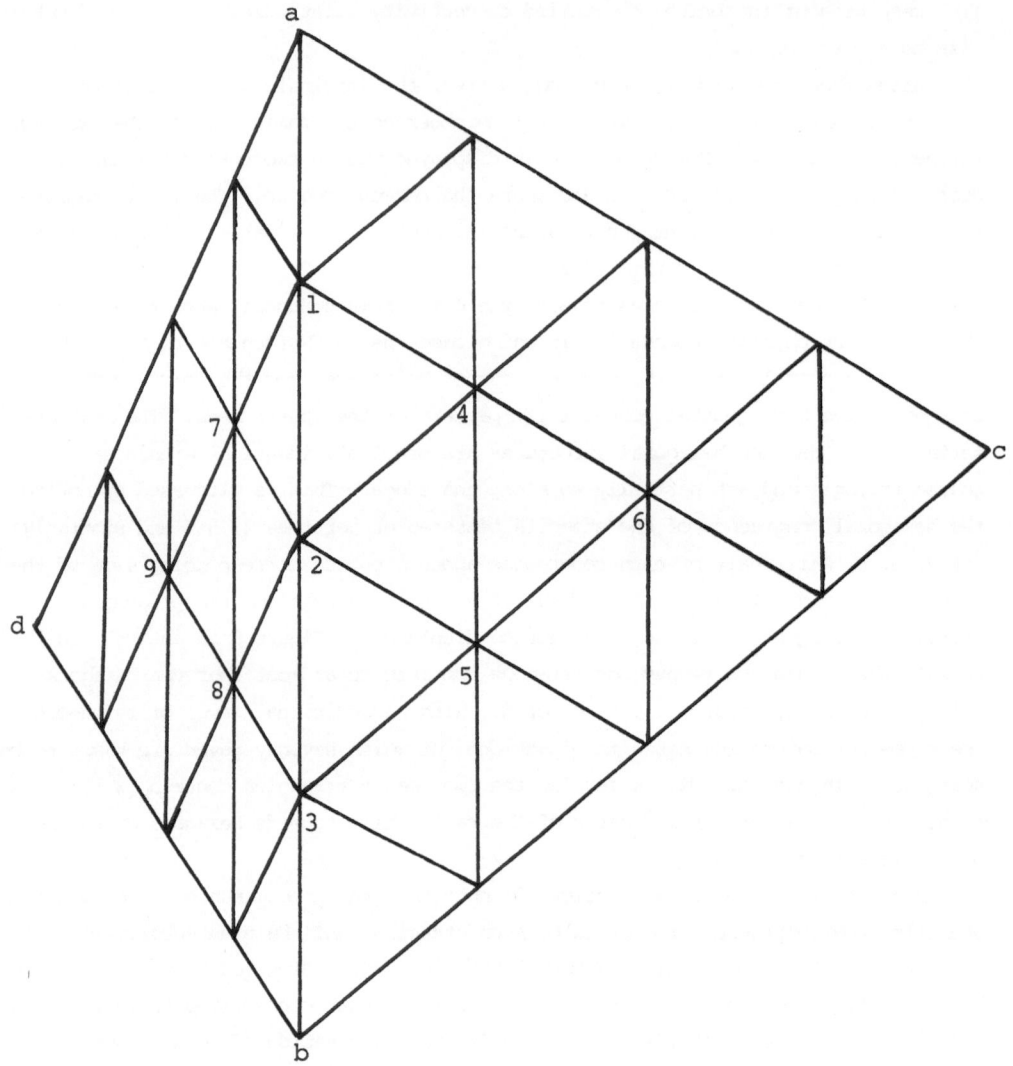

<u>Figure 8</u>: A two level refinement of the coarse mesh by
linear connections to successive edge midpoints

by Löhner et al. [11] in this conference, a further coarsening of the mesh for multigrid would seem possible. This would maintain the local efficiency while relieving the coarseness restriction.

While the above highly regular local structure is an ideal setting for multigrid methods, it is also less desirable for the solution adaptive simulations. There, a smoothly varying mesh is preferred so that numerical operators which are coupled by neighboring points do not abruptly change from point to point when the solution is smooth. This is needed for accuracy in such situations. To some extent, the desired smoothness is lost on the edges such as ab in Figure 8. As a consequence, there is then the motivation to allow for a completely free movement of the fine mesh while retaining its regular connectivity pattern. The movement can be applied everywhere or only in certain regions. Here, the integrity of straight edges for the coarse triangles may or may not be enforced. In the adaptive context, the basic motion is supplied by the center of mass expression of Eq. 3 which may be restricted by projection when straight edges are to be maintained. With these various levels of movement for the fine mesh, there is some motion relative to the most regular fine mesh which is also moving only because of the coarse mesh. As a consequence, a considerable degree of flexibility is available with a smoothly varying mesh that is locally regular in structure but is less rigid in geometry to various extents. In an intuitive sense, the basic pattern of the mesh is maintained while various parts of it are allowed to deform for adaptive purposes. Because of the connectivity pattern, multigrid methods are still applicable but may not be as efficient as in the highly regular case. As basic references on multigrid methods, it is worth noting Brandt [12] for elliptic equations and Ni [13] for hyperbolic equations since a number of other studies have been inspired by these works.

While the various locally regular structures described above provide many favorable attributes, they come with a price when the coarse moving mesh is to be restructured. A simple diagonal flip now requires that the pertinent fine mesh data also be transferred to a new fine mesh. In the most regular case, the new points are known analytically and the transfer is relatively simple. In the progression towards completely free movement for all points, the task entails progressively more work, although not an impossible amount.

## AN APPLICATION

To illustrate the basic features of the adaptive triangular mesh strategy, we take a brief look at an application to the study of plasma equilibrium. This is considered by solving the Grad-Shafranov equation to obtain cross-sectional equilibrium configurations for toroidal devices such as the TORUS II tokamak at Columbia University.

As a first step, the monitor surface must be suitably defined. In the early

stages, the solution itself, namely the magnetic flux function, was used mainly because of the positive results from the PEST codes developed at Princeton. There,
the contours of constant flux $\psi$ formed one family of coordinate curves which automatically isolated the magnetic axis and conformed with the plasma boundary. With
the flux $\psi$ also being the sought solution, its use as a coordinate seemed to be optimal. Unfortunately, however, the flux function varied too mildly at the plasma
boundary to cause its detection with a general adaptive strategy. The previous use
in PEST was by contrast a more analytically taylored method which could distinguish
this boundary in the more formal sense of selecting a constant $\psi$ value, typically 0.
As a consequence of the mild variations, the adaptive trials with flux surfaces often failed to converge upon any reasonable solution mainly because the plasma boundary could not be accurately represented.

With this failure came a second and deeper view into the physical situation so
that a more appropriate monitor surface could be prescribed. While it was desired
to adequately represent the magnetic axis, it was of critical importance to accurately represent the plasma boundary. From a physical viewpoint, there is a rapid
change in the toroidal current in crossing this boundary: the change is from a substantial current in the plasma to no current in the vacuum. Fluid mechanical analo-
gies to this behavior are pressure variations for shock waves and velocity variations for boundary layers. As a consequence, the toroidal current $J_T$ is a salient
quantity which should reliably distinguish the plasma boundary by its rapid variations. To also detect the magnetic axis, another salient quantity must be established. This is simply done by noting that the magnetic axis appears at the location where $\nabla \psi$ vanishes. To create a rapidly varying function at that location, the
gradient magnitude must appear as the variable in an expression which rapidly increases upon approaching zero. In a simple fashion, this was accomplished with
$(|\nabla \xi|^2 + \varepsilon)^{-1}$ where some small positive number $\varepsilon$ was inserted to yield a bounded
function. Upon normalization, the two salient quantities appear as

$$q_1 = \frac{J_T}{\max\{J_T\}}$$

$$q_2 = \frac{(|\nabla \xi|^2 + \varepsilon)^{-1}}{\max\{(|\nabla \psi|^2 + \varepsilon)\}}$$

[9]

and vary from 0 to 1. Noting that the magnetic axis and the plasma boundaries appear at distinct locations, the rapid variations in $q_1$ and $q_2$ are not conflicting
with each other. As a consequence, the monitor surface can be safely taken as the
linear combination

$$Q = a_1 q_1 + a_2 q_2$$

[10]

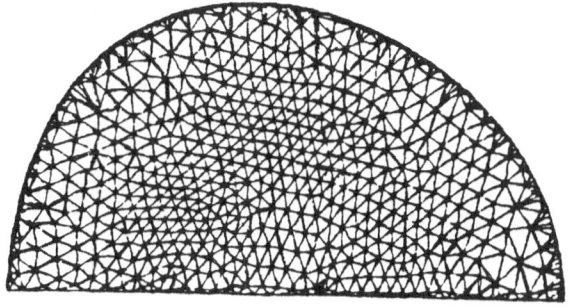

(a)

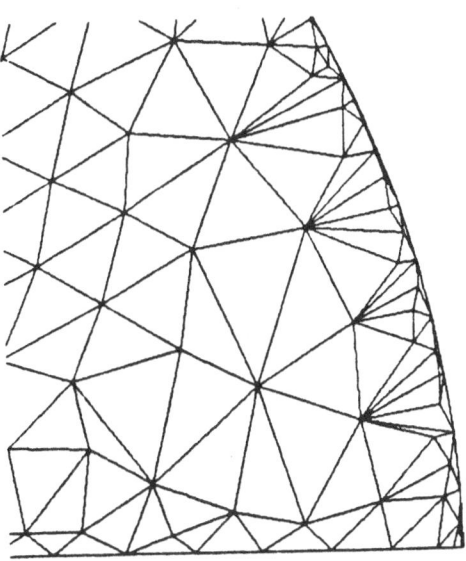

(b)

Figure 9: The final converged mesh for the Solove'v equilibrium
viewed (a) globally and (b) locally at the right-hand
corner under magnification

The coefficients $a_1$ and $a_2$ determine both the relative importance of quantities and the scaling with respect to the physical region.

In the test cases, typically chosen coefficients were $a_1 = 1$ and $a_2 = 2$. A remarkable feature occurred with the cases examined by using the new monitor surface: the simulation always converged to something meaningful! This was clearly the opposite to our previous experience using the flux surface. To affirm the meaningfulness of the results, extensive comparisons were made with other known studies and analytic solutions. The validity of the results were clearly established and provided some assurance in cases where no comparable results were available.

To witness the interplay between the mesh and the solution, two cases shall be displayed here in the form of the adapted mesh together with the current contours. In the first case, we examine the analytically known Solove'v equilibrium solution for a fixed plasma boundary. The converged mesh is displayed in Figure 9 both globally and locally. The converged current contours are displayed in Figure 10. The

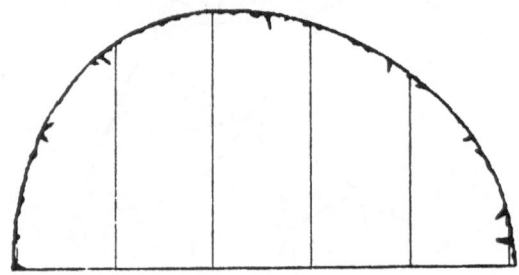

Figure 10:  The final current contours for Solove'v equilibrium

current is analytically known to be non-zero only in the plasma region where it varies linearly in the horizontal cross-wise direction. Observing Figure 10, the linear variation is certainly well represented by the vertical contours. More significantly, the rapid drop to zero at the boundary is also accurately represented by the exceedingly thin band of contours which were made possible by the adaptive motion that clustered points at the boundary. The boundary clustering is clearly evident in mesh when viewed globally in Figure 9a and under magnification in Figure 9b for the local section at the right-hand corner. In addition, the clustering of the magnetic axis is seen to occur at the expected location in Figure 9a which is near the center of the axis of symmetry. Altogether, the desired adaptivity for the mesh has been obtained and has resulted in an accurate solution which represents a substantial improvement over a parallel nonadaptive solution that greatly enlarged the current boundary thickness.

Regardless of the improvement from adaptivity, a more careful scutiny of Figure

10 leads to the observation of small spikes in the current contours that point in from the plasma boundary. The spikes appear to be of random orientation and to be triangular in shape. Upon further observation of the underlying mesh in Figure 9, the structure near the boundary does not appear to be good. This is particularly evident in the magnified portion displayed in Figure 9b where three nodes just adjacent to the boundary resolution not only have ten connections but also have their resulting small angles directed towards the boundary. In the somewhat random event that one of the connections strikes the boundary, the solution there is immediately propagated inward. This then is the most likely cause of the observed spikes in the current. It is also my principle motivating factor to propose the strategies for the adaptive restructuring and the control over the number of connections to each node. I have presented these strategies in the earlier sections. It should be anticipated that the proposed improvements in mesh structure will remove the spikes.

In a similar vein, the structural improvements will also enhance the solution quality in our second test case. Unlike the first, the second requires a determination of the plasma boundary. As a free boundary problem, the treatment may be either by fitting the evolving boundary with a mesh only within the plasma or by capturing the boundary with a mesh over the entire vacuum chamber which contains the plasma. The most noteworthy parallel in fluid mechanical simulations is the distinction between shock fitting and capturing. In our case, we chose the capturing strategy. The iteration for the solution was started from the initial current profile which is displayed in Figure 11 and is relative to the appropriately adapted mesh shown in Figure 12.

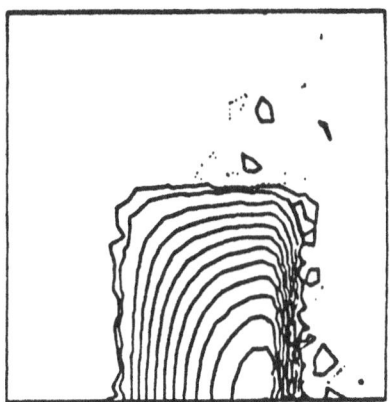

Figure 11:  Initial current profile

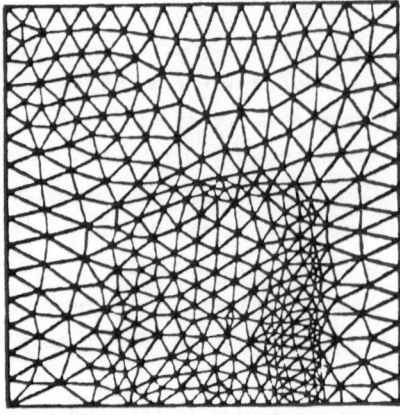

Figure 12: Mesh adapted to initial current profile

In keeping with the capturing strategy, the mesh covers the entire vacuum chamber
which is taken to be rectangular in cross-section. In this setting, the plasma
boundary appears as a rapid change in current which must be accurately represented
by the mesh. The mesh clustering here was accomplished with the adaptive movement
scheme of Eq. 3 by incorporating the approximate mean curvature of Eq. 6. The
boundary resolution is seen to vary in intensity in correspondence with the density
of constant current contours. Under a closer inspection of the side with the highest density, two bands of resolution are witnessed. These bands are the result of
curvature clustering. The outermost one resolves the rapid change in surface direction as the transition from no current to substantial current is made. The inner-
most one resolves the change as the steep climb in current levels off. The climb,
being a large but mildly varying gradient effect, does not need the same level of
resolution. Accordingly, the mesh thins out somewhat between the two bands.

Upon leaving the initial state and iterating towards convergence, the adaptive
movement scheme was used in an interlaced fashion with the solution procedure to
track and resolve the evolving solution. The converged current density is displayed
in Figure 13 while the corresponding adapted mesh is shown in Figure 14. The converged solution variable $\psi$ is displayed in Figure 15. An inspection of the results
shows that the current contours arrived at an equilibrium configuration where the
plasma boundary became more rounded, smaller, and sharply defined. The sharpness is
evident from the tight packing of current contours that had evolved from the initial
partial covering in Figure 11 to a complete one in Figure 13. In correspondence,
the curvature effect is now carried around the complete boundary as can be seen by

Figure 13: Converged current density contours in free boundary equilibria

Figure 14: Converged mesh in free-boundary equilibria

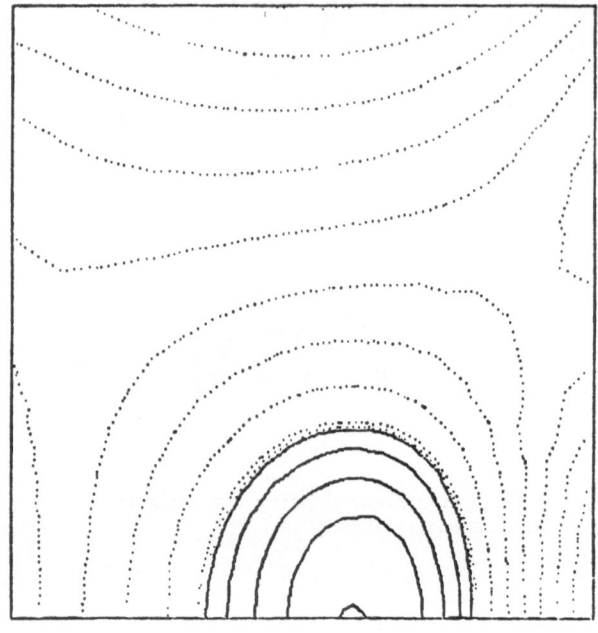

Figure 15: Converged flux contours in free-boundary equilibria

the full double band of resolution. Since the capturing strategy smears disturbances according to the number of connections to reach nearby points, this resolution
causes that smearing to occur over a short spatial distance. The result is a more
sharply defined solution as can be witnessed by the final current and flux contours.
The sharpness represents one of the distinctive advantages of adaptivity.

Aside from the adaptive resolution, certain irregularities are observed. These
are locally triangular in shape and are a result of the mesh connectivity pattern as
were the earlier spikes in the fixed boundary case. This represents further motivation to improve the mesh in the manner that I suggested earlier. As in the earlier
case, several nodes with far too many connections appear just to the right of the
outer band of resolution. As before, the consequent small angles are directed at
the band. In addition, improvement is also needed for the axis of symmetry where
the current contours enter on the left with a sudden departure from orthogonality.
However, this can be corrected by considering a reflected mesh upon which we then
flip diagonals to go across the symmetry line. This would add some more points
there due to the truncation, but the new connections would then enter orthogonally.

The discussion here has clearly focused upon the interplay between the mesh and the solution for an equilibrium configuration. Specific technical details have been purposely omitted. A good overview of these details is given by Erlebacher and Eiseman [9]. A more thorough discussion is presented by Erlebacher [6].

As a final note on the equilibrium problem, we observe that the fully power of a general connectivity triangular mesh was under-utilized. That was a consequence of the relatively simple boundaries in the test cases. Further cases can readily be considered although this was not done. For example, additional clustering can be included for resistive effects and, moreover, dynamic topological changes can be followed in such phenomena as the formation of islands.

CONCLUSION WITH A SENSE OF PERSPECTIVE

The simulation of physical phenomena with dynamically changing configurations was first considered in a serious sense in the context of pure Lagrangian hydro-dynamics. To avoid the diffusive rezoning interpolations, the emphasis was put on establishing a considerable freedom from the rigid mesh structures typically placed over the physical region. This provided the impetus to use a general connectivity triangularly mesh in which reconnections between points could be dynamically made to provide a good mesh structure while maintaining local conservation properties of the fluid. The chief restructuring mechanism was a simple diagonal flip for a quadri-lateral stencil formed by two adjacent triangles. Several criteria for such flips were established and used along with point additions at triangle barycenters and edge midpoints.

Under the restrictions imposed by conservation properties, there was some dif-ficulty in extending the applicability from hydrodynamics to a wider range of simu-lations even within the realm of fluid mechanics, let alone numerous other possibil-ities. At the same time, the more general problem class was mainly being examined with meshes that were not allowed to dynamically restructure and, at best, were able to provide additional resolution by adding points in a regular and structured man-ner. When the addition was done locally, the boundaries of the refinement would in-variably possess points that were not fully connected to all possible neighbors. The consequence was an internal artificial boundary in need of special treatment.

To address the widest possible range of applications, adaptive movement was es-tablished within the context of a general connectivity triangular mesh together with the earlier dynamic restructuring employed in Lagrangian hydrodynamics. The requi-site generality to enhance the range of applications came with the development of monitor surfaces. Adaptive movement was established with a local center of mass formula so that points could be pushed into positions that could more accurately see the monitor surface. A slightly simpler restructuring criterion than those previ-ously used was also developed for triangles in the physical plane.

In the present reporting, the adaptive setting was described with a discussion of the monitor surfaces and the movement formula. Based upon observations of computed results such as those presented in the last section, the need for some improvement in mesh quality was evident even though the basic resolution was well in place. This brought me to suggest new strategies to address the observed problems which caused the solution to have some undesirable properties. The new strategies included adaptive restructuring, fission and fusion operators, and control over nodal connectivity. The essence of adaptive restructuring was to determine the mesh connectivity from solution properties, a part of which was to provide automatic alignment with rapid solution variations by using curvature. Cellular fission and fusion were established to split a cell into two and to form one cell from two. The local coupling with the adaptive movement formula provided a gradual transition of cell sizes and a reasonably smooth means to change the number of mesh points. The use of movement also ensured that cellular fission and fusion were inverse operators in a strict sense. Their application to control the number of connections to each node is necessary to provide the opportunity to push towards nearly equilateral triangles throughout the mesh and thereby enhance the accuracy of the various numerical operators.

In addition to the strategies to improve the structure of the general connectivity triangular mesh, locally regular structures were presented as a means to enhance solution efficiency. This provides a cheaply expandable data base and an ideal setting for the application of multigrid methods for convergence acceleration. The idea is simply to consider the general connectivity mesh as a course mesh in which a regular fine mesh is embedded while maintaining full connections for all nodes.

## REFERENCES

1. Steger, J.L., Dougherty, F.C., Benek, J.A. 1983 "A Chimera Grid Scheme," Advances in Grid Generation, ed. by Ghia and Ghia, ASME FED-Vol. 5, pp. 59-70.
2. Eiseman, P.R. 1985 "Grid Generation for Fluid Mechanics Computations," Annual Review of Fluid Mechanics, Vol. 17, pp. 487-522.
3. Eiseman, P.R. 1983 "Alternating Direction Adaptive Grid Generation," Proceedings of the 6th AIAA Comput. Fluid Dyn. Conf., Danvers, Massachusetts, pp. 339-348; in press for AIAA Journal.
4. Dwyer, H.A. 1985 "Grid Adaptation for Problems in Fluid Dynamics," AIAA Journal, Vol. 22, No. 12, pp. 1705-1712.
5. Erelebacher, G. 1985 "Finite Difference Operators on Unstructured Triangular Meshes," Free-Lagrangian Methods Conference, Hilton Head Island, South Carolina.
6. Erlebacher, G. 1984 "Solution Adaptive Triangular Meshes with Application to Plasma Equilibrium," Ph.D. Thesis, Department of Applied Physics and Nuclear Engineering, Columbia University, 208 pp.
7. Crowley, W.P. 1970 "FLAG: A Free-Lagrange Method for Numerically Simulating Hydrodynamic Flows in Two Dimensions," Proceedings of the 2nd International Conference on Numerical Methods in Fluid Dynamics, ed. by M. Holt, Lecture Notes in Physics 8, Springer-Verlag, pp. 37-43.
8. Fritts, M.J., Boris, J.P. 1979 "The Lagrangian Solution of Transient Problems in

Hydrodynamics Using a Triangular Mesh," J. Computational Physics, Vol. 31, pp. 173-215.

9. Erelebacher, G., Eiseman, P.R. 1984 "Adaptive Triangular Mesh Generation," Presented at the 17th AIAA Fluid Dynamics, Plasma Dynamics, Lasers Conference, Snowmass, Colorado, Paper AIAA-84-1607; submitted to AIAA Journal.

10. Crowley, W.P. 1985 "Free Lagrange Methods for Compressible Flows," Free-Lagrangian Methods Conference, Hilton Head Island, South Carolina.

11. Löhner, R., Morgan, K., Peraire, J., Zienkiewicz, O.C. 1985 "Recent Developments in FEM-CFD," Free-Lagrangian Methods Conference, Hilton Head Island, South Carolina.

12. Brandt, A. 1977 "Multi-Level Adaptive Solutions to Boundary Value Problems," Math. of Computation, Vol. 31, No. 138, pp. 333-390.

13. Ni, R.H. 1982 "A Multiple-Grid Scheme for Solving the Euler Equations," AIAA Journal, Vol. 20, No. 11, pp. 1565-1571.

RECENT DEVELOPMENTS IN FEM-CFD

R. Löhner, K. Morgan, J. Peraire and O.C. Zienkiewicz
Institute for Numerical Methods in Engineering,
University of Wales, Swansea SA2 8PP, U.K.

## 1.  INTRODUCTION

We review the current status of CFD with respect to unstructured grids employing finite element methods and Eulerian frames.  Starting from our aims, we develop in subsequent sections algorithms suitable for the computation of large 3-D problems past arbitrary geometries This paper is presented to an audience which may not be completely familiar with 'Eulerian-CFD', but mainly interested in Free-Lagrangian Methods.  Therefore, its aim is not to give an in-depth explanation of all ideas, but to put forward concepts in a unified setting.

## 2.  AIMS

The aim of the ongoing research is to develop the capability of simulating the flowfields past arbitrary bodies in 2/3-D.  As a typical example we show the military aeroplane depicted in figure 1, although we also envisage the computation of flowfields past civil aircraft, re-entry vehicles, cars, lorries and trains.  The evaluation of the flowfields past these bodies are of major interest during their design phase and therefore we must seek not only to minimise actual run-times, but also mesh generation and result evaluation times.

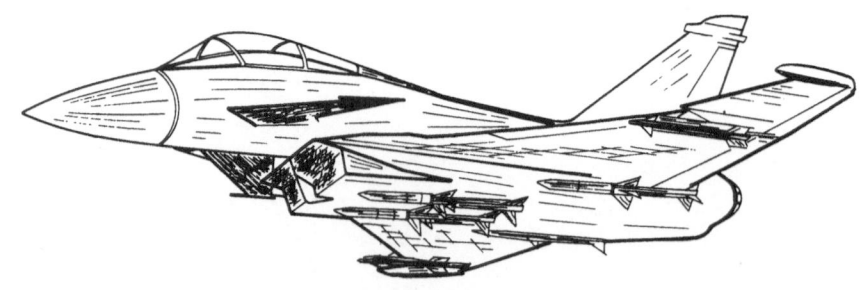

FIGURE 1    A TYPICAL 3-D CONFIGURATION

## 3. CONSEQUENCES

Having stated these aims, the following consequences follow immediately from them.


### 3.1 Accurate representation of any geometry

The numerical method must be able to cope with <u>any geometry</u>. This implies that <u>unstructured grids</u> must be employed.  By 'unstructured' we mean that the domain has been discretised in such a way that it cannot be mapped into a square/cube.  In other words, the distribution of points/elements is unordered.  As an example of such a grid, see figure 8 .  The degree of unstructuredness may vary from sets of structured subgrids/subdomains joined together to discretise a complicated domain (macro-element discretisation) to completely unstructured grids.  We adopted the latter approach, as it simplifies adaptive mesh refinement considerably.

We enumerate the advantages and disadvantages of unstructured grids.


### 3.1.1 Advantages

- Any geometry can be modelled without difficulties, particularly if triangular (2-D) or tetrahedral (3-D) elements are employed.

- The mesh generation algorithm/procedure has been <u>decoupled</u> from the solver.  Thus any mesh generator may be employed.

- The generation of grids starting from solid modellers or any similar CAD-CAM representation of objects is greatly facilitated.  Therefore, a considerable reduction in man-hours for grid generation and software development is achieved.

- As the grid is already unstructured, adaptive mesh refinement can easily be implemented.


### 3.1.2 Disadvantages

- As we are constantly operating on sets of unstructured data, CPU-times increase.  Vector/scalar speedup ratios may be as high as 25 on a two-pipe CYBER-205 in full precision, but the MEGAFLOP-numbers achieved are still a factor of 2.5-10 away from the optimally possible rates.  Heavy use is made of

GATHER/SCATTER operations, which are not as efficient as straight DO-loops.

- As the grids are unstructured, the concept of 'lines' or 'planes' vanishes. Therefore, no optimal implicit procedures such as ADI may be employed.

- For the same reason, no line/surface/zebra-relaxation [8] can be used in conjunction with multigrid-methods. The same applies to other orderings, like red-black Gauss-Seidel relaxation.

- One-dimensional concepts vanish: by this we mean that for a given point in the field, there does not exist a point 'to the left' or 'to the right'. The surrounding points are distributed randomly. This implies that we can only discretise operators (like $\underline{\nabla}$, $\underline{\nabla}^2$, etc.), but cannot apply stencils (like $(-\frac{1}{2}, 0, \frac{1}{2})$). In particular, the by now classic theory of limitors for TVD schemes [14]-[17] cannot be translated to unstructured grids.

## 3.2 Large Problems

The problems under consideration are big. In order to compute, with an acceptable degree of accuracy, the inviscid flow past a wing approximately $10^5$ gridpoints are needed. It is estimated that the simulation of viscous flow past an aeroplane will require as many as $4.10^7$ gridpoints. This poses formidable problems to any hardware, and any possible reduction of CPU and storage requirements must be pursued. This implies that:

- The algorithm under consideration must be vectorisable to a high degree. By this we mean that at least 95% of all operations must be performed in vector-mode (otherwise only a moderate increase in speed will be realised).

- As no optimal implicit schemes can be employed, only explicit schemes can be considered. This poses not as big a constraint as one may imagine, as implicit procedures tend to be inaccurate above the Courant number limit, and the problems under consideration may be of transient nature (separation, buffeting, vortex-streets etc.).

- Any conceivable effort must be attempted to achieve acceptable accuracy with as little a number of gridpoints as possible. As the solution is unknown a priori, automatic adaptive mesh refinement procedures will have to be considered.

- For steady state problems, local timesteps coupled with multigrid methods (MGMs) provide the fastest known convergence rates [6]-[12]. As the grids under consideration are unstructured, the nestedness of coarse/fine grids of classical multigrids will have to be abandoned. Instead, sets of non-nested grids are considered, resulting in unstructured MGMs.

- For transient problems, domain splitting will have to be employed, in order to advance efficiently in time on grids which have large variations in element size.

## 4. ALGORITHMS

The equations governing compressible fluid flow may be written in conservation form as

$$\underline{u}_{,t} + \underline{F}^i_{,i} = \underline{0}. \tag{1}$$

This equation is first discretised in space in a Lax-Wendroff manner [14], resulting in

$$\Delta \underline{u} = -\Delta t \ \underline{F}^j_{,j} + \frac{\Delta t^2}{2} (\underline{\underline{A}}^j \ \underline{F}^i_{,i})_{,j} \ , \quad \underline{\underline{A}}^i = \frac{\partial \underline{F}^i}{\partial \underline{u}} \tag{2}$$

and is then discretised in space by a straightforward Galerkin weighted residual procedure [30]. The justification for employing as weighting and test functions the same polynomials may be found in [1], [2]. As element type we have chosen linear triangles and tetrahedrons in 2/3-D, as these greatly simplify mesh generation and adaptive mesh refinement. After spacial discretisation we arrive at a system of algebraic equations of the form:

$$\underline{\underline{M}} . \Delta \hat{\underline{u}} = \hat{\underline{R}} \ , \tag{3}$$

where $\underline{\underline{M}}$ is not diagonal. This, seemingly, is against the stated explicitness which ought to be pursued (see section 3.2). However, the condition number of the $\underline{\underline{M}}$-matrix in (3) is always bounded, and therefore with very few iterations (we usually employ 3) of the form

$$\underline{\underline{M}}_\ell \cdot (\Delta \underline{u}_i - \Delta \hat{\underline{u}}_{i-1}) = \hat{\underline{R}} - \underline{\underline{M}} \cdot \Delta \hat{\underline{u}}_{i-1},$$

$$\Delta \hat{\underline{u}}_o = \underline{0}, \qquad i=1\text{-niter}$$

(4)

equation (3) is solved. $\underline{\underline{M}}_\ell$ denotes the 'lumped' diagonal mass-matrix, obtained as $M_\ell^{ii} = \sum_j M^{ij}$. Observe that only one evaluation of $\underline{R}$ is needed, as this remains fixed throughout the iteration (4). In order to demonstrate the need and effectiveness of the consistent mass matrix $\underline{\underline{M}}$ for transient problems we consider the classic example of the rotating cone, given by equation (1) with

$$\underline{u} = \phi, \quad \underline{F}^1 = -y \cdot \phi \quad \underline{F}^2 = +y \cdot \phi.$$

(5)

In figure 2 the initial configuration and the solutions obtained after one rotation for niter=1 (lumped mass similar to Lax-Wendroff) and niter=3 (consistent mass) are compared, and the superiority of the consistent mass is readily seen.

The right hand side vector of equations (2,3) may be constructed in an alternative fashion, which avoids the Jacobian $\underline{\underline{A}}^i$, by rewriting (2) as a Taylor expansion of the form

$$\Delta \underline{u} = - \Delta t \ (\underline{F}^j (\underline{u} + \delta \underline{u}))_{,j} = - \Delta t \ \underline{F}^j_{,j}\Big|_{\underline{u}} - \Delta t \left( \frac{\partial \underline{F}^j}{\partial \underline{u}}\Big|_{\underline{u}} \cdot \delta \underline{u} \right)_{,j} - \text{HOT}$$

$$= - \Delta t. \ \underline{F}^j_{,j}\Big|_{\underline{u}} - \Delta t. \ \left( \underline{\underline{A}}^j\Big|_{\underline{u}} \cdot \delta \underline{u} \right)_{,j} + \text{HOT}$$

(6)

Therefore if

$$\delta \underline{u} = - \frac{\Delta t}{2} \ \underline{F}^i_{,i}$$

(7)

is chosen, the following two-step scheme is seen to be equivalent to equation (2):

$$\underline{u}^{n+\frac{1}{2}} = \underline{u}^n - \frac{\Delta t}{2} \cdot \underline{F}^i_{,i} \qquad\qquad + \text{GALERKIN}$$

$$\Delta \underline{u} = \underline{u}^{n+1} - \underline{u}^n = - \Delta t. \ \underline{F}^j_{,j}\Big|^{n+\frac{1}{2}} \qquad + \text{GALERKIN}$$

(8)

We remark that the numerical equivalence between (8) and (2,4) is only regained if at the half-step $n+\frac{1}{2}$ <u>constant</u> shape functions are employed [5]. The avoidance of the $\underline{\underline{A}}^i$ matrices not only reduces CPU-times on

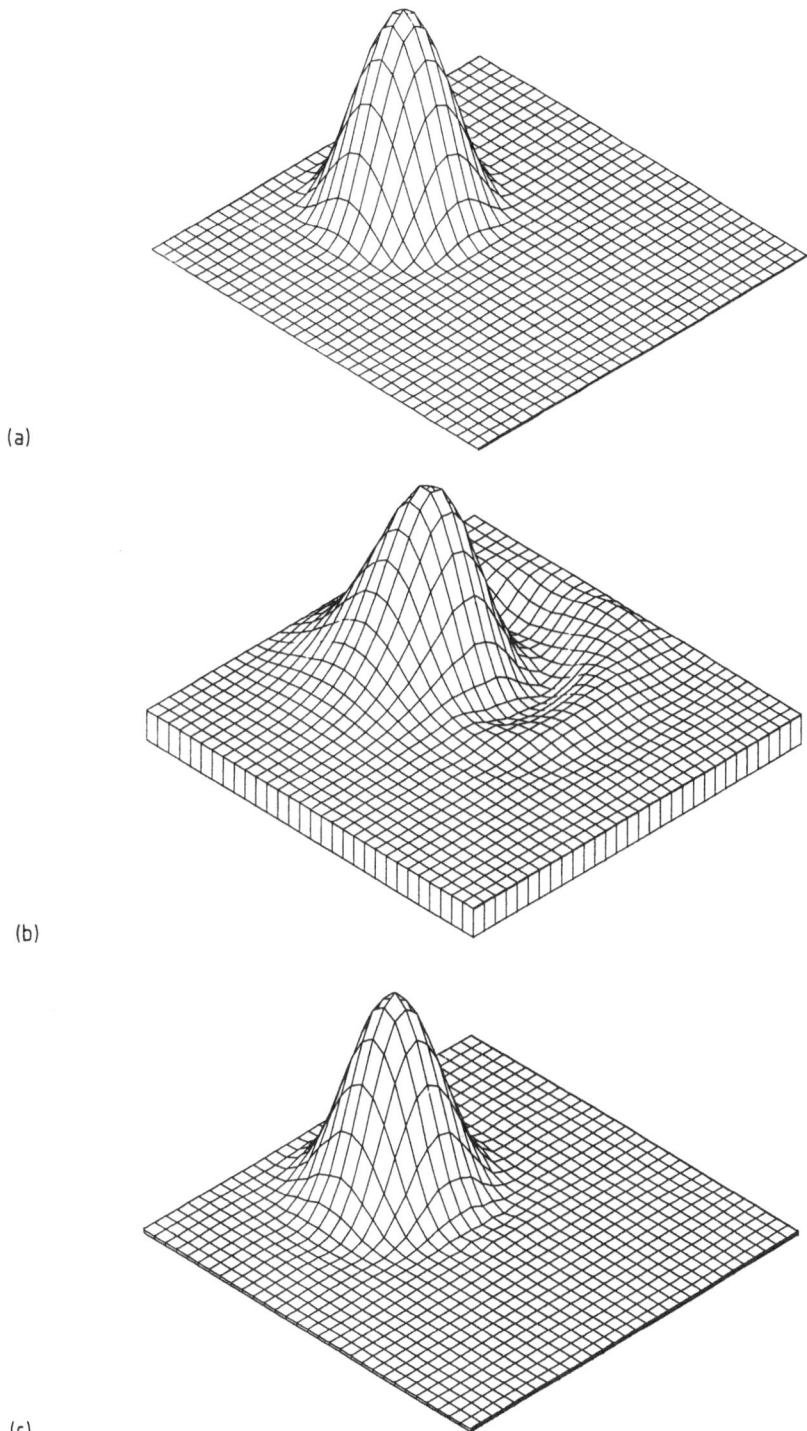

FIGURE 2  PURE ROTATION OF A CONE   (a) INITIAL STATE (b) AFTER ONE
ROTATION, LUMPED MASS (c) AFTER ONE ROTATION, CONSISTENT
MASS

a scalar machine by a factor of 5 in 3-D, but also greatly facilitates
vectorization. At the same time, for real gases, where no analytic
expression exists for the Jacobian, the expected reduction in CPU-
times is even more impressive. CPU-timings obtained for the 3-D
Euler-code for a problem of moderate size (605 nodes, 2400 elements)
are given in Table 1, showing that a high degree of vectorization has
been achieved.

TABLE 1                                      3-D

| | | | |
|---|---|---|---|
| CYBER-205 | (vectorizer off) | $202.0 \; 10^{-5}$ | |
| " | ( " on ) | $12.6 \; 10^{-5}$ | |
| CRAY-XMP[*] | ( " off) | $136.4 \; 10^{-5}$ | |
| " | ( " on ) | $45.1 \; 10^{-5}$ | |

* no hardware GATHER/SCATTER was given


For problems involving strong shocks, we employ a variant of the
Lapidus smoothing [3], given by (for the details, see [4]):

$$\underline{u}_s^{n+1} - \underline{u}^{n+1} = \Delta t \, \frac{\partial}{\partial \ell} \, k^{\ell\ell} \, \frac{\partial \underline{u}^{n+1}}{\partial \ell} \, , \tag{9}$$

where the direction $\ell$ is given by

$$\underline{\ell} = \frac{\mathrm{grad}\,|\underline{v}|}{|\mathrm{grad}\,|\underline{v}||}, \quad \underline{v} : \text{velocity vector}, \tag{10}$$

and $k^{\ell\ell}$ is of the form

$$k^{\ell\ell} = c_\nu \cdot h_e^2 \left| \frac{\partial\,(\underline{v}\cdot\underline{\ell})}{\partial \ell} \right| , \quad \begin{array}{l} h_e : \text{element size} \\ c_\nu : \text{Lapidus constant (1-2)} \end{array} \tag{11}$$

Observe that the smoothing acts only at places where $\frac{\partial\,(\underline{v}\cdot\underline{\ell})}{\partial \ell} \neq 0$, so
that contact discontinuities, slip-discontinuities and boundary layers
are not smeared.



5. ADAPTIVE MESH REFINEMENT

As the problems to be solved are very large, storage and CPU-
requirements can become excessive. Therefore, we explored adaptive
refinement procedures in [5], [21], [26], [27]. In order to be able
to refine adaptively, we must define the following three steps:

5.1 Aim

Given an error indicator/estimator: what is to be achieved? In
our case the problems are of hyperbolic type, implying that disturbances

or error can propagate through the domain.  Therefore, we are forced
to require that the maximum error over the whole domain under consideration be as small as possible:

$$\max_{\Omega} \; \|u-u^h\| \; \rightarrow \; \min. \tag{12}$$

This leads to the requirement that the error ought to be evenly distributed throughout the domain

$$\|u-u^h\|_{e\ell.} \; = \; \text{const.} \tag{13}$$

## 5.2 Error Indicator

Here, we simply use a priori (interpolation) estimates of the
form

$$\|u-u^h\|_0 \; \leq \; c.h^{\ell}|u|_{\ell}, \tag{14}$$

with $\ell=2$ ($\ell=1$ also works in most cases and has been advocated by other
authors ([23]).   Observe that a free constant c is included in (14):
this however disappears, as (14) has to be combined with (13).  Other
possible error indicators/estimators include the residuals of the
original equation (1) in discrete form, or, for inviscid transonic
flows, the variation of entropy [28].  Combining (13), (14), and substituting $u^h$ for u on the right hand side of (14) (which implies that
the current mesh must be fine enough to provide information for (14)),
we arrive at the requirement that

$$h^{\ell}|u^h|_{\ell} \; = \; \text{constant over each element} \tag{15}$$

## 5.3 Strategy

One can now seek to satisfy (15) either by mesh movement [18-22]
or by mesh enrichment [23-28,5].  Currently we are favouring mesh
enrichment for steady state problems, as in this way nearly all CPU
and storage requirements are spent on 'desired features' (discontin-
uities/boundary layers).

We are currently studying the following strategy for steady-state
problems in which both collapsible features (shocks) and 'smooth-high
gradient' (boundary layers) features are present:

- for the first mesh:  obtain $u^h$ at nodes and $e_E = h^{\ell}|u^h|_{\ell}$ in
  the elements, and refine all elements for which

$$e_E > a \max_{\text{elements}} e_E ; \tag{16}$$

- for all subsequent meshes obtain $u^h$, $e_E$ as before;  compare $e_E$
  with the error observed previously;  if the error decreases,
  regard it as 'smooth', if not as 'collapsible'.  For the smooth

features, see if the error (the constant c in equation (16) can be established quite accurately) is less than a preset tolerance: if not, refine, otherwise leave untouched.

- for the collapsible features, see if the element size is less than a preset length; if not, refine, if yes, skip. In this way, the only user-supplied parameters are an error tolerance for the smooth regions and an element size/length for the collapsible features.

At present, a major problem for fully transient adaptive refinement /derefinement via mesh enrichment is the de-refinement process. We are studying possibilities of 'eating' up 'idle' elements as described in [21], but so far no definitive strategy has emerged.

## 6.  DOMAIN SPLITTING/LOCAL TIMESTEPPING

Any mesh obtained from adaptive refinement processes will show large variations in mesh size. As the basic advancement algorithm is explicit, if a constant timestep is taken over the whole grid, CPU-efficiency will be poor and inaccuracies may be introduced in regions of low local Courant numbers. In order to alleviate this while still advancing time accurately the solution, domain splitting was introduced in [29]. The basic idea here is to overlap regions in which timesteps of different magnitude are taken, so that the errors introduced at inter-domain boundaries due to splitting are of the same order as the errors of the basic algorithm. It was found [29] that overlapping two layers of elements was sufficient to reduce the errors due to splitting to the same level as the basic scheme. For steady state problems, local timesteps may also be employed.

## 7.  MULTIGRID METHODS

In order to accelerate further the convergence rate for steady state problems, multigrid methods have often been employed in conjunction with FDMs/FEMs [6]-[12]. The common feature of these methods so far has been the 'nestedness' of the grids: coarser grids were obtained by deleting lines/planes of the finer grids. As the fine grid was assumed as unstructured (in order to model accurately the domain of interest), it cannot be obtained by 'enriching' coarser gris by lines/planes. Therefore, use is made of a series of unnested grids. This implies that:
- The fine grid still models accurately the domain.
- The generation of the coarser grids is simplified. For large problems grid generation will always be automatic: for this reason, once the fine grid has been generated (and all the

information to generate it has been stored), the coarser grids
may be generated without further work.
- Moving or deforming grids on the finest level may be employed
  in conjunction with fixed (or uniformly moving) 'background'
  grids, which are only used as a means to accelerate convergence.

In order to achieve good performances for unstructured multigrid
processes, the following developments were necessary:

## 7.1  Search Algorithms of Optimal Order

As the grids are not nested, search routines have to be developed
in order to interpolate residuals/increments across grids.  Given two
grids G1 and G2 with NELE1, NELE2, NPOI1, NPOI2 elements/points
respectively, we have to find which points of G2 lie in element K of
G1.  Obviously, this may be accomplished in 0(NPOI1*NPOI2) operations,
which is far too much.  An optimal-order algorithm is obtained if an
imaginary, structured, third grid G3 is introduced (see figure 3 ).
We then proceed as follows:

- find in which cell of G3 each point of G2 lies (0(NPOI2));

- find which cells of G3 each element of G1 covers (0(NELE1));

- for all the cells each element of G1 covers, see if the
  points found in the first step are inside/outside.

In this way, we have 0(NPOI2+NELE1) operations, which is acceptable.


## 7.2  Variable Timestepping Jacobi-Smoothers for Elliptic Problems

As only Jacobi procedures will allow full vectorization on
unstructured grids, these must be pursued.  However, if a constant
under-relaxation factor is employed, the smoothing rates are very
low [8].  Therefore, more suitable combinations of varying under-
relaxation factors (or timesteps) must be employed.  As an example
we consider the 1-D case for the Laplace-operator

$$\frac{d^2 \phi}{dx^2} = 0 \tag{17}$$

For each Jacobi sweep, the reduction factor of the Fourier modes is
given by

$$g = 1 - \Delta t (1 - \cos \frac{\pi}{K}) \tag{18}$$

In Figure 4  , the performances for the four-stage schemes
$\Delta t$ = (0.8, 0.8, 0.8, 0.8) and $\Delta t$ = (0.5, 2.0, 0.7, 0.9) are compared,
and, as can be seen, a much higher smoothing rate for the phase-range
$\frac{\pi}{3}$ - $\pi$ is achieved by varying $\Delta t$.  In general, the $\Delta t$-sequence is
obtained (heuristically) by the formula:

$$\Delta t_j = 2/(1+\cos(\pi(j-1)/1)) \qquad j=1...1 \qquad (19)$$

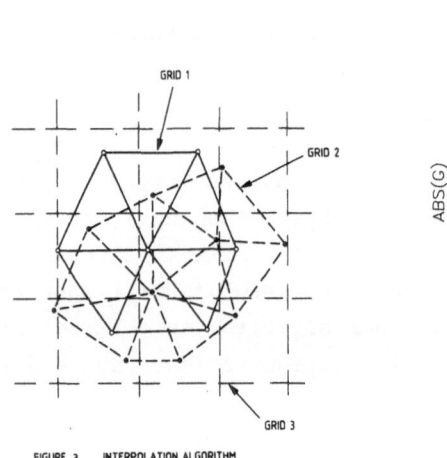

FIGURE 3    INTERPOLATION ALGORITHM

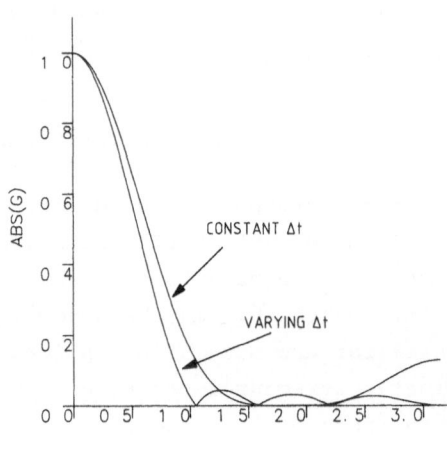

FIGURE 4  COMPARISON OF SMOOTHING RATES FOR JACOBI SMOOTHERS

## 7.3  Transport Concepts for Hyperbolics

Although the concept of smoothing can also be carried over to hyperbolic systems of PDEs [11] - mainly by devising new timestepping algorithms -, another concept can also be employed [9,10,12]: this is the one of transport. The basic idea is to approximate, on the coarser grids, in the formula (see equation (2))

$$\Delta \underline{u}_{2h} = (\underline{1} - \frac{\Delta t}{2} \partial_i \underline{A}^i)(- \Delta t \underline{F}^j_{,j}) \qquad (20)$$

the last term on the right-hand side by

$$-\Delta t_{2h} \underline{F}^j_{,j}\Big|_{2h} \cong \frac{\Delta t_{2h}}{\Delta t_h} \Delta \underline{u}_h, \qquad (21)$$

thus having

$$\underline{u}_{2h} = (\underline{1} - \frac{\Delta t}{2} \cdot \partial_i \underline{A}^i)(\frac{\Delta t_{2h}}{\Delta t_h} \cdot \Delta \underline{u}_h) \qquad (22)$$

Again, the use of the Jacobians can be avoided by setting [10]:

$$\underline{\underline{A}}^i \cdot \Delta\underline{u} = \underline{F}^i(\underline{u} + \Delta\underline{u}) - \underline{F}^i(\underline{u}). \tag{23}$$

This process has been shown to be both robust and effective, and due to its simplicity is preferred over traditional (smoothing) multi-grid methods [6-8, 11].

Examples

We demonstrate the capabilities developed on several examples.

1. Transient supersonic flow in a channel: the geometry of this example is that chosen by Woodward and Colella [31] as a general test example for numerical methods for hyperbolic PDEs. It consists of a uniform supersonic flow in a wind tunnel, into which a 'step' is suddenly introduced. The discretisation of the domain, as well as the initial conditions (not the same as in [31]) are shown in Figure 5 together with the contour plots of the density at several stages. Although no special treatment at the root of the expansion was employed, the results show the high accuracy of the method.

2. Regular shock reflexion of a wall: this example, proposed by Harten [32] and also studied by Colella [33] is depicted in Figure 6 together with the boundary conditions and initial values. Use is made of the adaptive refinement strategy and the domain splitting technique, described earlier, and the discretisations and solutions obtained are shown in Figure 6 ; the improvements in accuracy can be readily observed.

3. Viscous flow past a protruberance: this example, which simulates the flow past space-shuttle tiles, demonstrates again how problems with narrow 'regions of interest' may be economically solved with adaptive refinement and domain splitting. We only show here the final mesh and the solution obtained (Figure 7 ).

4. Potential flow past a cylinder: the domain of interest, together with the boundary conditions and the discretisation is shown in Figure 8 . One can also see the sequence of grids (no nesting), and the solution which was obtained after the equivalent of 30 iterations on the fine grid (accurate to 5 digits).

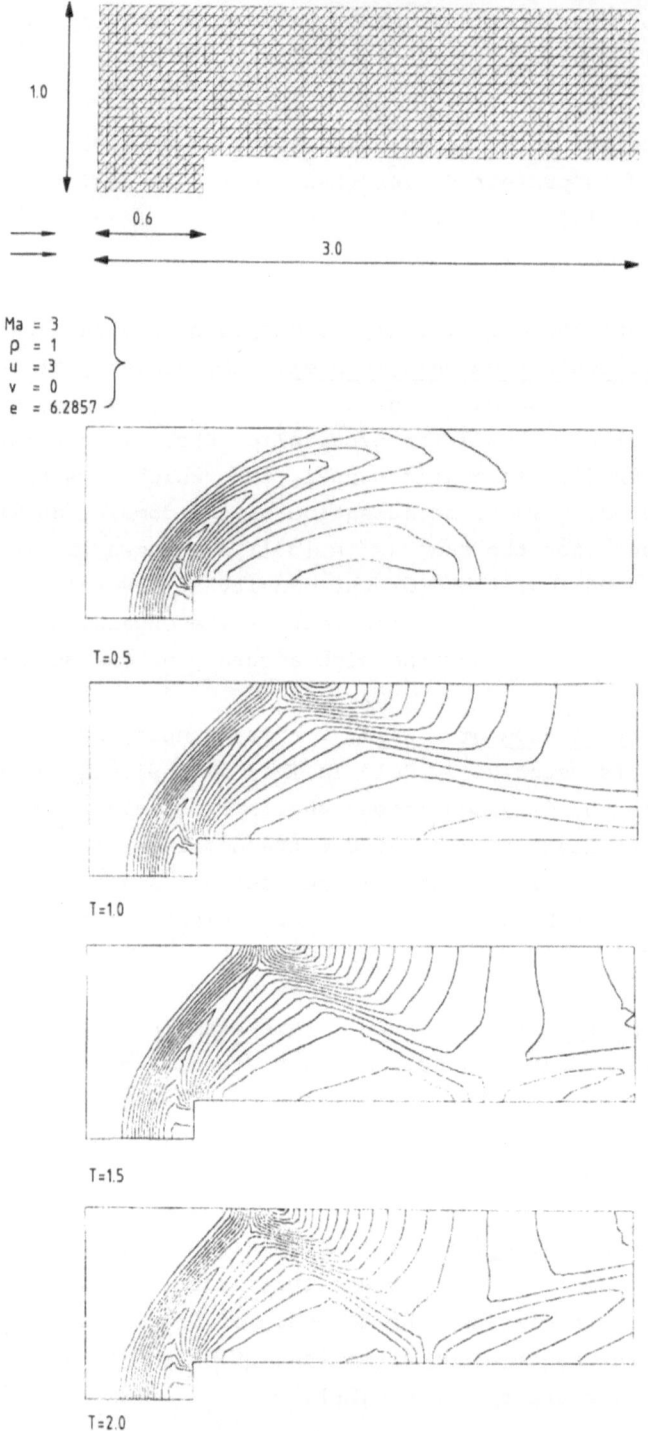

FIGURE 5   TRANSIENT SUPERSONIC FLOW IN A CHANNEL  PROBLEM
SPECIFICATION AND DENSITY CONTOURS AT DIFFERENT TIMES

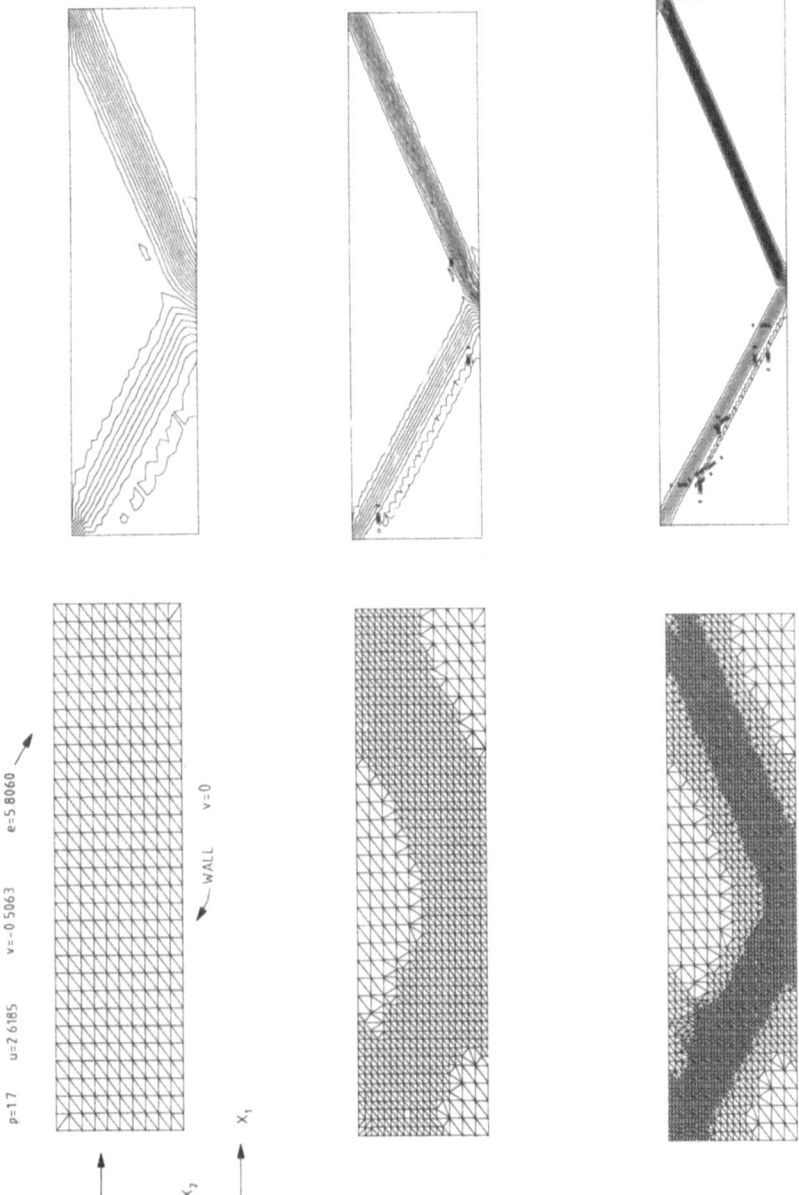

FIGURE 6    HARTEN-COLELLA PROBLEM

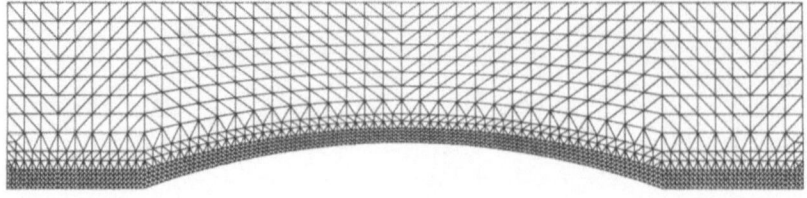

NASA-DOME 2

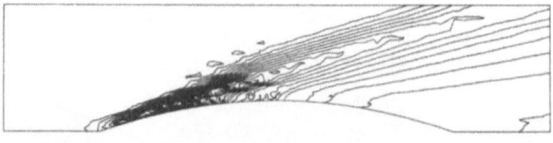

NASA-DOME 2
PRESSURE

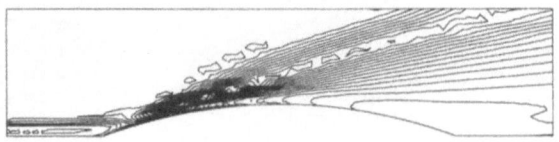

NASA-DOME 2
DENSITY

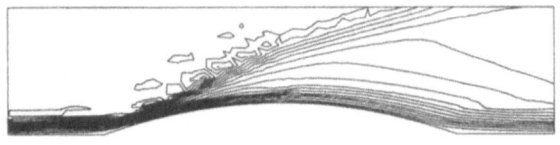

NASA-DOME 2
MACH

FIGURE 7    SIMULATION OF LAMINAR SUPERSONIC
FLOW PAST A SHUTTLE - TILE

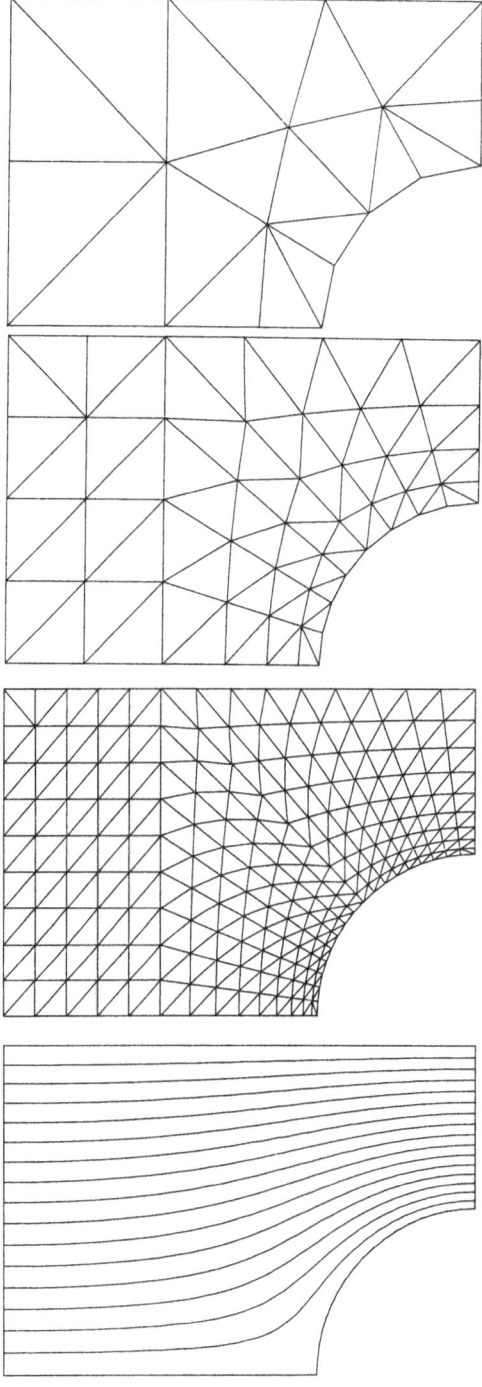

FIGURE 8    POTENTIAL FLOW PAST A CYLINDER
($\nabla^2 \psi = 0$) SEQUENCE OF GRIDS
AND SOLUTION

5.  <u>Burger's equation</u>:  this last example is used to show the performance of multigrid-methods for hyperbolic PDEs.   We consider the
solution of the inviscid Burger's equation, starting from the
initial configuration shown in Figure 9 .   We use 5 levels of
coarsening, and obtain a reduction in work of about 2.1. Although
this seems disappointing, others [10] have also reported similar
results.

ACKNOWLEDGEMENTS

The authors would like to thank the Aerothermal Loads Branch of
the NASA Langley Research Center for partial support of this research
under Grant No. NAGW-478, and especially A.R. Wieting and K.S. Bey
for this continued interest and encouragement.   J. Peraire ack-
nowledges the support given by the 'Ministerio de Educacion y
Ciencia' of Spain.

REFERENCES

1.  J. DONEA
    Int. J. Num. Meth. Eng. 20, 101-120 (1984).

2.  R. LÖHNER, K. MORGAN and O.C. ZIENKIEWICZ
    Int. J. Num. Meth. Fluids 4, 1043-1063 (1984).

3.  A. LAPIDUS
    J. Comp. Phys. 2, 154-177 (1967).

4.  R. LÖHNER, K. MORGAN and J. PERAIRE
    A simple extension of the artificial viscosity due to Lapidus
    to Multidimensional Problems, Comm. Appl. Num. Meth. To appear
    (1985).

5.  R. LÖHNER, K. MORGAN and O.C. ZIENKIEWICZ
    An adaptive finite element procedure for high speed compressible
    flows, Comp. Meth. Appl. Mech. Eng. To appear (1985).

6.  A. BRANDT
    ICASE - Rep. 79-8 (1979).

7.  A. BRANDT
    ICASE - Rep. 79-19 (1979).

8.  A. BRANDT
    Multigrid-Guide 1984.  To appear.

9.  R.H. NI
    AIAA Paper 81-1025 (1981).

10.  G.M. JOHNSON
     NASA TM 82843 (1982).

11.  M.J. BERGER and A. JAMESON
     Lecture Notes in Physics 218, 97-97 (1985).

12. CH. KOECK and J.J. CHATTOT
    Lecture Notes in Physics 218, 308-313 (1985).

13. R. LÖHNER and K. MORGAN
    The unstructured multigrid method for elliptic problems.
    In preparation (1985).

14. J.P. BORIS and D.L. BOOK
    J. Comp. Phys. 11, 38-69 (1973).

15. P.L. ROE
    pp. 2192-57 in Numerical Methods for Fluid Dynamics (K.W.
    Morton and M.J. Baines eds.), Academic Press (1982).

16. P.K. SWEBY
    Proc. AMS-SIAM Summer Seminar, La Jolla, California (1983).

17. A. HARTEN
    J. Comp. Phys. 49, 357-393 (1983).

18. P.A. GNOFFO
    AIAA J. 21(9), 1249-1254 (1983).

19. A.R. DIAZ, N. KIKUCHI and J.R. TAYLOR
    Comp. Meth. Appl. Mech. Eng. 41,29-45 (1983).

20. N. KIKUCHI and A.R. DIAZ
    pp. 41-47 of Proc. Fifth Int. Symp. on Finite Elements and
    Flow Problems (G.F. Carey and J.T. Oden eds.), TICOM (1984).

21. R. LÖHNER, K. MORGAN and O.C. ZIENKIEWICZ
    in Proc. ARFEC Conf. on Adaptive Refinement and Error
    Estimates in Finite Element Computations, Lisbon, June 1984.

22. K. NAKAHASHI and G.S. DEIWERT
    AIAA Paper 85-0486 (1985).

23. J.F. DANNENHOFFER and J.R. BARON
    AIAA Paper 84-0005 (1984).

25. M.J. BERGER and J. OLIGER
    J. Comp. Phys. 53, 484-512 (1984).

26. O.C. ZIENKIEWICZ, R. LÖHNER and K. MORGAN
    High speed inviscid compressible flow by the FEM, Proc. of
    the MAFELAP Conf. May 1984 (J. Whiteman ed.), to appear.

27. R. LÖHNER, K. MORGAN and O.C. ZIENKIEWICZ
    pp. 388-392 in Lecture Notes in Physics 218, Springer (1985).

28. B. PALMERIO
    Self-adaptive FEM algorithms for the Euler equations,
    INRIA Rep. No. 338 (1984).

29. R. LÖHNER, K. MORGAN and O.C. ZIENKIEWICZ
    An adaptive finite element procedure for high speed compressible
    flows, Comp. Meth. Appl. Mech. Eng. to appear (1985).

30. O.C. ZIENKIEWICZ
    The Finite Element Method, McGraw-Hill (1977).

31.  P. WOODWARD and P. COLELLA
     J. Comp. Phys. 54, 115-173 (1984).

32.  A. HARTEN
     NASA Ames Rep. NCA225 (1982).

33.  P. COLELLA
     LBL-17023. Preprint (1984).

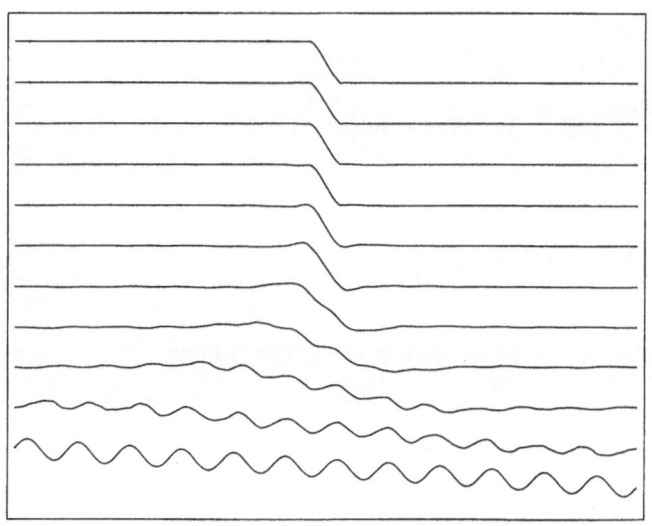

FIGURE 9    SOLUTION OF THE INVISCID BURGER'S EQU.
            VIA MULTIGRID-METHODS (128 ELEMENTS ,
            5 LEVELS , 1 PICTURE EVERY TWO CYCLES)

A CONSERVATIVE FREE LAGRANGE CODE

A.H.Armstrong

Atomic Weapons Research Establishment,
Ministry of Defence,
Aldermaston, England

CONTENTS

1  SUMMARY

    A Free Lagrange code for two-dimensional hydrodynamics, designed
specifically to facilitate later generalisation to three-dimensional
applications.

    The program sets up a network of triangular meshes filling a
convex region within specified boundaries. The nodes of the network
are Lagrangean in the region actually occupied by fluid, and
Eulerian elsewhere. Fluid nodes are preserved throughout the
subsequent hydrodynamic motion. The mesh lines connecting pairs of
nodes are changed automatically whenever appropriate.

## 2 INTRODUCTION

We adopt the normal Lagrangean concept of following the history of particles of the fluid (nodes of the network) throughout its motion. The entire "calculation region", occupying a fixed convex region of two-dimensional space, is covered also by a background Eulerian triangular grid. The nodes of this Eulerian grid are visible and effective only in those parts of the calculation region not actually occupied at that time by the fluid.

The fluid has the effect of a cloud or clouds moving across a starlit sky, obscuring and revealing the stars (background Eulerian nodes) as time passes.

The function of the Eulerian nodes is simply to keep account of the movement of the fluid boundaries, facilitating treatment of such phenomena as jet impingement, void formation and closure, etc.

The rates of change in density, energy and velocity of each fluid node during each time step are calculated from contour integrals round the boundary of the small area associated with that node. (If the area were to shrink to zero these contour integrals would give the exact equivalent of the usual partial differential equations of motion of hydrodynamics.)

Since the quantities are defined uniquely on each interface, and all exchanges, of area, energy and momentum, are made bilaterally between pairs of neighbouring nodes, all these quantities are exactly conserved.

The neighbours may change between one time step and the next.

## 3 NODES, LINES and MESHES

Our method is based on three interdependent concepts:

nodes: all the fluid's hydrodynamic properties are held by the nodes. Each node holds an updated list (in cyclic order) of all lines through it.

lines: each line joins two "neighbouring" nodes. Two lines may have a common end, but otherwise they may not intersect. Lines may be created or removed at the start of any time step, following certain precisely defined rules. Each line holds an updated list of its two nodes and its two ( or exceptionally one) mesh.

meshes: these are triangles and have an almost purely topological function. Each mesh holds an updated list (in cyclic order) of the nodes at its vertices, and the line opposite each vertex. A mesh has no physical properties, other than the "sign" of its area, which must be positive, and the position of its "centroid". The meshes form a "network" which covers the entire calculation region simply and completely, i.e. no overlapping and no gaps.

## 4 NODES: LOGICAL PROPERTIES

A node lies entirely inside a region of the calculational area: never on an interface or boundary.

A node, identified by an invariant positive integer, may be either one of two types:

(a) <u>real node</u>: this represents an actual particle of fluid, and is therefore Lagrangean. It is neither created nor destroyed at any time after the start of the hydrodynamic calculation.

(b) <u>virtual node</u>: this is an Eulerian node located in space (or spaces) available to the fluid, but not actually occupied by it. Together with the real Lagrangean nodes, they fully occupy the entire calculation region. One of their functions is to provide mutual logical awareness of two impinging parts of the fluid, as in void closure. Virtual nodes will also be referred to as "space nodes".

(Virtual nodes are not capable of motion, but can be obscured or revealed by the passage of any time step, according to defined rules.)

Virtual nodes must provide at least a fringe, with a non-concave outer boundary, round all parts of the fluid boundary, at all times during the duration of the problem.

Each node has a "material number" and a "region number", both of which are non-negative integers.

For a real node the "material number" identifies the Equation of State of that node. For virtual nodes the material number is zero.

The "region number" is used in setting up the mesh initially, each region, separated by internal boundaries, being populated with nodes in turn. The "region number" is also used subsequently for identifying the pressure region in which a virtual node lies.

Two nodes are "neighbours" when, and only when, they are connected by a "line" forming part of the triangular network. These lines may change between one time step and the next, in order to ensure that logical "neighbours" are also physical neighbours, in spite of shear motion, etc.

Each node (numbered N) is thus the logical centre of a "star of lines". This star consists of an ordered cyclic sequence of lines(L1, L2, L3, etc.):

$$LOFN(N,1) = L1$$
$$LOFN(N,2) = L2$$
$$LOFN(N,3) = L3$$
$$etc.$$

means "in the <u>l</u>ines <u>of</u> <u>n</u>ode N, the first is L1, the second is L2, etc.".

The number of lines in the star is called the "content". The content of a star is usually about 6, but it may be more or less.

Each line in the star has the node N (called the "star node") as one of its ends. The node at the other end of any line in the star is a "neighbour" of the node N.

Two non-zero consecutive lines in the star partially enclose a mesh, and the nodes at their other ends are neighbours.

If the number zero occurs in the star, it means that the lines whose numbers precede and follow the zero do not enclose a mesh, i.e. the nodes at their other ends are not neighbours. Two consecutive zeros are not permitted. After the initial setting up of the network, zeros may occur only in stars of such virtual nodes as are on the external boundary of the calculation region.


## 5 NODES: PHYSICAL PROPERTIES

A node is located at a point, but a real node is deemed to occupy a surrounding "nodal area", which abuts but does not overlap the areas occupied by the neighbouring nodes.

Each real node is deemed to possess certain physical properties:

nodal constant

mass (invariant during the hydrodynamic phase)    $m$

nodal primary variables

specific area (reciprocal density)                $a$
specific total energy (work)                      $e$
specific momentum vector (velocity vector)        $\underline{w} = (u,v)$
   position vector                 $\underline{z} = (x,y)$

nodal secondary variables

specific internal energy (total minus kinetic)    $T = e - \underline{w}.\underline{w}/2$
pressure                                          $p = p(a,\overline{T})$

The nodal area is sometimes more conveniently expressed through its "radius" $r$, i.e. the radius of a circle with equal area.

The specific internal energy of each real node is determined by subtracting its specific kinetic energy (defined through its specific momentum vector) from its specific total energy. The specific internal energy is, in fact, the "temperature" of the node, although it must be remembered that this temperature is measured on a non-linear scale, which is different for each material.

The pressure of each real node is determined from its "temperature" and its specific area (reciprocal density), using its equation of state, which may be different for each material. Since a real node (likewise a virtual node) never lies on an interface there is never any ambiguity about its equation of state, nor any question of 'mixed' equations of state. In calculating the resulting hydrodynamic motion, the nodal pressure is incremented by a non-negative pseudo-viscous pressure term depending quadratically on the rate of change of nodal area.

The pressure of a virtual node is specified (as a function of position and time) in the boundary conditions of the problem.

The change in position vector of each real node in each time step is determined by the product of the time step and the mean of the specific momentum vector at the start and the end of the time step.

# 6 BOUNDARIES

The outer boundary encloses the entire computation area, and has space regions on one side, "forbidden area" on the other. An inner boundary separates regions: either a "real region" containing "real nodes" from a "space region" containing "space nodes", or two real regions.

We recognise two types of inner boundary:

a): a "defined boundary" (hereafter called a "wall"), formed by a continuous sequence of straight lines and circular arcs, whose initial position and whose subsequent motion as a function of time are defined as part of the problem specification; or

b): a "free boundary", which may either separate two real regions, or a real region and a space region, for which the initial position, and in the latter case the pressure aa a function of position and time subsequently, are part of the problem specification.

No nodes of any kind are located on any boundary.

The fluid velocity at a wall is constrained by the requirement that the normal component of fluid velocity relative to that of the wall shall be zero.

In the initial setting up of the network each "free boundary" is regarded as a "wall". Subsequently the position of the free boundary is determined by the positions of the nodal boundaries of the nearest real nodes.


# 7 LINES

Each line (numbered L) joins two nodes (N1 and N2), one at each end. Each line separates two "back to back" meshes (M1 and M2) unless the line lies on the outer boundary, in which case one of these meshes is null.

$$NMOL(L,1) = N1 + N2$$
$$NMOL(L,2) = N1 - N2$$
$$NMOL(L,3) = M1 + M2$$
$$NMOL(L,4) = M1 - M2$$
$$NMOL(L,5) = 0 \text{ or a positive integer.}$$

NMOL(L,5) indicates the identifying number of the wall sector which this line intersects; if there is no such intersection this flag is zero.

The line is a part of the edge of the network if, and only if,

$$NMOL(L,3) = NMOL(L,4).$$

Each line has an "interpoint". In the case of a line joining two real nodes the position of the interpoint is the weighted mean of the positions of the two nodes.

In the "hydrodynamic phase", and for nodes in the same region, the weights are unity, in order to foster smoothness of the density and pressure profiles. For real nodes separated by a region interface "free boundary", the weights are the reciprocals of the nodal radii.

(In the "settling phase", which is a once and for all phase preceding the hydrodynamic calculation, the weights within a region are the nodal masses, and the nodal masses are revised to fit the resulting nodal areas and the specified densities. This weighting is gradually reduced to unity during successive cycles of the "settling phase". In this "settling phase" the nodal positions are also adjusted to coincide with the centres of mass of the nodal areas. This smooths out local inequalities of nodal masses within a region. In the "settling phase", the interpoints for lines crossing "free boundaries" are simply the intersections with the defined initial positions of these free boundaries.)

The velocity of the interpoint is the similarly weighted mean of the velocities of the two nodes. The interpoint pressure is similarly computed as a weighted mean, but additionally weighted (in the case of a line crossing a free boundary between two real regions) by the reciprocals of the respective nodal densities.

The interpoint velocity of a line crossing a "pressure boundary" (and therefore joining a real and a virtual node) is set equal to the velocity of the real node, plus an additional velocity component (called the "pressure differential effect" or p.d.e.) representing the particle velocity behind a normal shock wave caused by the difference in pressures at the two nodes. The interpoint position on such a line is obtained by integrating the interpoint velocity. The interpoint pressure is that of the space node.

For a line crossing a "wall" the interpoint is the perpendicular projection of the real node on the wall. The interpoint pressure is that of the real node, plus a "pseudo-viscous" component depending on the velocity normal and relative to the wall (called the v.d.e.).

If at least one of the two nodes is a virtual node the interpoint is said to be "forced". The properties of forced interpoints are summarised below. (Here we use R for a real node, S for a virtual node except when the line crosses a "wall", in which case W is used.)

| NODES | INTERPOINTS | | |
|-------|-------------|-------------|-----------|
| | Pressure | Velocity | Position |
| R/S | S | R ( + p.d.e.) | Calc. from vel.* |
| R/W | R ( + v.d.e.) | Projection of R on to the wall | |

* This interpoint is initially set at a distance from the real node equal to the nodal radius of the real node (and if the length of the line is less than this nodal radius, the space node must be deleted.)

At the end of each time step the calculated new position of the interpoint will not (in the case of a line crossing a "pressure boundary") lie exactly on the line joining the calculated new positions of the nodes: the new position of the interpoint is moved to the foot of the perpendicular on to the new line.

8 MESHES

Each mesh (numbered M) is a triangle with three nodes (N1,N2,N3) as its vertices, and opposite each vertex is a line L1,L2,L3 respectively.

$$NLOM(M,1 - 3) = N1,N2,N3$$
$$NLOM(M,4 - 6) = L1,L2,L3.$$

Each mesh has a "centroid", its location, pressure and velocity being (in the case of three real nodes) the "circum-offcentre". We define this as a slight generalisation of the well known circumcentre; it is the point such that the squares of its distances from the three nodes differ by the same differences as the squares of the nodal radii of the nodes (assumed reduced adiabatically to the same pressure.)

In the presence of "forced" interpoints, the "centroid" properties are as shown below.

| NODES | | CENTROID | |
|-------|----------|----------|----------|
| | Pressure | Velocity | Position |
| R/R/S | S | Mean of R/S, R/S | |
| R/S/S | Mean of S & S | Mean of R/S  R/S | |
| R/W/S | S | Projection of R/S on to wall | |
| R/R/W | Mean of R, R | Mean of R/W, R/W | |
| R/W/W | R | Mean of R/W, R/W | |

It will be noticed that this has the result that when a mesh is crossed by a boundary, the boundary conditions determine the position or the pressure, as appropriate, at the forced interpoints and at the centroid.

Each mesh is divided into three quadrilateral "sub-meshes" by the lines joining the "centroid" to the "interpoints" on the sides of the mesh. Obviously, the area of the mesh is equal to the algebraic sum of the areas of its sub-meshes. (Exceptionally the "centroid" may lie outside the mesh.)

The area of a sub-mesh pertains conceptually to the area of the node standing at one corner of the sub-mesh. Each non-boundary node is surrounded by its sub-meshes.

## 9 NODAL INTERFACES AND REGIONAL BOUNDARIES

If at least one of the two nodes on a line is real, then this line is also the common side of two "back to back" meshes, each of which has a "centroid" defined. The "interface" between the nodes at the ends of the common line consists of the joins from the "interpoint" on that line to the "centroids" of the two meshes. These joins are called the "arms" of the interface.

(Thus material boundaries are defined at all stages of the calculation in the normal course of defining nodal interfaces.)

The "interface arm normal vector" has magnitude equal to the length of the arm and direction normal to it, in the sense that makes an acute angle with the direction of the transfer.

The pressure and the velocity characterising an interface arm are defined to be the means of those of the mesh centroid and of the line interpoint, determined as in the previous section.

(Various other definitions have been tried, and eventually discarded. The simplest definition, the line interpoint value, was found to yield motions which were too erratic, involving only two nodal values. At the opposite extreme, a definition was tried which involved surface fitting to all the "neighbours" of all the four nodes involved in the two "back to back" meshes sharing the line. This procedure produced excessive smoothing of local nodal irregularities: the continuously curved surface fits could provide highly unrealistic interface values in cases where a real discontinuity, of pressure for instance, was moving across the mesh. The method now used provides interface values which are weighted means of, and necessarily bounded by, the nodal values at the four nodes involved in the two "back to back" meshes.).

## 10 CONSERVATIONAL CHANGES IN PHYSICAL PROPERTIES

The instantaneous rates of change in specific momentum, area, total energy and position in a continuous fluid during a time step dt can, in theory, be calculated from the local space derivatives by the equations:

$$\dot{\underline{w}} = - \, a.\mathrm{grad} \; p \tag{1.1}$$

$$\dot{a} = \quad a.\mathrm{div} \; \underline{w} \tag{1.2}$$

$$\dot{e} = - \, a.\mathrm{div}(p.\underline{w}) = - \, p.\dot{a} + \underline{w}.\dot{\underline{w}} \tag{1.3}$$

$$\dot{\underline{z}} = \quad \underline{w} \tag{1.4}$$

$$\dot{I} = - \, p.\dot{a} \tag{1.5}$$

In applying approximations of equations (1.1-4), however, with derivatives replaced by finite differences, there would be no intrinsic conservation of area, momentum or energy, although mass would be automatically conserved invariant in our Lagrangean scheme by its association with a real node.

Conservation of quantities within the interior of the fluid can be achieved most simply by ensuring that all exchanges are bilateral, i.e. whatever a star node gains, one of its neighbours loses, and vice versa. This is easily achieved, once we recognise that we are concerned, not with points and infinitesimals, but with small but finite regions of space (nodal areas) separated by common nodal interfaces.

The appropriate approach is to use contour integrals along the interfaces separating nodal areas as the quantities of intrinsic interest, not regarding these as providing approximations to partial spatial derivatives at the nodal centre itself, as is usually done. Conditions can be defined on these nodal interfaces in terms which apply equally to the nodal area on each side of the interface.

Each of the equations (1.1, 2, 3) is the limiting form, as the area shrinks to a point, of an integral over the area inside a closed contour enclosing the point. There is an equivalent equation related to it by Green's Theorem, namely:

(Instantaneous, point)                    (Instantaneous; finite space)

$$\dot{\underline{w}} = - \ a.grad \ p \quad (1.1) \qquad m.\dot{\underline{w}} = - \int p.\underline{n}.ds \ = \ \sum p_i.\underline{n}_i \quad (2.1)$$

$$\dot{a} = \quad a.div \ \underline{w} \quad (1.2) \qquad m.\dot{a} = \quad \int \underline{w}.\underline{n}.ds \ = \ \sum \underline{w}_i.\underline{n}_i \quad (2.2)$$

$$\dot{e} = - \ a.div(p.\underline{w}) \quad (1.3) \qquad m.\dot{e} = - \int p.\underline{w}.\underline{n}.ds \ = \ \sum p_i.\underline{w}_i.\underline{n}_i \quad (2.3)$$

where

$$\underline{n}_i = \int \underline{n}.ds$$

Here $\underline{n}$ is the unit outward normal to the closed integration contour, which follows the interfaces between the star node and its neighbours.

Each right-hand side is to be evaluated as the sum of the contributions due to each of the arms of those interfaces. For each interface, the values of $\underline{w}$, p, $\underline{n}$ and ds have already been defined uniquely in the previous section. Each of these values is unchanged (except for a change of sign) if the role of "star node" and "neighbour node" is interchanged. Transfer rates calculated in this way are strictly conservational.

Although a calculation based on equations (2) would ensure strict global conservation of area, as well as momentum and energy, it would not ensure that the allocation of area to nodes remained consistent with the nodal locations through a long succession of time steps. Each a would be obtained by integrating da/dt, and each $\underline{z}$ by twice integrating dw/dt (= $d^2z/dt^2$). The results eventually could become incompatible if the small errors were to accumulate in a systematic manner.

We can ensure the compatibility of the changes in the values of a with changes in the values of $\underline{z}$ if we deal, not with instantaneous rates of change with time, but rather with increments over small but finite time intervals $\delta t$.

For exact conservation, therefore, we replace equations (2) by

(finite time; finite space)

$$\delta \underline{w} = \delta t \sum p_i \cdot \underline{n}_i / m \qquad (3.1)$$

$$\delta a = \sum \delta a_i \qquad (3.2)$$

$$\delta e = \sum p_i \cdot \delta a_i \qquad (3.3)$$

where the values $p_i$, $\underline{n}_i$ on the right-hand side should be averaged over the time step, rather than taken at a particular instant. The quantities $\delta a_i$ are calculated directly from the trapezoidal areas swept out by that segment of the nodal interface.

For each time step we make a "predictive" computation, followed by a "corrective" computation. In the "predictive" phase, equations (2) are used, with the right-hand sides taking the values at the start of the time step, since at this stage that is all we know.

At this "predictive" stage we check that the predicted increments in $\underline{w}$, a, e, are not unacceptably large in comparison with the "start" values. In particular, we ensure that the specific area and the specific internal energy would not be reduced at any node by an excessive factor. We also check that $\delta t$ is small enough to satisfy a Courant condition at every node. If these criteria are not satisfied, $\delta t$ is reduced until they are. Notice that no question arises at this stage of "aborting" the time step, but merely of reducing its duration.

Next, the calculated increments for the primary properties of each node are added to the "start" values to give the predicted "end" values of $\underline{w}$, (and hence of $\underline{z}$), of a, and of e. Furthermore, the predicted "end" position, velocity and pressure of each line interpoint is calculated. The predicted "end" positions, velocities and pressures of the mesh centroids are then calculated.

Applying now equation (3.1), and using the mean of "start" and predicted "end" (i.e. "mid-step") values on the right-hand side, gives the corrected "end" values of w (and hence of z). The corrected "end" position of the interpoint on each line and hence the centroid of each mesh are now calculated in the manner previously specified.

At this point the ABORT criterion is applied. If any "inverted" meshes are present, the time step is now ABORTed (with a shortening of the time step) before any REMESHing is carried out. From this point onwards the time-step cannot be ABORTed. The final stage of the time step computation now proceeds.

The exact area $\delta a_i$ of the trapezium swept out by each interface arm during the time interval is now calculated, and this represents the increment of area transferred between the corresponding pair of nodes across that particular interface arm during the current time step.

Applying equations (3.2, 3), using "mid-step" values of $p_i$ on the right-hand side, gives corrected "end" values of a which are compatible with the corrected "end" values of $\underline{z}$, and the corresponding corrected "end" values of e.

REMESHing is now carried out, according to the criteria described in the following section.

Subsequently we re-calculate the "end" values of the nodal radii, internal energies, pressures, and hence the interpoint and centroid velocities and pressures, for use in updating the "start" values for the next time step.

The computation scheme is set out diagrammatically in Fig. 1. The columns represent successive procedures in the computation of one time step. The variable being computed is indicated by a CAPITAL LETTER (indicating the sub-routine), using as data the variables indicated by the symbol @. This form of diagram has been found useful in showing the dependence of variables upon other variables, the occasions when they are used (@), and when they were last updated. It also imposes a necessary self-discipline upon this particular code writer.

## 11 REMESH CRITERIA & PROCEDURE

We test the network in turn against three criteria, which are, in order,

(a) each line for "cyclicity" of the quadrilateral with this line as diagonal;
(b) each line for possible need to SPAWN a space node;
(c) each line for possible need to DISCARD a space node.

The order is important, since unsatisfactory "cyclicity" needs treatment before lines are considered for SPAWNing or DISCARDing.

Fritts and Boris (Ref.1) have shown that the preferable diagonal for a quadrilateral is that which joins the opposite pair of angles whose sum exceeds 180 degrees.

It will be noticed that, if the "centroids" of two back-to-back meshes coincide when the diagonal is swapped, the nodal areas will be unaffected by this procedure. This can be achieved, for example, by using the circumcentre for the "centroid", combined with the Fritts and Boris criterion for diagonal swapping. (Similarly, if the "centroid" is chosen in any other way such that the two centroids of two back-to-back meshes move along a common line and instantaneously coincide, a criterion for "diagonal swapping" can be devised that will ensure nodal area conservation.)

It is therefore convenient to have a measure of the angle A at any vertex of a mesh. Such a measure , which involves less computation than the use of inverse trigonometric functions, is the quantity

$f(A) = 1 - \cos A \cdot \text{abs}( \cos A )$.

This increases monotonically from 0 to 1 to 2 as A increases from 0 to 90 to 180 degrees, and its calculation does not involve even a square root. (After it had been devised by the writer, he discovered that this measure had been previously introduced by W.P.Crowley, LLNL USA.)

This function also has the useful property that f(A) + f(B) is less than or greater than 2, depending on whether A + B is less than or greater than 180 degrees.

We accordingly replace one diagonal of a quadrilateral by the other if, and only if, the change is necessary to ensure that the two angles joined by the diagonal shall be equal to or greater than 180 degrees.

The "line stars" of the four nodes involved, the nodes and lines associated with the two meshes, and the nodes and meshes of the four lines involved are appropiately up-dated.

Each faulty line of the network is corrected in numerical order in this way as it is detected, and the check then cycles from the first line. The recycling goes on until complete.

In a given space region (of which a void is a particular case) the number of space nodes inserted initially is governed by the area of the region. As the calculation proceeds, however, the size of this region may alter greatly, or even vanish, and new space regions may arise, so that it is necessary to be able to SPAWN or DISCARD space nodes at the end of any time step. It is necessary to specify a criterion and a procedure.

(Some of these routines may also be invoked in real regions when setting up the network - see Section 15.)

To decide whether SPAWN is appropriate, we find each line in turn which has a mesh on one side with space nodes only, but a real node in the other back-to-back mesh. Then an "image" space node is tentatively inserted in the mesh containing the real node, and tested to see whether it is too close to any of the existing nodes. If not, this new space node is incorporated into the network by joining it to each of the three nodes of the mesh in which it is located. (This method, though not the simplest, has the merit that new nodes fit into the regular pattern of the existing nodes.)

To decide whether DISCARD is appropriate, each line joining a real node and a space node across a "free surface" is tested to see whether its length is less than the nodal radius of the real node. If so, the space node is discarded, and all the non-redundant lines in its line star are incorporated into the line star of the real node.

## 12 CREATION OF VOIDS

When a new space region or "void" arises, this will occur in a region of real nodes, and will be indicated by the falling of the local pressure below a preset minimum (e.g. vapour pressure or zero). We therefore need to detect any line such that the real node at each end satisfies the "vaporisation pressure" criterion. We then invoke the "spawning" procedure to insert a space node at the interpoint and name this as a new region.

(The coding of the void creation facility is obviously a little tedious, but logically straightforward. It has therefore been left as a simple exercise for the professional programmer.)

# 13 TIME STEP CONTROL

When ABORT is invoked the time step is reduced by a factor of 2, and the calculation of the time step is recommenced.

To compensate for the reduction of the time interval in case of ABORT, we cause the time interval to be increased (by the square root of 2) for the start of the next time step, subject to the requirement that the maximum increase or decrease of the specific area or "temperature" associated with any node (in the time step just computed) be not far different from unity, and the additional requirement that the time step shall not exceed that permitted by a Courant condition, evaluated at the most unfavourable node. These rules constitute, in effect, automatic time interval control.

# 14 NODE BUNCHING

There may be a tendency, due to the chance accumulation of local statistical scatter in nodal velocities, for some nodal pairs to approach one another more closely than the local density values would justify. The position of a node is not exactly at the "areal centre", however this be defined, of its nodal area. If the displacement is marked, and in opposite directions for neighbouring nodes, these nodes will appear bunched.

An obvious remedy to adopt would be to "adjust" the pressure between each pair of neighbouring nodes, so that it was raised when the nodes were too close together, and lowered when they were too far apart. This remedy is effective in preventing node bunching, but it must be implemented with discretion in order to avoid a quite fundamental drawback.

The essential difference between an inviscid fluid on the one hand, and a viscid fluid or a solid on the other hand, is that in the first case the interaction between neighbouring particles can be entirely represented by a <u>scalar</u> pressure, i.e. a quantity with no directional properties whatever, whereas for a viscous medium there is a stress tensor.

Pressure "adjustments" between node pairs may violate this fundamental requirement of an inviscid fluid, by making the pressure different in different directions at a given point in the fluid. In our first implementation of this idea, the pressure was scaled according to the ratio of the distance separating neighbouring nodes to the sum of their effective nodal radii. Results obtained appeared plausible at first, but closer examination revealed features of the flow pattern which are characteristic of a highly viscous fluid or even a deformable solid. For example, in the zone of flattened meshes behind a plane shock wave, the "pressure" takes quite different values perpendicular to, and parallel to, the shock front.

And again, the problem of a jet impinging on a flat plate (Section 16.1), solved using this "pressure adjustment" device, gave a plausible shape for the free surface, but it was found that the jet suffered slowing down for many jet calibres before the point of impact with the plate. With an inviscid fluid jet this would not happen.

That particular implementation, therefore, has been discarded after extensive trial.

The implementation now incorporated avoids the above pitfall. For each star node one determines the effective "areal centre" of the nodal area, as the mass centre of the polygon bounded by the interfaces between the star node and its neighbour nodes. This "areal centre" will in general not coincide with the actual location of the node.

The mesh centroid pressure, calculated by the rules already described, is now divided by the ratio of the area of the triangle formed by the nodes themselves, to that formed by their "areal centres".

This has the effect of superimposing on the large scale inter-nodal pressure distribution a local intra-node pressure differential, which is always nudging the node towards its "areal centre": precisely the desired effect. Moreover, conservation is still retained. It does not have the undesirable effect of producing higher pressures along the short axis, and lower pressures along the long axis, of an elongated nodal area.

(Time has not permitted an investigation into whether this "anti-node bunching" device is really necessary with the present version of the code: it was introduced during an earlier stage of development, when it certainly was necessary.)


15 SETTING UP THE NETWORK

To set up a calculation one normally wishes to fill specified regions, each with a certain uniform material.

The regions are defined by specifying, in order, all the boundaries comprising first the interfaces between the regions and, secondly, the outer boundary of the calculation zone. These boundaries are temporarily regarded as walls, and they are placed in a numbered sequence such that access to the next region is obtained by overriding the walls in succession.

The scheme can be envisaged as a set of upright walls standing on a plane surface. Any given wall is of uniform height, but each wall is a little higher than its predecessor in the sequence. One imagines liquid being poured into Region 1: it is prevented from escaping by the surrounding walls. As the liquid continues to be poured, its level rises until it overflows all the walls separating Region 1 from Region 2, so that now Regions 1 & 2 are both occupied.

We need to specify a set of nodes with convenient positions in each region, and a legal network of meshes based upon them. A simple way of doing this is to construct, from some interior point of the first region (by a sort of triangular "crystal growth"), a system of equilateral triangular meshes, with nodes situated at all the mesh vertices. The mesh has to be sufficiently extensive to cover the regions so far flooded, as completely as possible without extending beyond the present boundary.

Continuing the process, we fill each additional region in turn until only the outer boundary of the whole calculation zone contains the extent of the fluid.

For each region it is required that the "material" filling it shall have certain specified properties. If the region is to contain "space nodes" we need only specify the position and pressure. If, on the other hand, the region is to contain "real nodes", the material must have a stipulated specific area (or density), a stipulated Equation of State, a velocity, and either a stipulated pressure or specific internal energy (i.e. temperature). (In Examples 3,4 we have assumed the initial velocity to be zero everywhere.)

If a region contains material of exceptionally high density it may be desirable to multiply the number of nodes in that region, either by the factor 3, by "spawning" a node at the centre of each mesh, or by 4 by "spawning" a node at the centre of each line, or by some intermediate ratio by allowing spawning in some parts of the region and not others. Subsequent "settling mode" motion will later redistribute the nodes in a more uniform manner throughout the region.

We locate all the line interpoints and mesh centroids in the manner described in Sections 7&8 and calculate the nodal area initially associated with each node. From the stipulated density of each region (whether uniform or not) we now calculate the total mass of each region. Dividing this by the number of nodes in that region, we find the (equal) masses of the nodes in that region.

(In some circumstances it may be desirable near a particular location to obtain greater definition by making the node masses smaller. The effect of the ensuing "settling mode" motion will then be to concentrate the lighter nodes in that area.)

During the ensuing "settling mode" iterations, we treat all boundaries as "walls" and move each real node to its "areal centre" and then recalculate its nodal area and hence mass, and iterate.

## 16 NUMERICAL EXAMPLES

### 16.1 EXAMPLE 1

A model used in development of an earlier version of this code was the numerical example presented by J.V.Dukowicz in a recent LASL report describing another Free Lagrange code [ref.2]. That code uses the Voronoi-Delaunay mesh, and makes conventional use of partial differential equations, but otherwise the two codes have some similarities.

Dukowicz's example treats a two-dimensional jet impinging normally on a flat plate. The fluid in the jet satisfies the Chaplygin equation of state, which can be written as

$$p = ( 1 - \tau/\tau_0)\ c_0^2/\tau_0$$

Here $p$, $\tau$ and $c$ are the pressure, specific volume and sonic speed, respectively, and the subscript zero refers to conditions far upstream in the jet. The undisturbed velocity of the jet is $c_0/2$, i.e. the upstream Mach number is 0.5.

The starting point of the problem is the instant when the jet first strikes the plate. After a sufficiently long time the configuration should approach a steady state, for which the form of the free surface is known analytically and is quoted by Dukowicz.

This example was run in 1981 with an earlier version of the present code (when, for example, explicit "wall nodes" were still being used), and using both a p.d.e. (i.e. non-conservational) formulation and also a finite element (conservational) formulation. Time has unfortunately not permitted repeating this example with the present version of the code. Fig. 2 shows the mesh and the pressure distribution at an advanced stage of the problem.

## 16.2 EXAMPLE 2

Another problem, attempted unsuccessfully several years ago with an earlier version of the code, was a "stirred tea-cup". The inner core of fluid is given an initial solid rotational motion, the outer annulus being stationary. One might expect to see Helmholtz waves developing on the interface. Once again, time has not permitted a repetition with the present version of the code.

## 16.3 EXAMPLE 3

The numerical example illustrated in Figs. 3-5 was used as the main development vehicle for this program.

A fluid piston of ideal gas (gamma = 2.0) is contained between two pressure boundaries and two straight parallel fixed walls. The pressure boundaries are initially straight lines, perpendicular to the walls. Fig. 3a shows the nodal connections in the initial configuration, Fig. 3b the nodal interfaces.

The upper pressure boundary is held at a constant pressure of 4 units. The material of the piston, together with the lower pressure boundary, are initially at a pressure of 1 unit and a density of 100 units.

A plane shock (theoretical speed: 0.2550 units) passes through the fluid until it strikes the lower pressure surface and is then reflected as a plane rarefaction wave through the piston, as the lower surface of the latter flies downwards. The shocked fluid has a theoretical speed of 0.1177 units and a density of 185.7 units.

With a more realistic equation of state, giving a liquid nature to the piston, the lower pressure boundary could be set at zero pressure throughout the calculation. The example was not intended, however, as a study in detailed equations of state.

The initial "beehive pattern" mesh has three-fold symmetry, as opposed to the two-fold symmetry available with rectangular meshes. As a result, the most drastic change in the orientation of the mesh is obtained by a rotation of one right-angle. The first configuration gave 119 real (i.e. Lagrangean) nodes, arranged in 14 rows of alternately 8 and 9 nodes per row. The 111 time steps involved a total of 482 "diagonal swaps".

The second configuration gave 112 real nodes, arranged in 25 rows of alternately 4 and 5 nodes per row. The 111 time steps involved a total of 314 "diagonal swaps".

This example involved one shock surface and two free surfaces. The total time was little different in the two configurations, the first giving a time of about 7.5 milliseconds per real node per time step on an IBM 3083 B.

Fig. 4 shows the flow field (first configuration) when about half the fluid has been shocked. Shown also are the isobars in the flow field, and on the right-hand side, the profiles of pressure and vertical velocity component (the ordinate of each point being the same as the ordinate in the flow field).

Fig. 5 shows the history of the position (with standard deviation) of the shock surface (identified by the highest values of the pseudo-viscous pressure term) and that of the upper free surface. The shock is seen to move at the theoretical speed. Although the interface ("piston") was a little sluggish from t = 0. to 0.5, it maintained the theoretical speed between t = 0.5 and t = 1.5. Fig. 5a refers to the first configuration, Fig. 5b to the second configuration.

16.4 EXAMPLE 4

An example of more current interest is Dr. Sod's shock-tube problem. To facilitate critical examination of the numerical results, however, we have modified the initial data slightly to give initial densities of 0.8741 and 0.10526 at the high and low pressure ends of the shock-tube. The pressures remain at 1.0 and 0.1 (gamma = 1.4) as specified by Dr. Sod, and the effect is that the analytical solution now gives a shock pressure of 0.3, an interface speed of exactly unity, and a shock speed exactly twice that of the interface.

Once again the problem was run with two alternative meshings, one being at right-angles to the other. The first gave 272 real nodes, arranged in 32 rows of 8 and 9 nodes alternately per row: 417 "diagonal swaps" occurred in 111 time steps. The second configuration gave 252 real nodes, arranged in 56 rows of 4 and 5 nodes alternately per row: 300 "diagonal swaps" occurred in 103 time steps. In this example the second arrangement gave results (Figs. 8 and 9b) appreciably smoother than the first (Figs. 7 and 9a).

Discounting an over-eager start during the first half unit of time, the interface ("piston") speed agrees well with the theoretical speed (1.0). The shock speed in the second configuration also agrees to within a few per cent with the theoretical speed (2.0).

Involving one shock only, this example took about 6 milliseconds per real node per time step on an IBM 3083 B.

REFERENCES

1. M.J.Fritts and J.P.Boris "The Lagrangean solution of transient problems in hydrodynamics using a triangular mesh." J.Comp. Phys. 31, 173 (1979).

2. J.K.Dukowicz "Lagrangean fluid dynamics using the Voronoi-Delaunay mesh" Los Alamos Scientific Laboratory, presented at the International Conference on Numerical Methods for Coupled Problems at Swansea, Wales (1981).

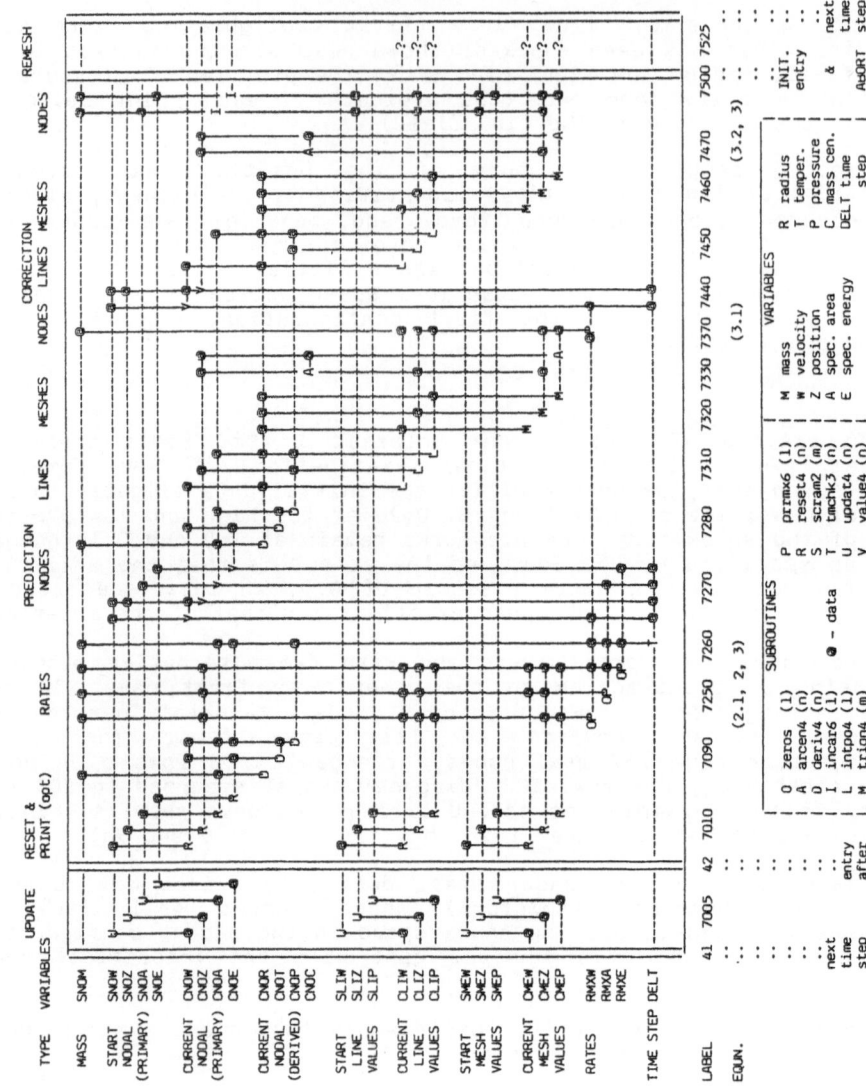

FIGURE I   HYSTEP COMPUTATIONAL SCHEME (1985)        0022E   22/02/85

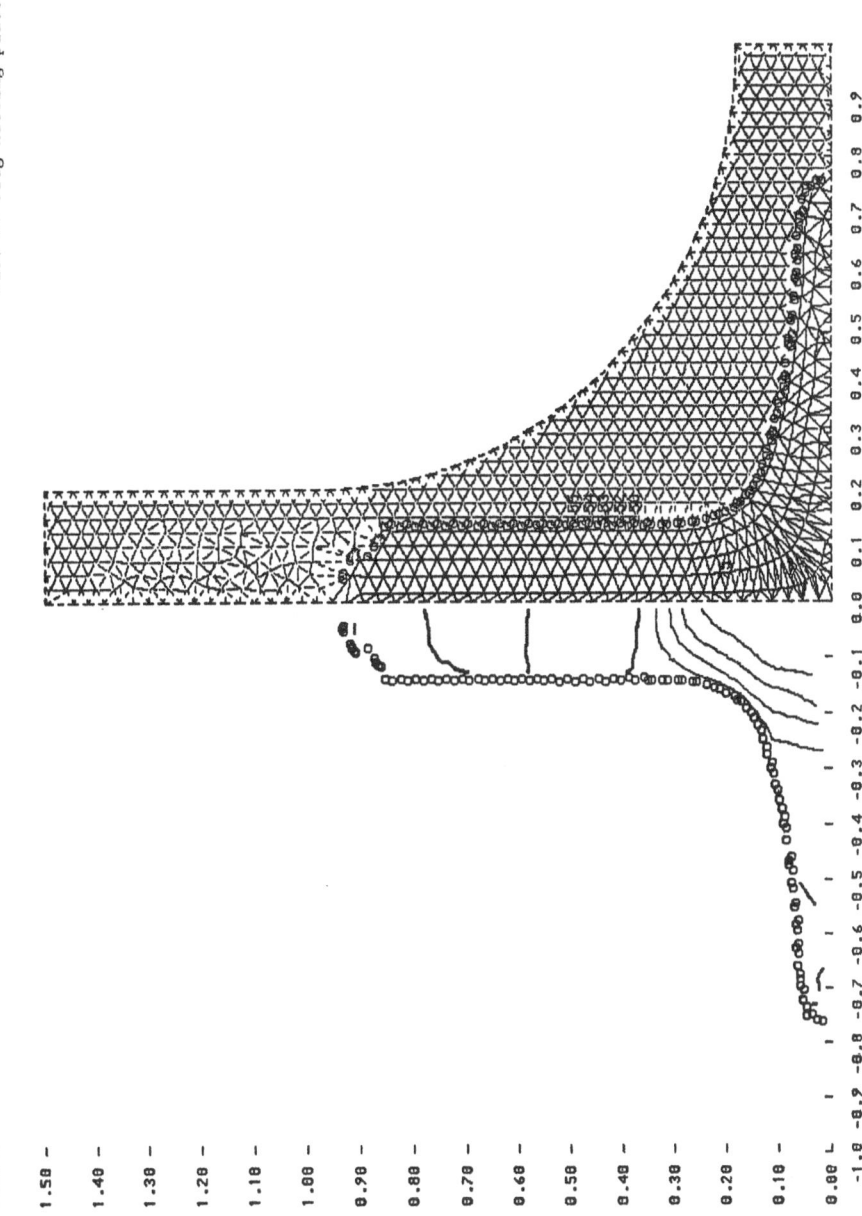

GAHAHYD     81/18/81     18.42.29

SPOOL    1  26

( 246)         248
STEP        8.8828
TIME        8.473

FIG. 2: Slug hitting plate

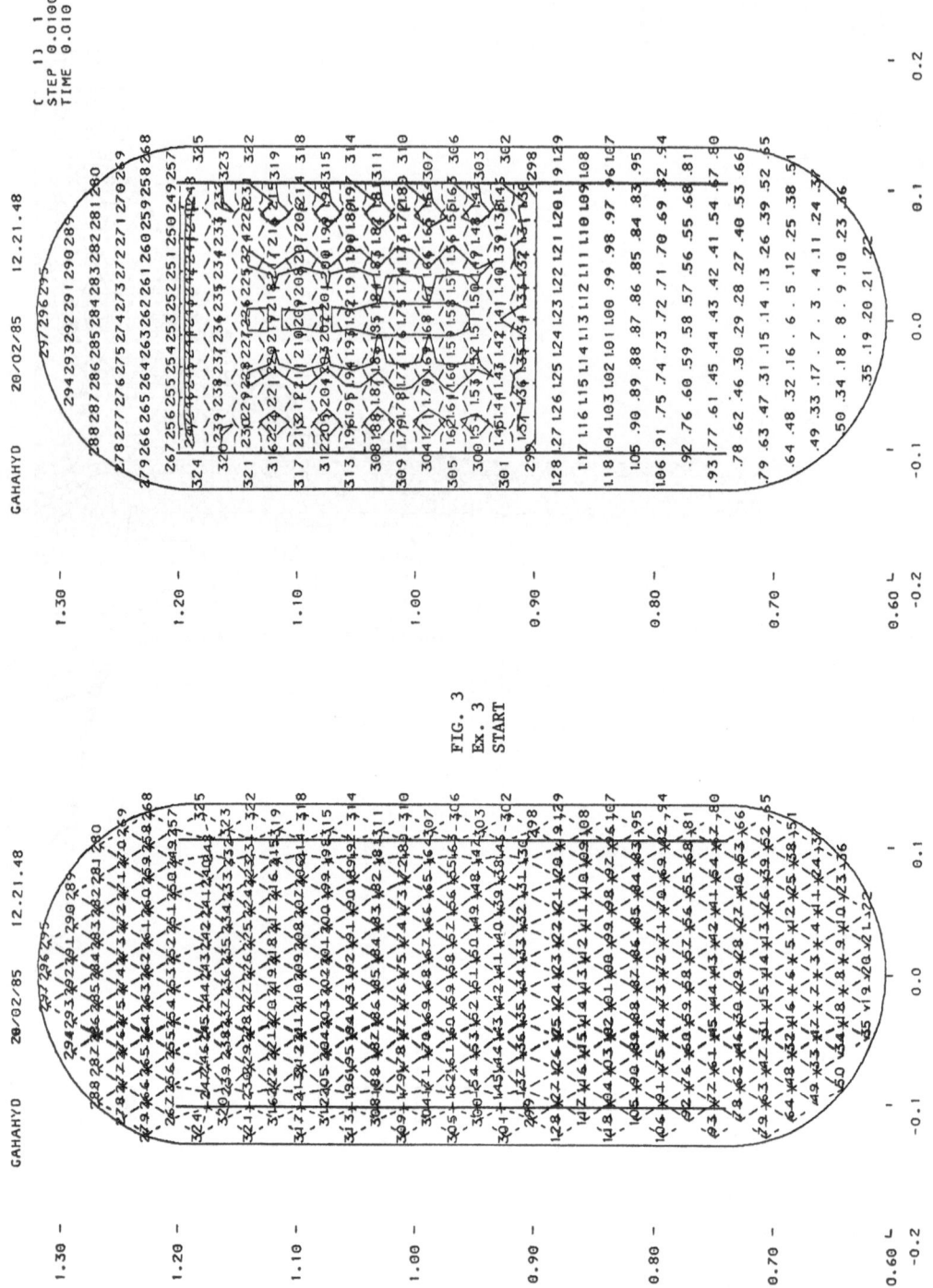

FIG. 3
Ex. 3
START

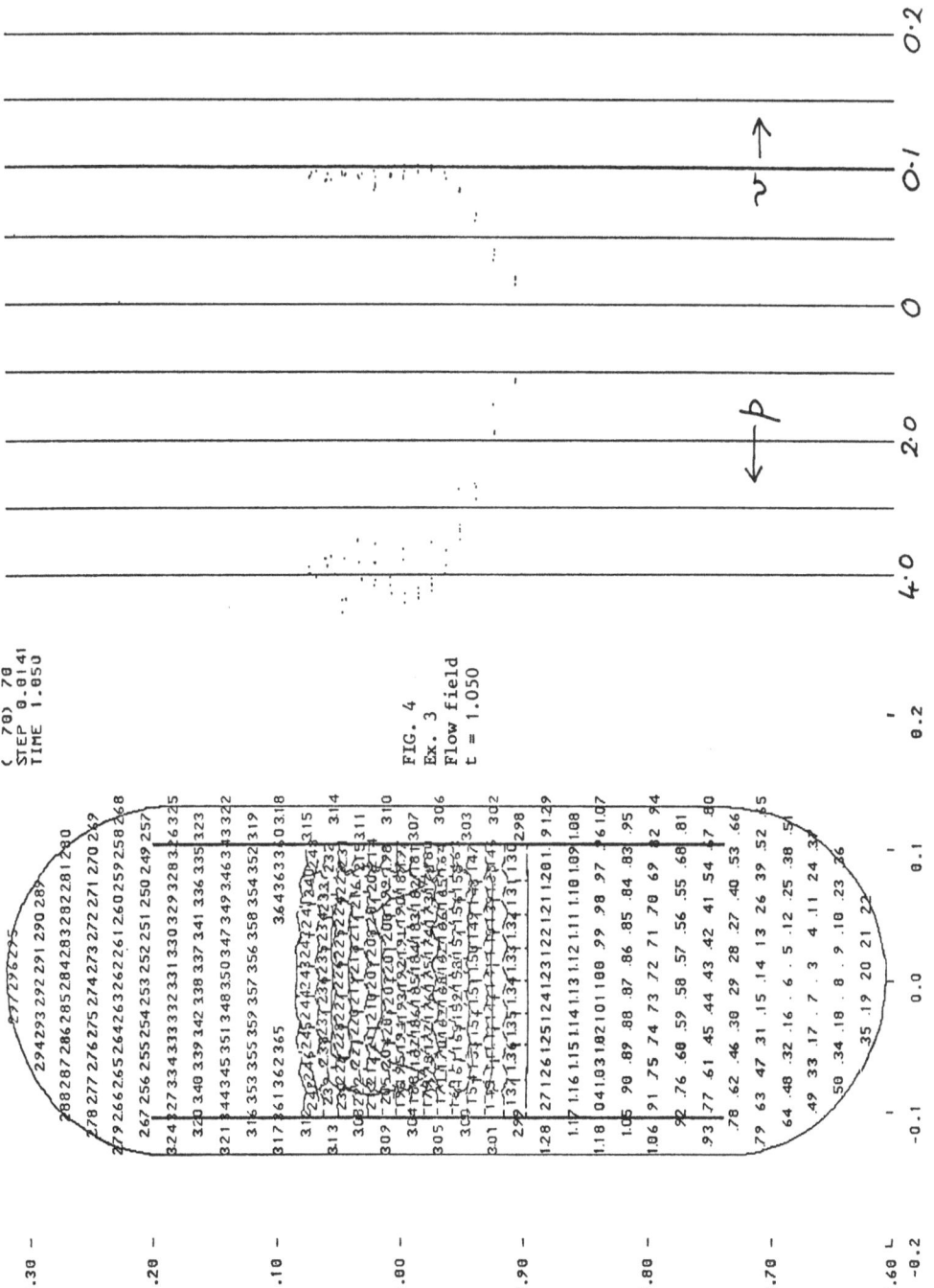

FIG. 4
Ex. 3
Flow field
t = 1.050

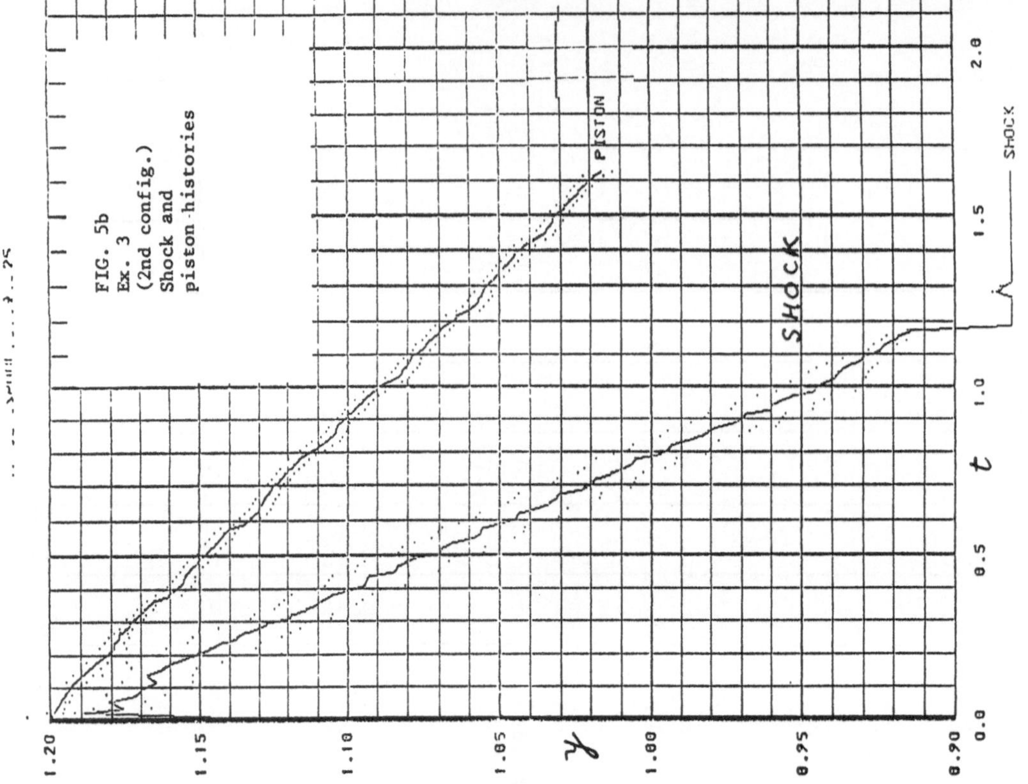

GAHAHYD   19/02/85   11.48.52

FIG. 5b
Ex. 3
(2nd config.)
Shock and
piston histories

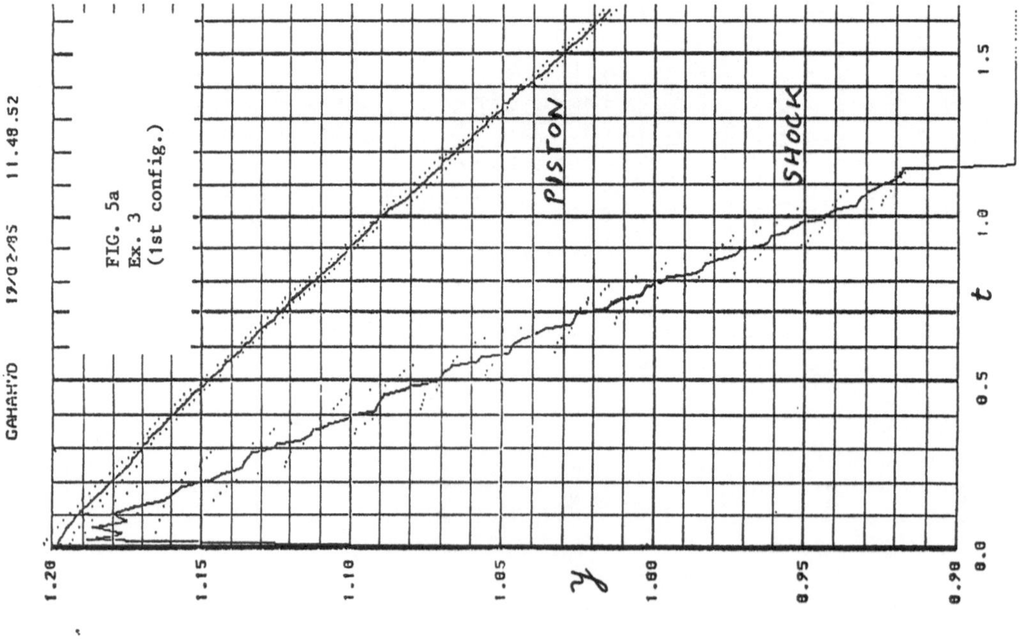

FIG. 5a
Ex. 3
(1st config.)

277

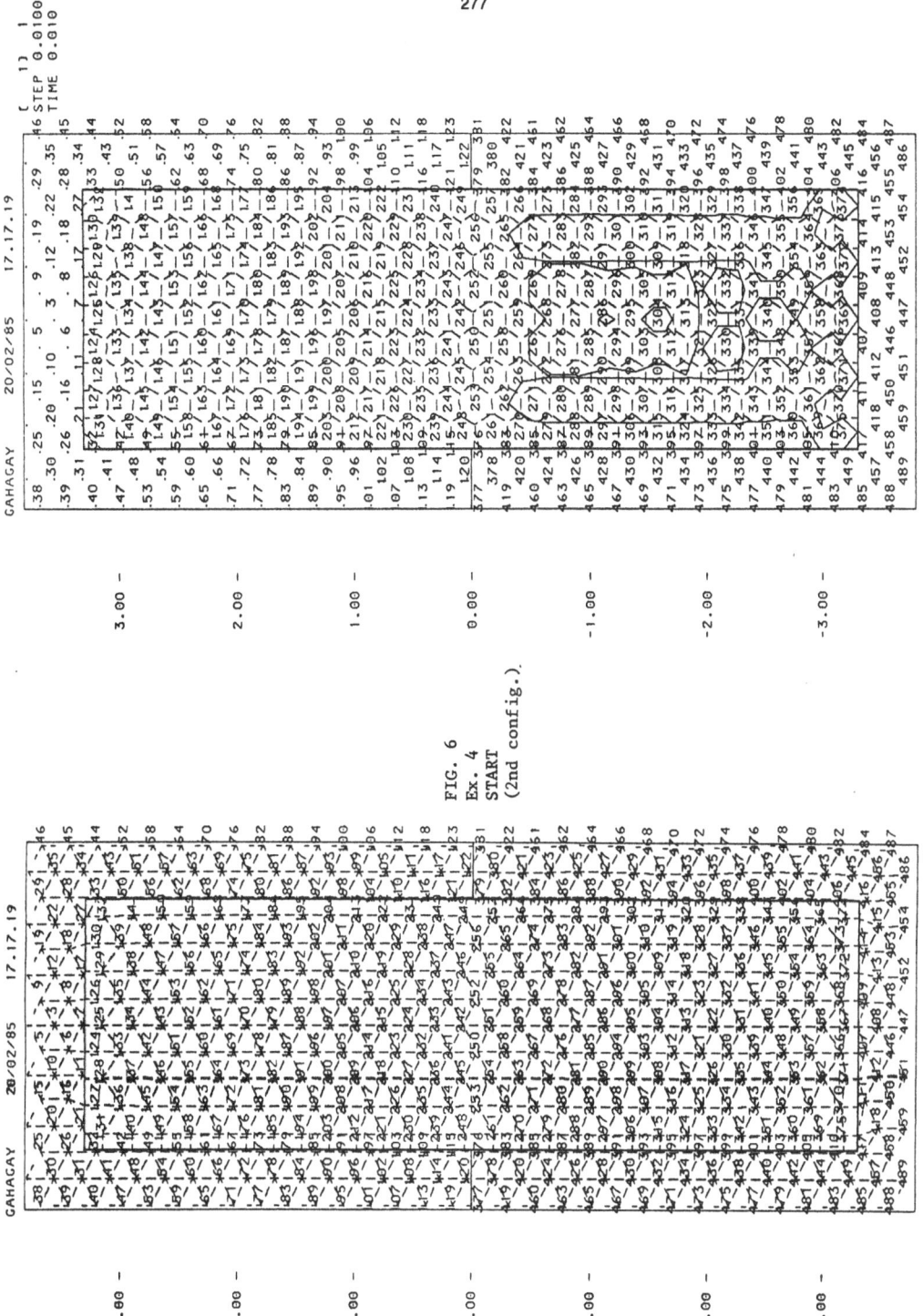

FIG. 6
Ex. 4
START
(2nd config.).

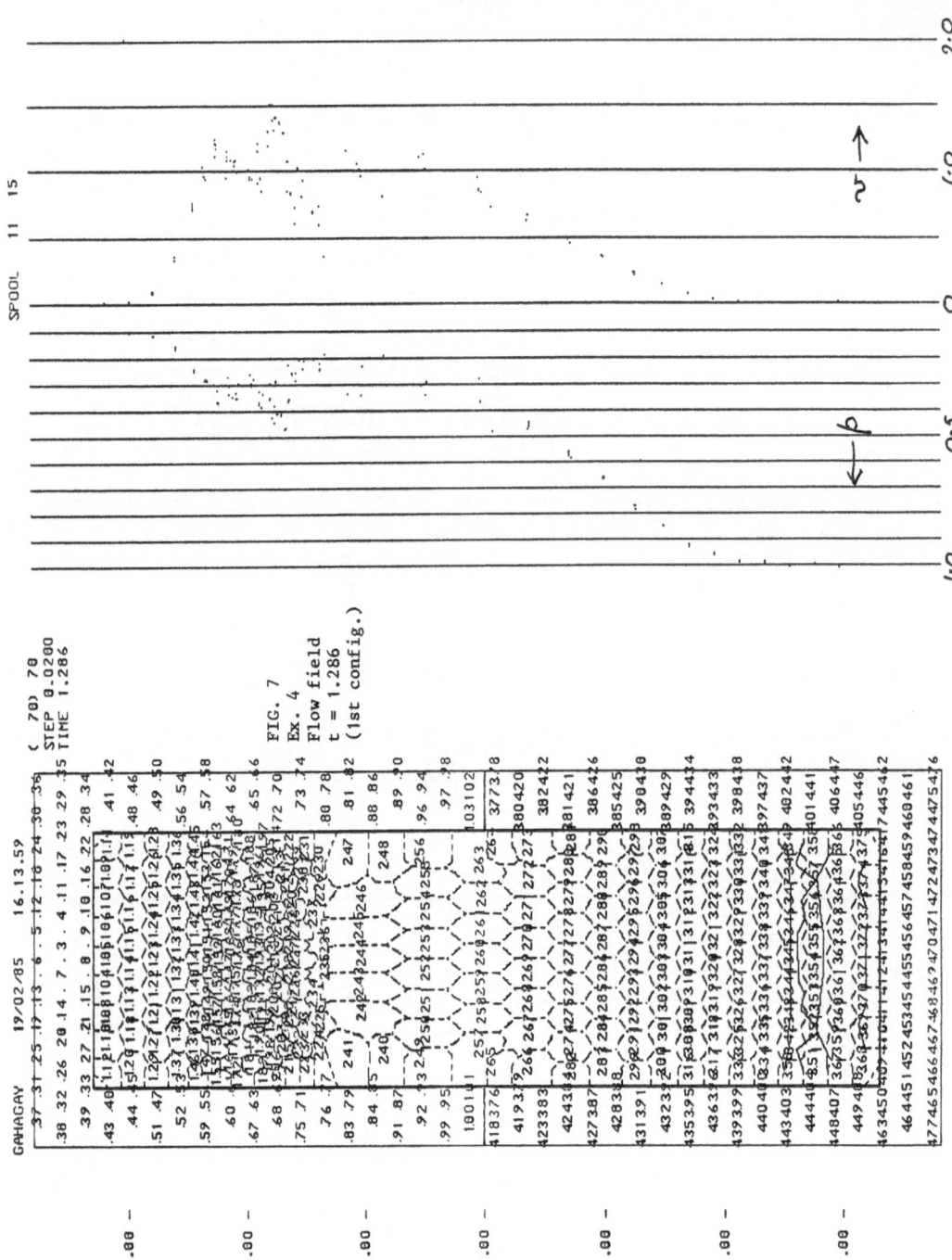

FIG. 7
Ex. 4
Flow field
t = 1.286 (1st config.)

279

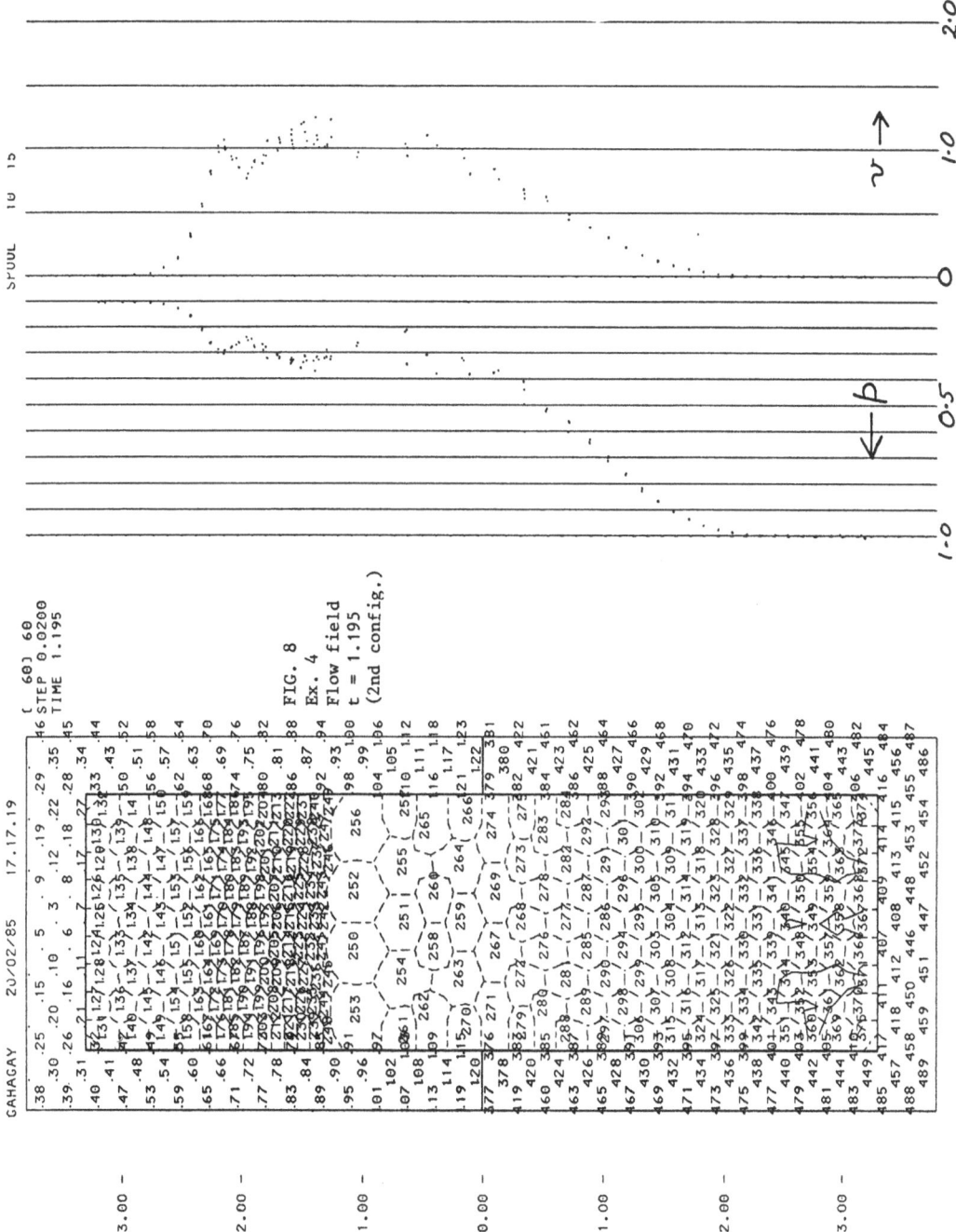

FIG. 8
Ex. 4
Flow field
t = 1.195
(2nd config.)

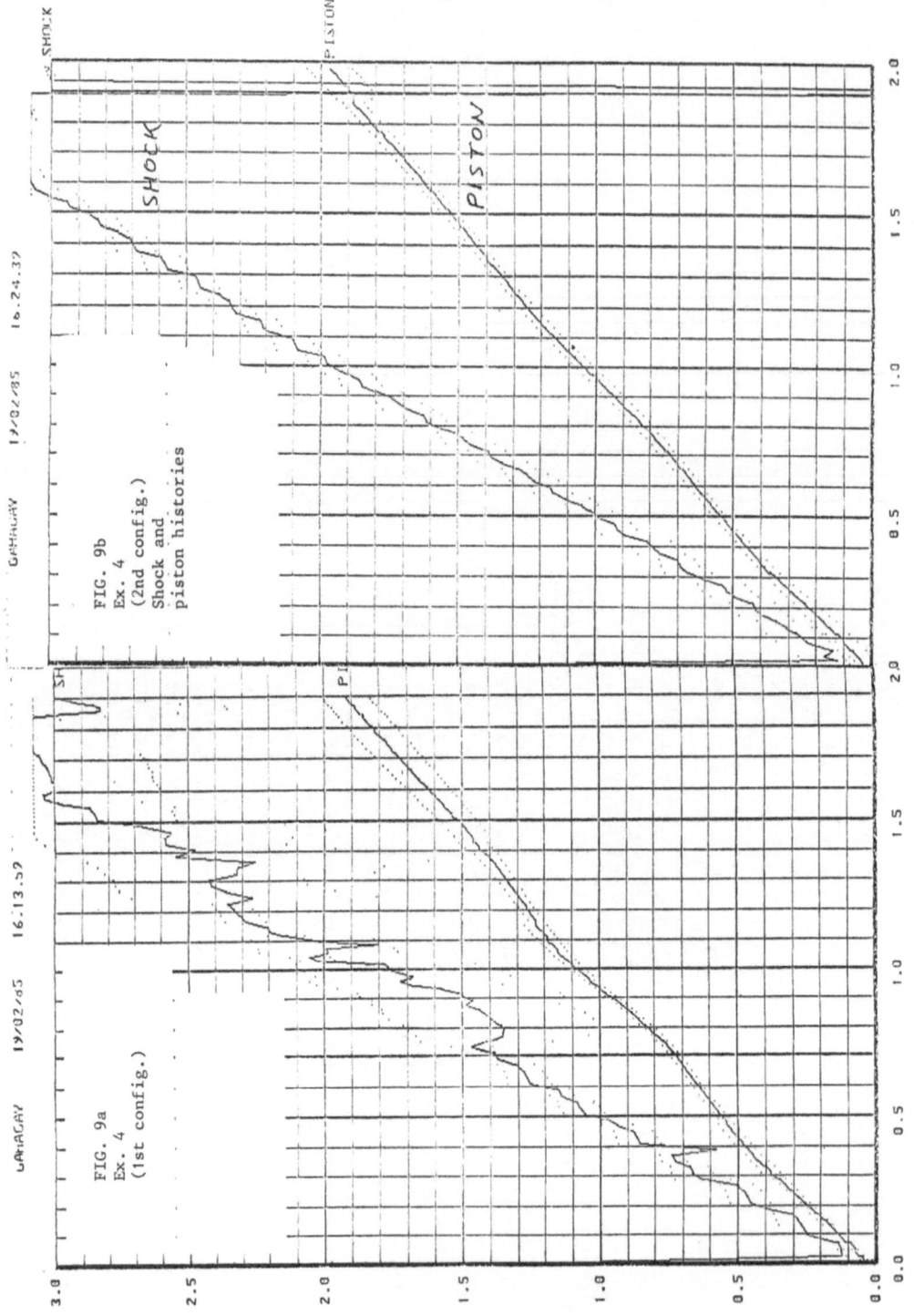

FIG. 9a
Ex. 4
(1st config.)

FIG. 9b
Ex. 4
(2nd config.)
Shock and
piston histories

# COMPRESSIBLE LAGRANGIAN HYDRODYNAMICS WITHOUT LAGRANGIAN CELLS

Robert A. Clark
Computational Physics
Group X-7, MS B257
Los Alamos National Laboratory
Los Alamos, New Mexico  89745

## I.  INTRODUCTION

The formulation normally used to calculate compressible Lagrangian hydrodynamics in two dimensions is the following. First define a two-dimensional mesh containing a set of Lagrangian cells. Assign each cell a fixed mass. Compute the acceleration of the mesh points and move the points. The volume of the cell changes with the motion of the points. The changes in cell density, energy, and pressure are computed from the changes in volume. Difficulties occur when there are large distortions in the flow that cause similar large distortions in the Lagrangian cells. The usual solution is to somehow adjust the mesh as the calculation proceeds. This involves either moving individual mesh points or actually re-connecting the mesh. In either case, it becomes necessary to re-map the mass from the old cells to the new. This necessarily produces some amount of undesirable numerical diffusion. When and how to adjust the mesh and how to accurately re-map the mass and other variables so as to minimize numerical diffusion are the problems.

One way to eliminate these problems is to abandon the idea of the Lagrangian cell since it is the distortion of the Lagrangian cell that is the cause of all the other problems. In the next section we will discuss how the conservation equations can be solved directly without resorting to Lagrangian cells. Next we will give some examples of calculations using this method. Finally, we will give details of the calculational method presently being used.

## II.  SOLVING THE CONSERVATION EQUATIONS

The equations we are trying to solve can be written

$$\frac{D}{Dt} \rho = -\rho \, \vec{\nabla} \cdot \vec{U} \qquad\qquad [2.1]$$

$$\frac{D}{Dt} \vec{U} = - \frac{1}{\rho} \vec{\nabla} P \qquad\qquad\qquad [2.2]$$

$$\frac{De}{Dt} = - \frac{P}{\rho} \vec{\nabla} \cdot \vec{U} \qquad\qquad\qquad [2.3]$$

$$P = P(\rho, e) \qquad\qquad\qquad [2.4]$$

where $\vec{U}$ represents the vector velocity, $\rho$ the density, e the specific internal energy and P the pressure of the fluid. Equation [2.1] expresses conservation of mass, [2.2] conservation of momentum and [2.3] conservation of energy. The Lagrangian time derivative, i.e., the derivative following the fluid, is indicated by $\frac{D}{Dt}$.

In a standard Lagrangian calculation only Eq. [2.2], the momentum equation is solved directly. The procedure is to integrate [2.2] over some region of space to arrive at the acceleration of each mesh point. The mesh points are then moved and the new cell volumes along with the fixed cell mass determine the new density, hence, indirectly solving Eq. [2.1]. The associated PdV work term updates the cell energy and indirectly solves Eq. [2.3] and the new pressure is obtained from the equation of state [2.4].

We propose the following: Instead of Lagrangian cells, we think of a set of Lagrangian points which are embedded in and move with the fluid. There is no mass associated with these points. They are just moving tracer points at which we will attempt to keep track of the velocity, density, energy, and pressure of the fluid. In our later example calculations we will show point positions at various times in the calculation. At each of these points, we know the density, energy and velocity of the fluid, but we do not associate any particular mass with the point.

Looking now at Eq. [2.1], we note that to approximate the time integral of the density change from time t to time t + δt we need an approximation to $\vec{\nabla} \cdot \vec{U}$ at that point. To solve Eq. [2.2], we need an approximation for $\vec{\nabla}P$ and for [2.3] we again need $\vec{\nabla} \cdot \vec{U}$. To obtain these, we select a set of "representative" neighbors. We then make a finite difference approximation to $\vec{\nabla}P$ and $\vec{\nabla} \cdot \vec{U}$, using these neighbors, and update $\rho$, $\vec{U}$ and e at each point. Each point is then moved the distance $\vec{U}$ $\delta t$ and one time step is completed.

At the next time step the selection of a set of "representative" neighbors may change, but this does not require any sort of re-mapping of variables. It only means that a different set of points will be used in the next finite difference approximation to $\vec{\nabla} \cdot \vec{U}$ and $\vec{\nabla}P$. Large distortions in the flow will produce frequent changes in neighbor selection, but since there are no cells to distort and no re-mapping to be done the calculation proceeds from cycle to cycle with no difficulty.

## III. SOME EXAMPLE CALCULATIONS

3.1  Here we will give three examples of calculations performed by the code HOBO using the free Lagrangian method described herein.  The first test problem is the implosion of a gaseous sphere.  The initial condition is a sphere of perfect gas with a gamma of 5/3.  The gas is divided into four regions as seen in Fig. 3.1.  Pressures are in megabars, density in gm/cc and dimensions in cm.

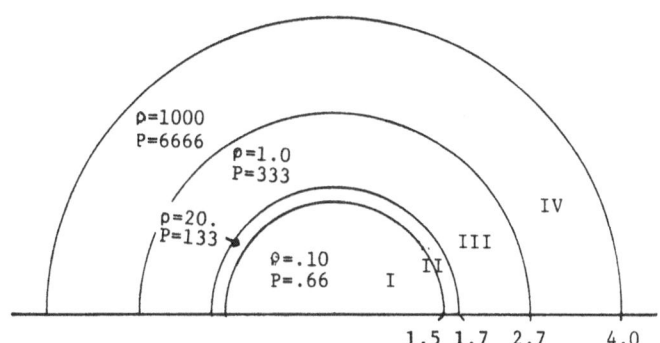

Fig. 3.1

The high pressure in region IV will drive a spherical implosion which will greatly compress region III, II, and particularly I.  There are two challenges to this problem, the first is to maintain a spherical ball while running the calculation in cylindrical (r,Z) geometry.  Six snapshots of region II are shown in Fig. 3.2.  Region I is interior to region II.  The minimum volume of region I occurs in the fifth snapshot after which region I begins to expand.  We ran 1000 calculational cycles with 73 points in the radial direction and 64 points covering 180° of angle. The left half of the snapshot is a reflection of the right half which was calculated.

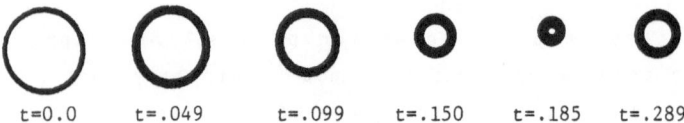

t=0.0    t=.049    t=.099    t=.150    t=.185    t=.289

Fig. 3.2

The second challenge is the accuracy of the solution. For comparison purposes we ran a standard one-dimensional Lagrangian code using 800 zones, 200 zones in each region. In Figs. 3.3.a, b, c, and d. We have plotted the average density and average specific internal energy in regions I and II as calculated by HOBO with 73 points in the radial direction and the one-dimensional Lagrangian calculation with 800 points. We feel the agreement to be quite good. One notable difference is the time at which minimum volume is reached. HOBO is

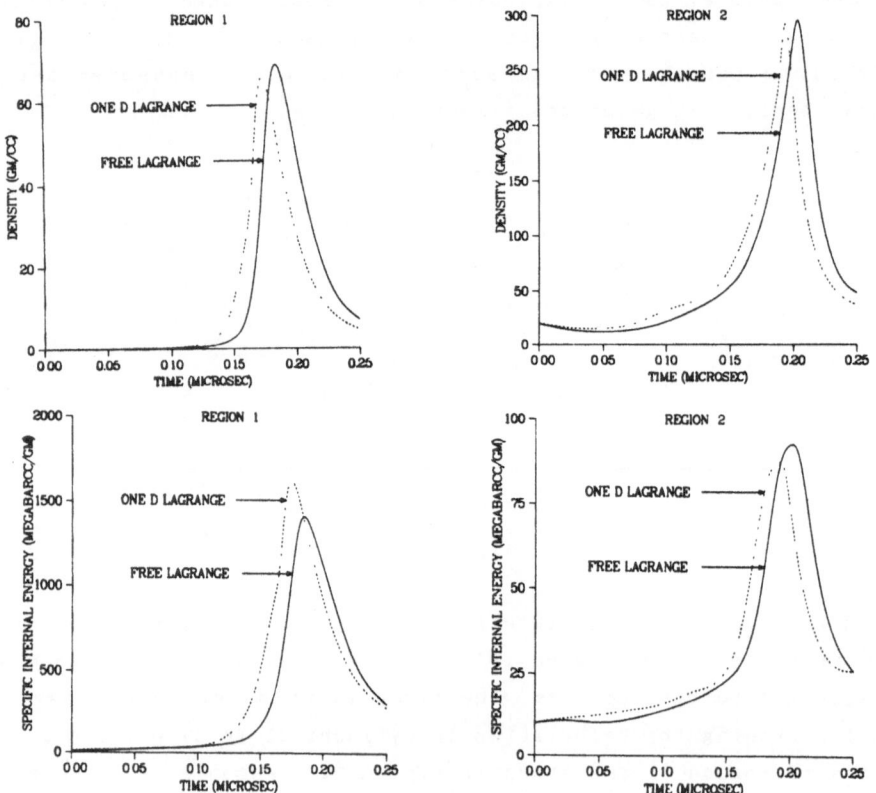

Figs. 3.3.a, b, c, and d

slow by about .0075 µsec or 4% of the problem time at that point.
Since average density and energy are integral quantities we have
plotted one of the variables as a function of radius in Fig. 3.4. We
chose radial velocity, but the agreement in all other variables is
very similar. The plots are from slightly different times to compensate for the time shift just mentioned. The 1D Lagrange plot is at
2.125 µsec and the HOBO plot is from 2.25 µsec. Apart from the in-
ability of the more coarsely zoned HOBO to resolve the shock front at
the radius 1.2 cm we feel the agreement is excellent. The time chosen
for the plot is late in the calculation when region II has expanded
almost back to its original volume.

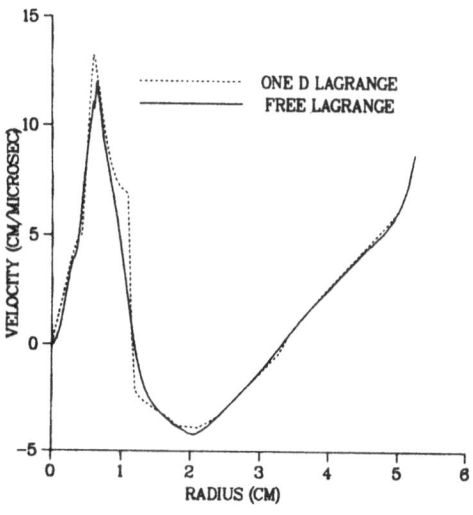

Fig. 3.4

3.2. For our second test problem we have chosen a Meshkov
instability[1] based on the geometry used in one of Meshkov's
experiments. The initial conditions are shown in Fig. 3.5. A piston
driven shock is driven through a region of air and then helium. The

| 33.8cm | 12.675cm | 16.9cm |
|---|---|---|
| air γ=1.4<br>P=1.8289<br>ρ=1.795x10⁻³<br>U=15.31 | air γ=1.4<br>P=1.01325<br>ρ=1.184x10⁻³<br>U=0. | He γ=1.63<br>P=1.01325<br>ρ=1.664x10⁻⁴<br>U=0. |

Fig. 3.5

air to helium density ratio is just over 7. There is an initial perturbation in the air -- He interface which grows with time after the
shock passes through the interface. In Fig. 3.6 we plot several snapshots of the Lagrangian point positions in the air (the He is not
plotted). For comparison purposes we ran the same problem on a two-
dimensional Eulerian code with the cell size similar to the point

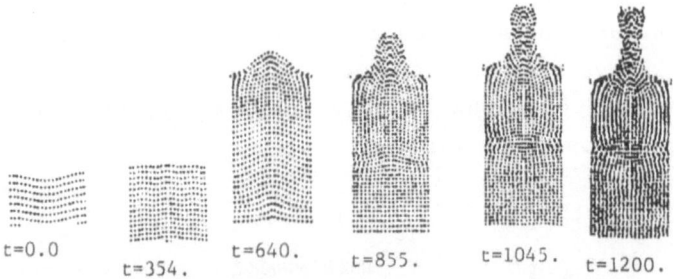

t=0.0  t=354.  t=640.  t=855.  t=1045.  t=1200.

Fig. 3.6

separation used in HOBO. In Figs. 3.7a and b, we compare the size of
the perturbation as it grows in time. In 3.7.a the initial perturbation, $\delta$, is .2 cm and in 3.7.b it is .4 cm in width. The agreement
between the two codes is excellent.

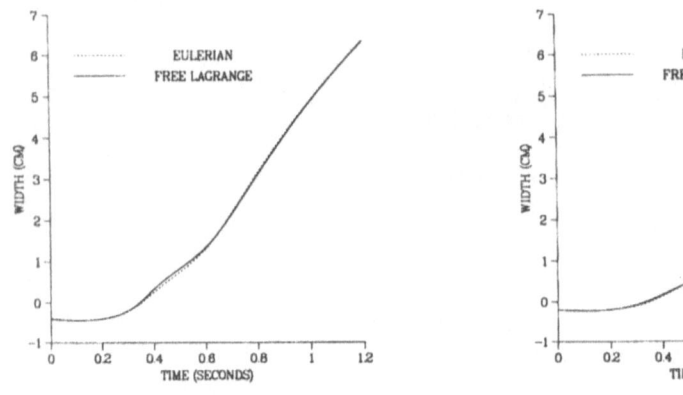

Fig. 3.7.a                    Fig. 3.7.b

3.3. Our third test problem is the penetration of a concrete plate by
a steel rod moving at an initial velocity of .2134 cm/μsec. The rod
is 9.066 cm in diameter and 45 cm in length. The concrete is 50 cm
thick. In Fig. 3.8 we show six snapshots of the rod penetrating the
concrete. Incompressible theory[2] predicts a constant time rate of

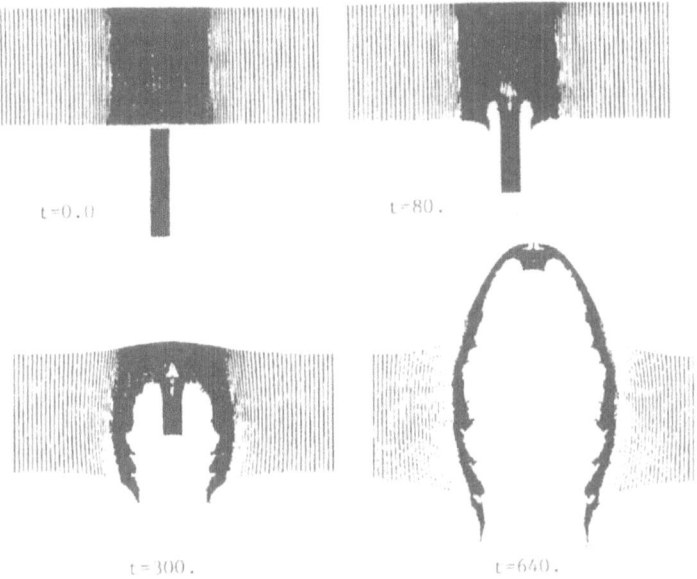

Fig. 3.8

change in the length of the steel rod. The sound speed in the rod is
.4545 cm/μsec and $(v/c)^2$ = .22, so this problem should not be too far
from the incompressible solution. As is shown in Fig. 3.9, the rod
length as a function of time matches the incompressible theory very
well. Calculations with a two-dimensional Eulerian code produced an
almost identical result.

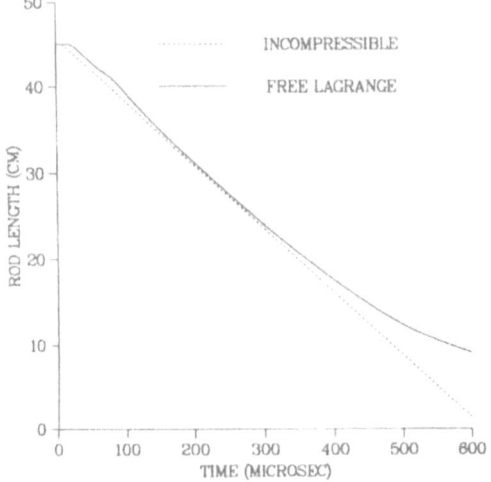

Fig. 3.9

# IV. THE FINITE DIFFERENCE SCHEME

## 4.1 The pressure gradient

We want to approximate $\vec{\nabla}P$ at the point k whose neighbors are the points $k_1$, $k_2$ ... $k_{nmax}$. Our neighbor selection guarantees at least three neighbors for each point, the average is six and there is no maximum number. Clearly there are many methods that could be used to approximate $\vec{\nabla}P$. The following was arrived at through much trial and error and appears to work very well.

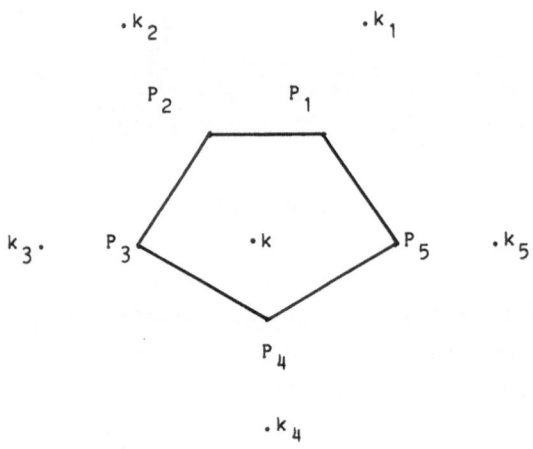

Fig. 4.1

Consider point k in Fig. 4.1 that has five neighbors. We construct a polygon with vertices midway between the point and each of its neighbors. The position of the nth vertex is $\vec{x}_n = 1/2 \ [\vec{x}(k) + \vec{x}(k_n)]$ and the vector from $\vec{x}_k$ to $\vec{x}_n$ is denoted by $\delta\vec{x}_n = \vec{x}_n - \vec{x}_k$. The pressure at the nth vertex, $P_n$, is a weighted average of $P(k)$ and $P(k_n)$ (to be described in section 4.3). We assume a linear pressure distribution along each edge of the polygon and integrate the pressure over the surface to get a force $\vec{F}$. We assume a constant density $\rho_k$ over the polygon to calculate a mass M. Then we have $\frac{D}{Dt} \vec{U} = \frac{F}{M}$. Now let $\vec{x}'_n = \vec{x}_k + \varepsilon\delta\vec{x}_n$ and the pressure at the new vertex is $P'_n = P_k + \varepsilon(P_n - P_k)$. Now F and M are functions of $\varepsilon$ and we compute

$$\lim_{\varepsilon \to 0} \frac{F(\varepsilon)}{M(\varepsilon)}.$$

The resulting expression for the pressure gradient is

$$
\vec{\nabla} P_k = \frac{\hat{x} \sum_n P_n (\delta y_{n-1} - \delta y_{n+1}) + \hat{y} \sum_n P_n (\delta x_{n+1} - \delta x_{n-1})}{\sum_n (\delta x_{n+1} \, \delta y_n - \delta y_{n+1} \, \delta x_n)} \qquad [4.1]
$$

where $\hat{x}$ and $\hat{y}$ are respectively the unit vectors in the x and y directions and $\delta \vec{x}_n = \delta x_n \, \hat{x} + \delta y_n \hat{y}$.

If the preceeding is done in cylindrical geometry, the result is identical for $\vec{\nabla} P$ with x and y replaced by r and z. It is of interest to note that if the $\lim_{\epsilon \to 0}$ is not taken, the result does not give a spherically symmetric pressure gradient in a spherically symmetric problem using cylindrical coordinates.

There is an easier way to arrive at Eq. [4.1] although the method just described is how we originally derived it. Since it takes only three points to describe a plane surface, each consecutive pair of neighbors along with the point k defines a pressure plane to first order. If we assign a weight to each of these approximations we have an approximation for $\vec{\nabla} P$. If the weighting function is the area of the triangle formed by the three points, the result is the same as Eq. [4.1]. We have tried other weighting functions, $\theta$ and $\sin\theta$ where $\theta$ is the angle between $\delta \vec{x}_n$ and $\delta \vec{x}_{n+1}$ both work fairly well, but area weighting appears to be best at this time.


### 4.2   The divergence of the velocity field

In cartesean coordinates we represent the velocity at the point k by $\vec{U}_k = u_k \hat{x} + v_k \hat{y}$. The divergence of the velocity field can be expressed as $\vec{\nabla} \cdot \vec{U} = \frac{1}{V} \frac{\partial V}{\partial t}$ where V is the specific volume of the fluid. Referring back to Fig. 4.1 the specific volume of the constructed polygon is proportion to the area of the polygon given by

$$
A = 1/2 \sum_n (x_{n+1} + x_n)(y_{n+1} - y_n)
$$

Hence we can write

$$\vec{\nabla} \cdot \vec{U} - \frac{1}{A} \frac{\partial A}{\partial t} = \frac{\sum_n (u_{n+1} + u_n)(y_{n+1} - y_n) + (x_{n+1} + x_n)(v_{n+1} - v_n)}{\sum_n x_{n+1} y_n - y_{n+1} x_n}$$

$$\vec{\nabla} \cdot \vec{U} - \frac{1}{A} \frac{\partial A}{\partial t} = \frac{\sum_n (u_{n+1} + u_n)(y_{n+1} - y_n) + (x_{n+1} + x_n)(v_{n+1} - v_n)}{\sum_n x_{n+1} y_n - y_{n+1} x_n}$$

[4.2]

Equation 4.2 can be derived directly from Eq. 4.1 by noting that 4.1 implies a definition for the operators $\frac{\partial}{\partial x}$ and $\frac{\partial}{\partial y}$ and when these are applied to $\vec{\nabla} \cdot \vec{U} = \frac{\partial u}{\partial x} + \frac{\partial v}{\partial y}$ Eq. [4.2] is obtained. Thus, we have in effect three ways of deriving the same finite difference approximation to the operators $\frac{\partial}{\partial x}$ and $\frac{\partial}{\partial y}$. In cylindrical coordinates we express the divergence of the velocity field as

$$\vec{\nabla} \cdot \vec{U} = \frac{1}{r} \frac{\partial}{\partial r}(ru) + \frac{\partial v}{\partial z} = \frac{u}{r} + \frac{\partial u}{\partial r} + \frac{\partial v}{\partial z}$$

where $\frac{\partial u}{\partial r} + \frac{\partial v}{\partial z}$ is calculated by Eq. [4.2] with x,y replaced by r,z.

### 4.3  The midpoint pressure and velocity

In 4.1 we use a pressure $P_n$ which is midway between points k and $k_n$. This is not a numerical average. Consider the one-dimensional problem depicted in Fig. 4.2.a.

What pressure should we use for $P_i^+ = P_{i+1}^-$? If we use the average, 1/2 $(P_1 + P_2)$ the acceleration at i+1 will be much greater than at i. However, we know that the velocity should be continuous across the discontinuity. Given equal zoning the boundary pressure which gives equal accelerations to points i and i+1 is $P_i^+ = (P_i \rho_{i+1} + P_{i+1}\rho_i)/(\rho_i + \rho_{i+1})$.

It can be shown that the resulting finite difference approximation $P_x$ = $(P_i^+ - P_i^-)/\delta x$ is second order accurate when the density is continuous.

Now consider the problem depicted in 4.2.b. Here we have a heavy material on the left moving into a very light material on the right. What should we use for $U_i^+$ = $U_{i+1}^-$? If we use the average, $1/2\ (U_i + U_{i+1})$, there will be a very large rate of compression in region 2 which is incorrect because region 1 is moving into a near vacuum. The quantity that should be continuous is pressure. The velocity which causes equal pressure increases at points i and i+1 is $U^+$ = $[(\rho c^2)_i\ u_i + (\rho c^2)_{i+1}\ u_{i+1}]/[(\rho c^2)_i + (\rho c^2)_{i+1}]$. This assumes the sound speed c is a constant. Again it can be shown that the resultant finite difference approximation to $U_x$ is second order accurate if $\rho c^2$ is continuous.

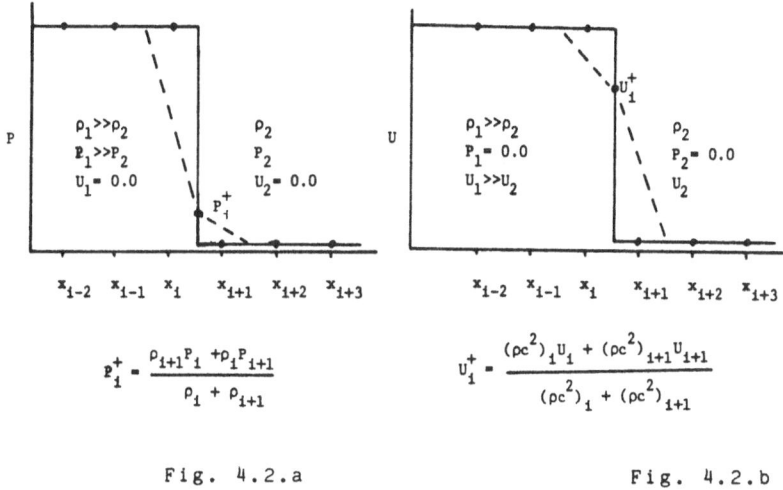

$$P_i^+ = \frac{\rho_{i+1}P_i + \rho_i P_{i+1}}{\rho_i + \rho_{i+1}}$$

$$U_i^+ = \frac{(\rho c^2)_i U_i + (\rho c^2)_{i+1} U_{i+1}}{(\rho c^2)_i + (\rho c^2)_{i+1}}$$

Fig. 4.2.a          Fig. 4.2.b

The midpoint pressure used in Eq. [4.1] are inverse density weighted and the midpoint velocities in Eq. [4.2] are $\rho c^2$ weighted.

### 4.4.    The artificial viscosity

An artificial viscosity, q, is added to the midpoint pressure in Eq. [4.1]. It is quadratic in form. Let $U_c$ be the closing rate between points k and $k_n$, i.e.

$$U_c = (\vec{U}_k - \vec{U}_{k_n}) \cdot \frac{(\vec{X}_k - \vec{X}_{k_n})}{\left| \vec{X}_k - \vec{X}_{k_n} \right|}$$

Then let $q_k = a^2 \rho_k U_c^2$ and $q_{k_n} = a^2 \rho_{k_n} U_c^2$. In the spirit of paragraph 4.3, we inverse density weight the two to get our expression for the midpoint q, i. e.,

$$q_n = 2a^2 U_c^2 / (1/\rho_k + 1/\rho_{k_n}) \quad [4.3]$$

In all of our example calculations in section 2 we used $a^2 = 5.76$. Now we must fold q into the internal energy equation in which we need to evaluate $(P + q)\vec{\nabla} \cdot \vec{U}$. Our approximation for $\vec{\nabla} \cdot \vec{U}$ is given by Eq. [4.2]. The q term is brought inside the summation so that

$$(P + q)\vec{\nabla} \cdot \vec{U} = \frac{\sum_n (P_k + q_n)u_n(y_{n-1} - y_{n+1}) + \sum_n (P_k + q_n)v_n(x_{n+1} - x_{n-1})}{\sum_n x_{n+1} y_n - y_{n+1} x_n}$$

$$[4.4]$$

## 4.5.  Prevention of density striations

The method so far described has one remaining difficulty.  By having all of ·the variables centered in space it becomes impossible to detect a sawtooth type wave as depicted in one dimension in Fig. 4.3.

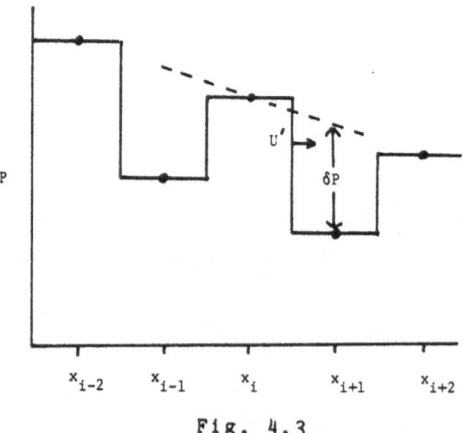

Fig.  4.3

If such a wave develops it cannot be detected by a centered difference scheme. To correct for this, we define an artificial velocity u' as depicted in Fig. 4.3. We use our calculated $\vec{\nabla}$ P to extrapolate from point k to point $k_n$ giving $P_{k_n}^{ext} = P_k + (\vec{X}_{k_n} - \vec{X}_k) \cdot \vec{\nabla} P_k$. If the pressure field is linear then $P_{k_n}^{ext} = P_{k_n}$. If they are not equal, there is a second derivative in the pressure field which we attempt to reduce. Physically what should happen is a velocity would be produced at the midpoint as indicated in 4.3, which would decompress point i and compress point i+1. This velocity must be proportioned to $\delta P = P_{k_n}^{ext} - P_{k_n}$. We chose to use u' = $b^2 \delta P / \rho c$. We then use $\rho c^2$ weighting between points k and $k_n$ to arrive at

$$u_n' = \frac{b^2 \delta P (c_k + c_{k_n})}{\rho_k c_k^2 + \rho_{k_n} c_{k_n}^2} \qquad [4.4]$$

$u_n'$ is added to $u_n$ in calculating $\vec{\nabla} \cdot \vec{U}$.

In our present calculations $b^2 = 1.44$. We further limit $|u_n'|$ to be less than 20% of the maximum of $(C_k, C_{k_n})$. In practice, u' is a very small term, but an absolutely necessary one. For example, in test problem 1, density striations of around 50% will occur without using u'. We note also that $\delta P$ is proportional to $\delta x^2 P_{xx}$ and thus is quadratic in nature. The similarity between q and u' is striking. The q is an aritificial pressure which smooths the velocity field while u' is an artificial velocity which smooths the pressure field.

### 4.6 Neighbor selection

The method requires a good selection of representative neighbors at each point in time. We have found out that the neighbors whose bisectors form the Vornoi polygon[3] around the point k are an excellant choice. The $k_{th}$ Vornoi polygon is defined as that region of space which is nearer point k than any other point.

### V. SUMMARY

The partial differential Eqs [2.1, 2.2, and 2.3], along with the equation of state 2.4, which describe the time evolution of compressible fluid flow can be solved without the use of a Lagrangian mesh. The method follows embedded fluid points and uses finite difference approximations to $\vec{\nabla}P$ and $\vec{\nabla} \cdot \vec{u}$ to update $\rho$, $\vec{u}$ and e. We have demonstrated that the method can accurately calculate highly distorted flows without difficulty. The finite difference approximations are not unique, improvements may be found in the near future. The neighbor selection is not unique, but the one being used at present appears to do an excellent job. The method could be directly extended to three dimensions. One drawback to the method is the failure to explicitly conserve mass, momentum and energy. In fact, at any given time, the mass is not defined. We must perform an auxiliary calculation by integrating the density field over space to obtain mass, energy and momentum. However, in all cases where we have done this, we have found the drift in these quantities to be no more than a few percent.

## References

1.  JETP, Vol. 44, No. 2, p. 424, (1976)

2.  G. Birkhoff, E. H. Farantonello, "Jets, Wakes and Cavities," Academic Press (1957)

3.  G. Vornoi, J. Reine Angew, Math., 134, 198 (1908)

MR. ARMSTRONG: I would like to make one comment on the last speaker's presentation. Bob Clark pointed out the Voronoi mesh would pick nearest neighbors, and drew three vertical lines of dots. I insist it's much easier to understand the Voronoi mesh if you understand the triangular network which must always underlie it. If you visualize the triangular network which would be associated with any Voronoi mesh, you then get planar triangles which must be altered because they contain practically 180 degree sums of vertex angles. Therefore, they would immediately have diagonal swapping. So you can see they wouldn't possibly contain several neighbors in the same column in this diagram for that reason. To me, it's much more awkward to apply Voronoi.

ANONYMOUS: I have one or two comments on the most recent two papers. One of the things Arthur Armstrong said earlier, is that he prefers to write down the discrete approximations to the conservation equations, rather than the differential equations themselves. With this approach you never have information about the canonical form of the mathematical problem to be solved, whether it's parabolic or whatever. This knowledge influences the type of method required for solving the problem. And that carries over into the last paper that we heard in which, it seems to me, that what is being solved is hyperbolic rather than centered parabolic. The last paper was more of a method of characteristics solution without the characteristics. The information should be drawn from a zone of influence rather than to the extent that the choice of nearest neighbors influences the solution. You are either passing information through some type of filter, or you are solving a problem that you may or may not realize.

MR. GLENN: I don't think so. He has a Courant condition in that code which is based upon the distance between nearest points; in fact, so did Armstrong.

MR. LÖHNER: It's not hidden; the time step is based upon the smallest distance between two points.

MR. ARMSTRONG: I don't think Bob Clark ever admitted he was using a Voronoi mesh, which, in fact, is quite different than the Voronoi mesh in today's version.

MR. TREASE: How many neighbors do you find; an average of six, minimum of three?

MR. ARMSTRONG: At any given instant, in any given calculation, it is not arbitrary, it is determinate. I pointed out if you had at least five neighbors, you have six points.

MR. LÖHNER: I don't think that using the difference equations, which are analogues of differential equations, has anything to do with changing the canonical form of the equations. It seems to me you would have the same form.

MR. ARMSTRONG: I think the last question has a point in saying it is very much easier in analytical terms to judge whether the system is hyperbolic or parabolic or elliptic by looking at the equations. Like I say, I don't think in terms of PDE's, but otherwise, of course.

I could shoot down my own method. It's occurred to me in the last few hours that I have come to a bit of a dead end in my method. If you look at the median method and the Voronoi method, the Voronoi method is the only one that gives you, I think, invariance of distribution of the area of the quadrilateral zone among the four nodes, when you flip the diagonal. It is the only thing that ensures that for you. On the other hand, it doesn't give you interfaces which are Lagrangian. You have to sacrifice one or the other. What I want and cannot get is interfaces which are Lagrangian in character and which don't lose invariance of area when I flip diagonals. To sum up, I don't know any method whatever which will give you strict invariance of area if you swap the nodes.

MR. TREASE: Yes. The median-based mesh is one of those that preserves interface invariance.

MR. CLARK: We made some comment about running some test calculations among the codes. Maybe there is a group of people who are interested in designing test problems.

MR. WINSLOW: What type of problems?

MR. CLARK: Pick out one of them. Marty, myself or Pat can be a focus for those comments. We should start thinking about a real test problem or two.

MR. FRITTS: I have a suggestion. What struck me during this meeting, first of all, is that we have a very large variety of methods here, and a large amount of interplay among them. We have done two things over the last several years. One is to reinvent each others methods, and in a lot of cases, ignore some of the advances that were made elsewhere. The other is that some real advances have been made in new ways of approaching or interpreting this area by people who haven't been in touch with the "mainstream". We have to communicate more. The comment I would make, however, is that despite all the activity that has gone on in the field, that progress is still a bit slow. This is in part due to the fact that the problem is extremely difficult. I suggest we all think about doing more than could normally be done in this situation, say, by having workshops. The suggestion I have is that since we all have computers, we should use them to set up central locations where we have test codes. People can work on a test code at the central location, and communicate much more directly with other people about the test problems. In that way we could accelerate our progress. Different people could advise the whole group on what they have done; for instance you could put a notice on the computer on what you have done recently. Here is my test run, run it if you want. Test it out and see what is going on. Everyone can look at it immediately and perhaps incorporate it into their codes.

MR. TREASE: I have to make a comment on that, Marty. I would like to see the test problems distributed.

MR. FRITTS: We are talking about test problems anyway. This is another way of distributing them. People keep working on them, and you can't wait for completed codes.

MR. ARMSTRONG: I would like to ask a question based on your suggestion, Marty. Are we really trying to find even greater accuracy on existing problems or are we trying to develop a code which will tackle problems which are much more outrageous and unexpected in terms of their behavior than the problems we have been able to tackle in the past.

MR. BORIS: My comment is just about that. What I have seen so far is that the irregular nature of the grid implies the accuracy is lower, and fluctuations are much higher. I think we must do better than we are doing in general before we can believe the answers. Of course, that is what we are trying to do. And what you say, is that we have to do better with test problems before we can convince ourselves about the code solutions for the situation which we know nothing about.

MR. TREASE: I guess we can all give the test problem idea some thought, maybe anyone with ideas can get back to one of the three of us: Pat, Marty or I.

MR. N. JOHNSON: One possibility is to set up a direct computer link just to send messages back and forth.

MR. BORIS: I would go for that.

MR. CLARK: Referring to the list Pat made last night (see Figure 4). Are the Free-Lagrange codes continuously rezoned codes? Are there implicit remapping features built into the remeshing? I kind of believe there are.

MR. WINSLOW: It's hard to believe.

MR. BORIS: I presume that everybody who switches neighbors and comes up with two new zones does some sort of rezoning. Rezoning leads to remapping variables. That is exactly one the things I am trying to avoid, one of the reasons I worked so hard...

MR. WINSLOW: It seems to me one of the very large differences between the Voronoi scheme and the Lagrangian scheme is the topological activity of the two different methods. People who have typically done Lagrangian codes in the past are using logically connected meshes which may move, but the mesh is topologically connected in a fixed way. The Free-Lagrange mesh has a lot more generality. Other than that, it seems to me that, the two methods are pretty much trying to do the same thing. As for the accuracy, I don't know. Jay has a point there. I don't think that schemes like FCT are accurate in comparison with the ALE scheme which seems to be fairly accurate.

ANONYMOUS: If you put points close enough, then first order accurate schemes give results that can be better than more accurate schemes with the points more widely separated.

MR. FRITTS: There is also the question of the value of the coefficient for the first order terms in addressing the accuracy. The first order term depends on the distance between a centroid and the vertex position. It is not the full mesh spacing. There is a large difference. Depending on how close you are, you could

AGENDA

- Voronoi vs Regular vs MLG vs ...
      Accuracy
      Mesh Properties (Nearest Neighbor Selection Schemes, Angle
      Criteria, etc.)
- 3D
- Collaboration
- F-L Node Centered vs ?
- Point Quantities vs Distributed
- Trigger Mechanisms
- F-L vs Continuous Rezone (ALE)
- Mesh Instabilities
- Boundaries (Mesh Changes Near Boundaries)
- Limitations?
- Improve Current Methods - Why?  How?

Figure 4.  Concensus Agenda Suggested for the Round-Table Discussions.

be totally second order accurate.

MR. EISEMAN: The question of accuracy leads you to comparing temporal accuracy. You do something to an object to approximate its shape, then the error bound you were talking about previously, the projection of physical space, is less relevant. In a way, I think first order might do a better job, in time differencing, than second order. In fact, I know it would.

MR. N. JOHNSON: I think I'll follow up what Bob Clark said. It's always been a disturbing fact to me in the Lagrangian method that when you do reconnect or redefine anything continually, you need to do remeshing, and interpolation of variables. If you do not do that, then you are destroying the idea of conservation. I would like to hear comments on that. I have no results presently.

MR. CROWLEY: I would like to comment on Bob's talk. I've seen Bob's work over the years, and I think he has really got something now. The thing that I think is impressive is that he's gotten it without conservation. I think that maybe Bob is not intending to conserve, but that he _is_ conserving. For example, he could be conserving because of the centering of the momentum. That is, the pressure between two nodes pushes both ways, and you are conserving momentum. I'll probably work on it on the airplane going home. I think he must be conserving energy pretty well, even though he doesn't explicitly set out to do that. Otherwise, he wouldn't get the jump conditions as well as he does in a shock problem.

MR. FRIEDMAN: I could possibly define mass, momentum and energy in a scheme such as Bob Clark's, so that it is conservative. In such a definition the possibility is, if it is a sufficiently reasonable one, that it is in fact a conservative scheme.

MR. CLARK: Let me make one other comment. I meant to make it at the end of my talk. It is relevant to conservation. Since I do use this auxiliary Voronoi mesh, I can make auxiliary calculations which are totally separate from what I described. I can use the Delaunay triangles on that mesh, I can plunk a mass in each triangle and start a new problem, and I can let them move just like most people would do, except there is no feedback from that auxiliary calculation into the original calculation. At any point in time, I can tell you exactly where the mass is, based on the velocity field that I have calculated. For people who are critically concerned with mass conservation, this auxiliary calculation is almost free.

MR. BORIS: What concerns me, with respect to your question again, is that something is going on here when you conserve at a point. You are forcing it to move in a certain way. You can't get past a certain number of zones.

MR. CLARK: I must comment. The points can move through each other deliberately in this code.

MR. BORIS: In your code, that's right. In other cases, I have a problem doing this. I have a weird shaped thing trying to get past another weird shaped thing, and I cannot allow a node to jump non-physically to the other side into one of the other zones. When you do what you do, the stiffness loosens up, but conservation goes away.

MR. WINSLOW: Would your code interchange mass points?

MR. BORIS: If their mass is conserved exactly. In that case we move the pointers of where those points are located in memory, but do not change anything else.

MR. WINSLOW: Usually when they get too close, you exchange them?

MR. BORIS: They get right through. But, for example, if you think of those points as repelling each other with a pressure force, and in the process think of two rows of points equally spaced and another moving between them. You see a washboard effect whereas the real problem would just have a single layer with no displacements generated at all. Trying to push it with all the points next to each other; I don't think that is physical. You conserve mass, momentum and energy with the Voronoi cells exactly. The problem is reminiscent of rolling a ball down a washboard. You do anything else, and you are allowing these restrictions on mass points to be constraints. It doesn't have to be this particular volume; it will attach to these side volumes, as it goes by, or it may reattach to this other volume. You have not forced that extra local strength, which is the source of stiffness. You may have to play a lot of games like Marty and I did in the incompressible case or in some other case. I think in some cases there is a trade-off there. It's a feeling, rather than anything I have ever been able to prove.

MR. KIRKPATRICK: What about the Q (artifical viscosity) that Bob Clark has used.

MR. BORIS: He is using that to help smooth out the variation between one zone and another. Maybe we are saying the same thing. I think Q is playing an important role.

MR. WINSLOW: Is it the tensor Q?

MR. CLARK: That is what I'm talking about there.

MR. N. JOHNSON: Maybe there is another way of looking at this. I think we all agree that in the mesh structuring, that Pat Crowley's is symmetric. If point "I" is connected to point "J", then point "J" is connected to point "I". You can also have equal and opposite forces across two surfaces, and it is not clear to me, we're all doing that. I'm curious to ask Bob if that is true in his code.

MR. CLARK: I didn't want to say at the time. It is not conservative. There is no way to be conservative even if you assigned a mass to each one of those points. The way my gradients are done does not insure that the same momentum is being imparted on both sides of the line. So, it is not explicitly conservative. In one dimension, you can conserve. But what happens in two dimensions is that the two points not only see each other, but they also see other points. Since this pattern up here can be completely different than this other pattern, those patterns are taken into account by weighting the pressure gradients, and that is why the contribution is not identical. In fact, I ran the code in an attempt to be conservative a year ago, and computed the contribution of K to K' and K' to K, and looked at the difference. For a conservation method, they should be the same. They are not. I adjusted them a little bit so they were, but that failed to prove anything. But, the contributions definitely are different.

MR. BORIS: You could, in principle, instead integrate over a mass, put in a specific volume on the other side of the equation to conserve momentum. If you unfold the way you were doing pressure gradients, you will get the same pressure to apply in both points. That is all that is really required to have momentum conservation.

MR. CLARK: When you do that, you again arrive at the remapping problem. If you want this point to have a mass associated with it, even if it is density times a volume, the mass will be changing and the mass will be shifting. If you want to conserve energy, then every time you shift mass around you must also shift energy around also.

MR. BORIS: And shift the same amount of momentum?

MR. CLARK: You are adding more and more complications to what is effectively a very simple code right now, and I have not done what you suggest. I'm not totally convinced that it is not worth doing. Lack of conservation does not affect the answers of any calculations I have done so far.

MR. ARMSTRONG: I would like to help you. If you are, in fact, off by three percent of your mass, you would have pretty good accuracy, wouldn't you.

MR. FRIEDMAN: I would be curious to hear other impressions about whether there are equal and opposite forces.

MR. ARMSTRONG: The difficulty seems to be: when you don't store explicit interfaces, like I do, then when a node gains or loses energy or momentum, it is impossible to tell which of its neighbors got what fraction of that energy or momentum. It seems to be very difficult to achieve conservation in any other way!

MR. BORIS: I would like to make a comment. The question you were asking is: have we done this acceleration match and if it works could you make it conserve? It works remarkably well in dealing with the inaccuracies you have shown, which is part of what we are doing. It is not the density jump, but the difference in the zone size when you do that acceleration matching in the test problem I used. There was a problem of uniform space zoning and then a problem of zones that varied in size by a factor of 100. We found in the one case that we couldn't do it at all, and in another case it eventually made no difference. That convinced me that for conservation this type of trick is extremely valuable. I contend that that is the bad part about having a large cell variation and why you were getting the results you were.

MR. CLARK: I will admit, going back to Pat, that I gave a rather extreme example of neighbors of one point being grossly different than those of another point. The fact is that this is usually not the case. It's a well posed problem. The zoning is consistent throughout; it does not change rapidly. At least it is not supposed to.

MR. BORIS: Even if it did, the trick may help.

MR. FRIEDMAN: One possible generalization of the Voronoi mesh might be; at an interface between disparate type of fluids, where the density is greatly different, you might imagine keeping a smaller cell for the denser fluid, where you have the same amount of mass associated with each. Now, you might want to keep that a little bit farther into the less dense material so you can imagine having different wavelengths on the Voronoi mesh, and in something like Clark's scheme, that would give you a different choice of neighbors. It would be very interesting to see if a somewhat different choice of neighbors had an effect on the solution at an interface separating disparate materials.

MR. ARMSTRONG: May I suggest trying that by using the generalized Voronoi mesh procedure you have mesh (numerical) instabilities and physical instabilities in the mesh in the same way.

MR. TREASE: I would like to ask Bob Clark a question. Have you tried any problems other than matching jump conditions? Did you try any problems where you had a terrible interface where your methods didn't work at all?

MR. CLARK: Other type of problems, the latest method? No! The kind of things that originally cause trouble are contact discontinuities or material discontinuities, where you had greatly different properties. I described how the intermediate compression or velocity being calculated took care of all of those problems. I don't have any problems I know of.

# NEAREST NEIGHBOR ALGORITHM

Ronald C. Kirkpatrick
University of California
Los Alamos National Laboratory
Los Alamos, New Mexico  87545

## INTRODUCTION

Various problems solved by finite difference methods require
construction of a Voronoi mesh for some distribution of points in
2-D or 3-D space.  For some of these problems, this construction
represents a significant, even limiting overhead.  Therefore, it
seems reasonable to endeavor to derive an efficient algorithm for
constructing a Voronoi mesh.  The initial presentation will be in
terms of a 2-D space, but later the extension to 3-D will be made.

The Voronoi mesh is a set of polyagonal cells surrounding each
point such that the sides are perpendicular disectors of the lines
between all neighboring pairs of points.  Thus, the apexes are the
centers of circles defined by adjacent triplets of points (see
Figure 1).  It can be shown that a random distribution of points
defines a Voronoi mesh that completely fills the space between the
points with polygons (no holes) and that associated with the
Voronoi mesh is a unique relation between the triplets of points
called Delaunay triangulation.  Changes in this unique relationship
due to relative motion of the points is often referred to as
reconnection.  Although this reconnection is abrupt, the corresponding changes in the Voronoi mesh are incremental (continuous for
continuous motion).  Points that are connected in the Delaunay
triangulation are referred to as nearest neighbors, and in 2-D
each point will typically have four to seven nearest neighbors.  A
list must be established for each point's nearest neighbors and it
must be periodically updated when the relative positions of the
points are changing.  Computation of the initial mesh is an
N-squared problem, but the work for subsequent meshes is greatly
reduced if the relative motion is restricted so that only the
neighbors of nearest neighbors must be tested for connection to the
point of interest.

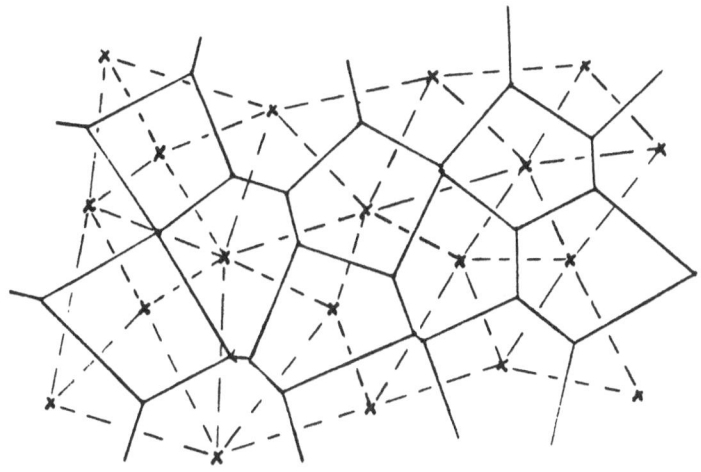

FIGURE 1


## THE ALGORITHM

Although Voronoi meshes have been successfully obtained by
construction, the most efficient method relies on simply finding
the apexes for the polygons (centers of circles in 2-D or for
polyhedrons in 3-D, the centers of spheres) and applying a simple
test.

The center of a circle may be found for a given triplet by applying
the equation for a circle:

$$(x - x_o)^2 + (y - y_o)^2 = R^2$$

$$x^2 + y^2 - 2xx_o - 2yy_o - R^2 + r_o^2 = 0$$

For three points the three equations formed by substituting each
$(x,y)$ point successively can be solved algebraically to get the
three coefficients $-2x_o$, $-2y_o$, and $R^2 + r_o^2$. The center will be
at the intersection of the perpendicular bisectors between the
three pairs of points of the triplet.

For a fourth point sharing one of the pairs of the (1st) triplet
that is common to another (2nd) triplet involving the fourth point,
it is connected to the third point if the apex for the 2nd triplet
lies inside the circle defined by the first triplet.  Then the
common pair cannot be connected and the third and fourth points
become the new common pair.  The old pair now constitutes the third
points for the new adjacent triplets.  This is reconnection (see
Figure 2).  This test corresponds to the sum of the opposite angles
test ($\gtrless$ 180°) but requires no trigonometry and is equally applicable to 3-D.  In addition it is equivalent to the minimum Z test
for Step 4 used by Tanemura et al. (1), but more efficient.

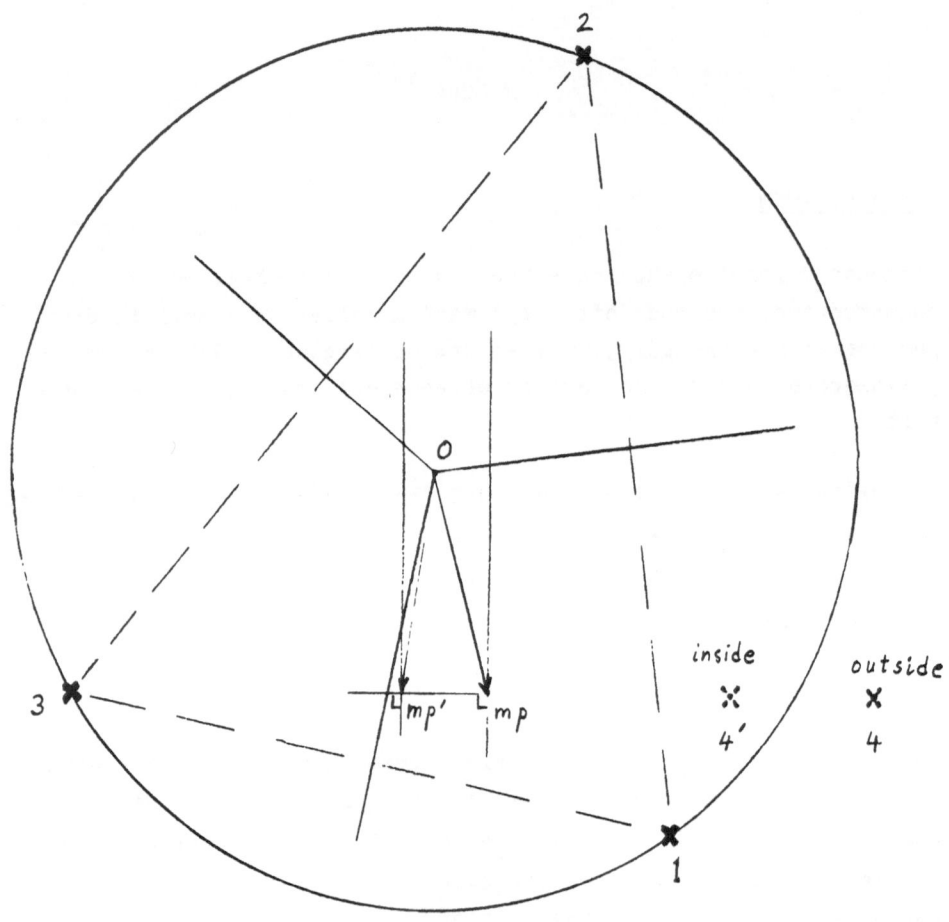

FIGURE 2

When reconnection occurs the nearest neighbor (NN) lists for all
four points must be updated.  NNs are deleted from two and NNs are
added to two.


## ESTABLISHING THE NEAREST NEIGHBOR LISTS

In establishing the NN list for each point, it seems advantageous
to first close some set of points about each point and then search
for additional NNs from among the remaining points.  This ensures
that all NNs for each point are found.  If for some point it is not
possible to geometrically close the set, then that point must be a
boundary point.  After finding all the NNs for all the points, the
NN list for each point can be ordered geometrically by simply
sorting through the NN lists of the NNs associated with each point
in turn.  This geometrically ordered list of NNs can then be used
to establish a geometrically ordered list of apexes for each point.

The algorithm for establishing an initial closed set of points
about a given point is to first find the closest neighbor.  Then
excepting the given point and its closest neighbor, find the
closest point to the mid point between the given point and its
closest neighbor for which the product

$$(\hat{r}_p - \hat{r}_o) \times (\hat{r}_n - \hat{r}_o) > 0,$$

where $\hat{r}_o$ is the given point, $\hat{r}_p$ is the (previous) closest
point, and $\hat{r}_n$ is the (next) closest point to the mid point (see
Figure 3).  Next, by excepting only the given point and the
previously found closest point to the mid point each time,
eventually the next point will be the initial closest point, thus
geometrically closing the set.  Generally speaking this set should
be a reasonable approximation to the NN list.

The search for additional NNs involves first getting the apexes
(circle centers) for the closed set, successively applying the
reconnection test for a fourth point to all the remaining points
(those not included in the closed set), and updating the NN lists
as necessary.  Using this process the NN lists for all points can
be established.  As stated, the algorithm is redundant, since each

time a new NN is found, that new NNs list can be added to as well,
and the new NN's partial list can be used as a starting set for the
NN's geometrically closed set. However, the order of the partial
list must be revised.

The procedure for ordering the NN lists is as follows: Start with
the given point and one of its NNs. The first is convenient. Then
from the NN list of this NN find an NN for which the given point is
one of its NNs. There are two possibilities. Next, for the NN
chosen (NNa), find one (NNb) for which the given point is one
of its NNs and it (NNb) has not yet been added to the ordered list.
Continue this process until both possible choices have already
been added to the ordered list, meaning that the ordered list is
complete (see Figure 4). Upon taking any two successive NNs from
the ordered list, the cross product of their directions will give
the geometric sense of the ordered list (clockwise or counter
clockwise). The ordering of the list is needed to properly
calculate the apexes for the Voronoi mesh about a given point.
However, because the process is straight forward and could be
applied point by point as needed when the apexes are being found,
it is not necessary to reorder all the lists or carry them in the
calculations that way.

## UPDATING THE NEAREST NEIGHBOR LISTS

Quite simply stated, for a given point about which the Voronoi mesh
has been established, if the perpendicular bisector of the line
between the given point and some other point not associated with
the mesh intersects the mesh (that is passes closer to the given
point than one of the apexes), then that other point becomes a new
NN. Equivalently, in order to be a new NN the other point must lie
inside one of the circles defined by the triplets for the mesh
about the given point.

For sufficiently restricted changes in the positions of the points,
it is only necessary to test the NNs of the given point's NNs to
see if they should be added to its NN list. Updating the
appropriate NN lists (four for 2-D) involves additions to each of
the lists for the new NN pair and deletions from each of the NN
lists for the other pair that used to be NNs.

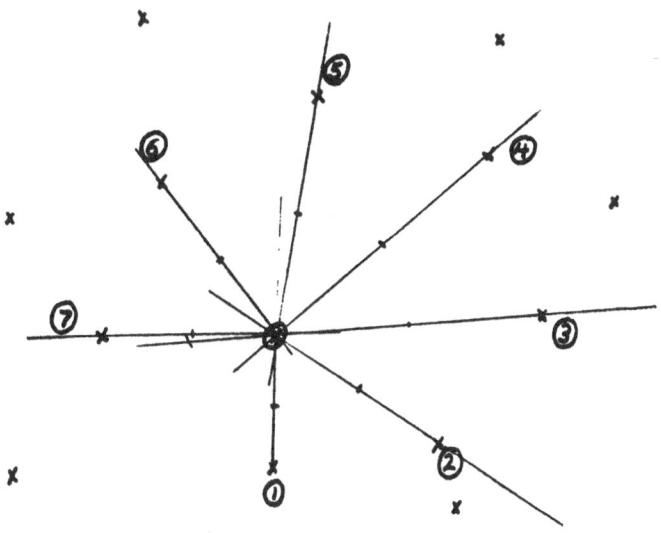

FIGURE 3

## THE ALGORITHM IN 3-D

Similar to the 2-D case, the center of a sphere associated with a
set of four points may be found by applying the equation of a
sphere:

$$(x - x_o)^2 + (y - y_o)^2 + (z - z_o)^2 = R^2$$

$$2xx_o + 2yy_o + 2zz_o + (R^2 - r_o^2) = x^2 + y^2 + z^2$$

$$
\begin{bmatrix}
1 & x_1 & y_1 & z_1 \\
1 & x_2 & y_2 & z_2 \\
1 & x_3 & y_3 & z_3 \\
1 & x_4 & y_4 & z_4
\end{bmatrix}
\begin{bmatrix}
R^2 - r_o^2 \\
2x_o \\
2y_o \\
2z_o
\end{bmatrix}
\begin{bmatrix}
x_1^2 + y_1^2 + z_1^2 \\
x_2^2 + y_2^2 + z_2^2 \\
x_3^2 + y_3^2 + z_3^2 \\
x_2^2 + y_4^2 + z_4^2
\end{bmatrix}
$$

or

$$M\ V = W$$

$$V = M^{-1}\ W$$

The vector V is most efficiently obtained by Gaussian elimination and back substitution and the first element $(R^2 - r_o^2)$ is not needed.

For a fifth point sharing one of the triplets of the four points for which the center (apex) was found, it does not change the connections of the four if it lies outside the sphere, but does if it lies inside. That is, if

$$(x - x_o^2) + (y - y_o^2) + (z - z_o^2) > R^2$$

no reconnection occurs. By continually reapplying this test, the first apex with its defining points (a quad) can be found. Each quad has four associated triplets. There is no angular sum rule for reconnection in 3-D.

The second nearest neighbor (2nd NN) for a triple will be that fifth point, as determined by the solution of the equations of the sphere. Finding the 2nd NN provides a new quad and three possible new triplets beyond which to search for their 2nd NNs.

By keeping a list (a stack) of the new triplets, choosing one for the next direction of search, and adding to the appropriate NN lists, it is possible to continually move through the points defining quads and apexes until a boundary is reached (no fifth point beyond the triplet) or a previously defined quad is encountered. When either of these two eventualities occurs, another triplet is taken from the stack and the process is continued. When there are no more triplets on the stack, all quads (w/ their apexes) have been defined. A previously defined quad may be detected by checking to see if the triplet's NN lists all contain the same NN which is not the fourth of the quad just previously found. Confirmation is obtained by checking to see if that same NN contains all three triplet points in its NN list.

All quads associated with a given point can be found by starting with that point, finding the 1st of its quads, then choosing from the triplet stack only those triplets which include the given point. In addition, by restricting the choice to a triplet that contains, in addition to the given point, also one of the previous triplet's points which is kept the same until a previously defined quad is encountered, then an ordering may be established for the quads. When no more undefined quads can be found that include the given point, then all of its NNs have been found. See Figure 5.

REFERENCE

1.  Tanemura, M., Ogawa, T., Ogita, N., J. Comp. Phys. 51, 191 (1983).

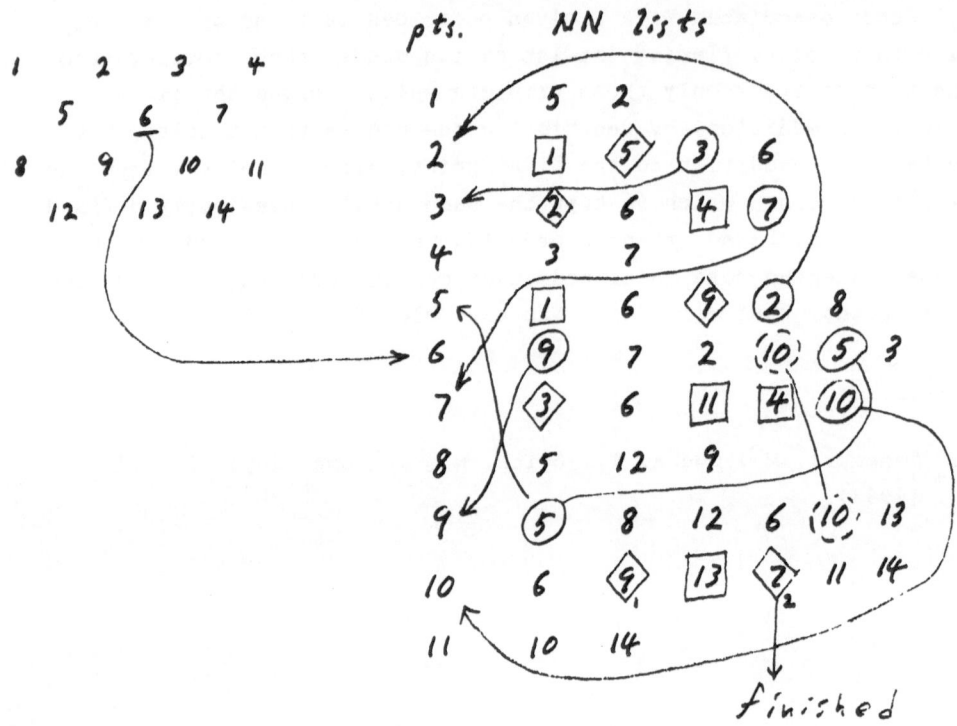

□ - not given pt. NN
◇ - already chosen
○ - next NN

FIGURE 4

311

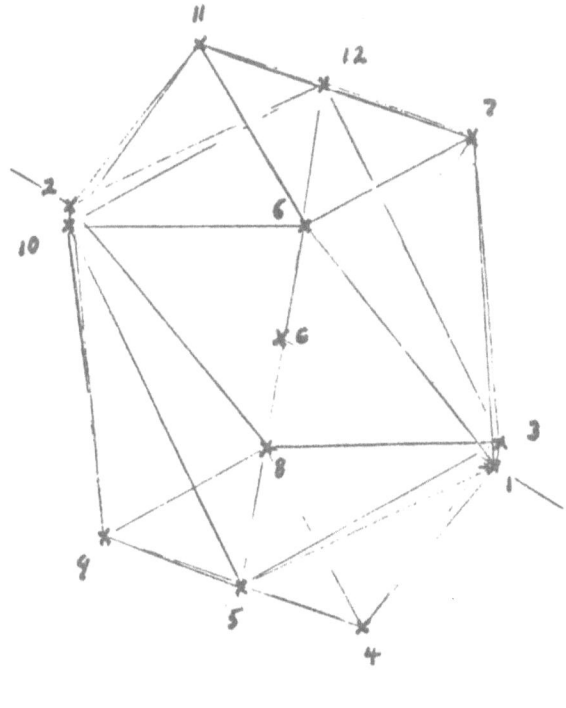

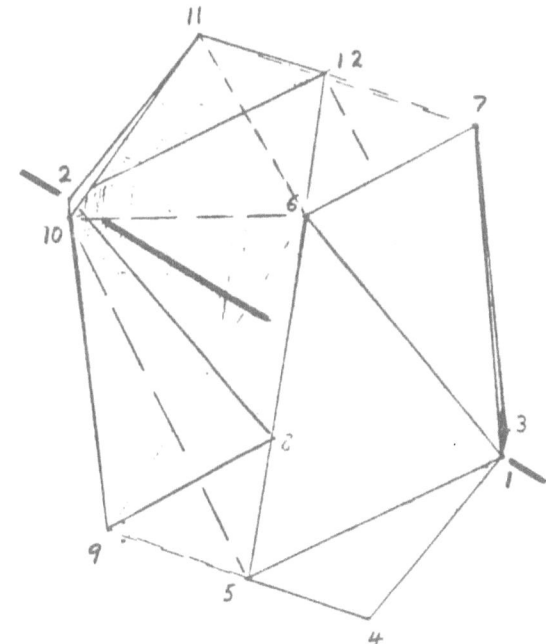

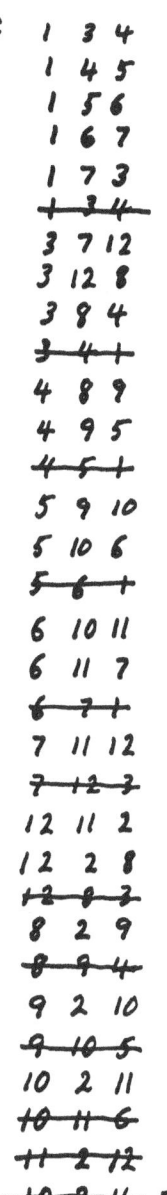

G
1 3 4
1 4 5
1 5 6
1 6 7
1 7 3
~~1 3 4~~
3 7 12
3 12 8
3 8 4
~~3 4 1~~
4 8 9
4 9 5
~~4 5 1~~
5 9 10
5 10 6
~~5 6 1~~
6 10 11
6 11 7
~~6 7 1~~
7 11 12
~~7 12 3~~
12 11 2
12 2 8
~~12 8 3~~
8 2 9
~~8 9 4~~
9 2 10
~~9 10 5~~
10 2 11
~~10 11 6~~
~~11 2 12~~
~~10 2 11~~ F

FIGURE 5

# LIST OF CONFERENCE ATTENDEES

Arthur Armstrong
Ministry of Defence (P.E.)
A.W.R.E., Aldermaston
Berkshire, U.K.

Armand V. Attia  (L-202)
Lawrence Livermore National Lab.
P.O. Box 808
Livermore, CA  94550

Jeffrey Augenbaum  (Code 611)
NASA/Goddard
Goddard Space Flight Center
Greenbelt, MD  20771

David Bailey  (L-477)
Lawrence Livermore National Lab.
P.O. Box 808
Livermore, CA  94550

Christoph Börgers
Courant Institute
251 Mercer Street
New York, NY  10012

J.P. Boris  (Code 4040)
Naval Research Laboratory
Washington, D.C.  20375

Daniel Carroll
Group X-7, MS B257
Los Alamos National Lab.
Los Alamos, NM  87545

Robert A. Clark
Group X-7, MS B257
Los Alamos National Lab.
Los Alamos, NM  87545

Robert E. Cooper  (L-13)
Lawrence Livermore National Lab.
P.O. Box 808
Livermore, CA  94550

Benoit Couet
Schlumberger-Doll Research
Old Quarry Road
Ridgefield, CT  06877-4108

Pat Crowley  (L-13)
Lawrence Livermore National Lab.
P.O. Box 808
Livermore, CA  94550

John Dukowicz
Group X-7, MS B216
Los Alamos National Lab.
Los Alamos, NM  87545

Peter R. Eiseman
Department of Applied Physics and NE
Columbia University
New York, NY  10027

Peter G. Eltgroth  (L-298)
Lawrence Livermore National Lab.
P.O. Box 808
Livermore, CA  94550

Gordon Erlebacher
M.S. 15G
NASA Langley Research Center
Hampton, VA  23665

Charlie Finan
Schlumberger-Doll Research
Old Quarry Road
Ridgefield, CT  06877-4108

Alex Friedman  (L-477)
Lawrence Livermore National Lab.
P.O. Box 808
Livermore, CA  94550

Martin Fritts
SAIC
134 Holiday Court, Suite 318
Annapolis, MD  21401

Lewis A. Glenn  (L-387)
Lawrence Livermore National Lab.
P.O. Box 808
Livermore, CA  94550

Norman L. Johnson
Group WX-4, MS G787
Los Alamos National Lab.
Los Alamos, NM  87545

Mark Johnson  (L-16)
Lawrence Livermore National Lab.
P.O. Box 808
Livermore, CA  94550

Ronald C. Kirkpatrick
Group X-2, MS B220
Los Alamos National Lab.
Los Alamos, NM  87545

Samuel G. Lambrakos  (Code 4040)
Naval Research Laboratory
Washington, D.C.  20375

Wen Ho Lee
Group X-7, MS B257
Los Alamos National Lab.
Los Alamos, NM  87545

Deyuan Li
Institute of Applied Physics
    and Computational Mathematics
P.O. Box 8009, Beijing
China

Rainald Löhner
Universtion College Swansea
Singleton Park
Swansea SA2 8PP, U.K.

Elaine Oran  (Code 4040)
Naval Research Laboratory
Washington, D.C.  20375

J. Ovadia
C.E.A. Limeil
B.P. 27
94190 Villeneuve St. Georges
France

Samuel Paolucci
Div. 8245
Sandia National Lab.
Livermore, CA  94550

Walter F. Reddall
The Aerospace Corp, M4/964
P.O. Box 92957
Los Angeles, CA  90009

Fred Rockenbach
Group X-7, MS B257
Los Alamos National Lab.
Los Alamos, NM  87545

Hans Ruppel
Group T-3, MS B216
Los Alamos National Lab.
Los Alamos, NM  87545

Jack R. Schendel
Group X-2, MS B220
Los Alamos National Lab.
Los Alamos, NM  87545

Richard W. Sharp  (L-477)
Lawrence Livermore National Lab.
P.O. Box 808
Livermore, CA  94550

Aleksei Shestakov  (L-477)
Lawrence Livermore National Lab.
P.O. Box 808
Livermore, CA  94550

Kurt Sinz  (L-202)
Lawrence Livermore National Lab.
P.O. Box 808
Livermore, CA  94550

Harold Trease
Group X-7, MS B257
Los Alamos National Lab.
Los Alamos, NM  87545

A.M. Winslow  (L-16)
Lawrence Livermore National Lab.
P.O. Box 808
Livermore, CA  94550

Zhilu Yu
Institute of Applied Physics
    and Computational Mathematics
P.O. Box 8009, Beijing
China

## Numerical and Physical Aspects of Aerodynamic Flows

Editor: **T. Cebeci**

1982. 302 figures. X, 636 pages. ISBN 3-540-11044-5

**Contents:** Numerical Fluid Dynamics. – Interactive Steady Boundary Layers. – Singularities in Unsteady Boundary Layers. – Transonic Flows. – Experimental Fluid Dynamics.

## Numerical and Physical Aspects of Aerodynamic Flows II

Editor: **T. Cebeci**

1984. 171 figures. XI, 416 pages. ISBN 3-540-12659-7

**Contents:** General Review. – Keynote Paper. – Navier-Stokes Methods. – Interactive Methods and Measurements: Subsonic Flows. – Interactive Methods: Transonic Flows. – Analytical Methods. – References.

## Turbulent Shear Flows

### Volume 1

Selected Papers from the First International Symposium on Turbulent Shear Flows, The Pennsylvania State University, University Park, Pennsylvania, USA, April 18–20, 1977
Editors: **F. Durst, B. E. Launder, F. W. Schmidt, J. H. Whitelaw**

1979. 256 figures, 4 tables. VI, 415 pages. ISBN 3-540-09041-X

### Volume 2

Selected Papers from the Second International Symposium on Turbulent Shear Flows, Imperial College, London, July 2–4, 1979
Editors: **L. J. S. Bradbury, F. Durst, B. E. Launder, F. W. Schmidt, J. H. Whitelaw**

1980. 310 figures, 12 tables. IX, 391 pages. ISBN 3-540-10067-9

### Volume 3

Selected Papers from the Third International Symposium on Turbulent Shear Flows, The University of California, Davis, September 9–11, 1981
Editors: **L. J. S. Bradbury, F. Durst, B. E. Launder, F. W. Schmidt, J. H. Whitelaw**

1982. 244 figures. VIII, 321 pages. ISBN 3-540-11817-9

### Volume 4

Selected Papers from the Fourth International Symposium on Turbulent Shear Flows, University of Karlsruhe, Karlsruhe, FRG, September 12–14, 1983
Editors: **L. J. S. Bradbury, F. Durst, B. E. Launder, F. W. Schmidt, J. H. Whitelaw**

1985. 286 figures. VIII, 397 pages. ISBN 3-540-13744-0

Springer-Verlag
Berlin
Heidelberg
New York
Tokyo

# Lecture Notes in Physics

# Selected Issues from
# Lecture Notes in Mathematics